Physics of Semiconductor Devices

WSPC—COSTED SERIES IN EMERGING TECHNOLOGY

Physics of Semiconductor Devices

PROCEEDINGS OF THE FOURTH INTERNATIONAL WORKSHOP

Madras (India) December 10–15 1987

Editors

S C JAIN

Former Director
Solid State Physics Laboratory
New Delhi

S RADHAKRISHNA

Professor – Department of Physics
Indian Institute of Technology
Madras

World Scientific
Singapore • New Jersey • Hong Kong

**COMMITTEE ON SCIENCE & TECHNOLOGY
IN DEVELOPING COUNTRIES**

Published by

World Scientific Publishing Co. Pte. Ltd.
P.O. Box 128, Farrer Road, Singapore 9128

U. S. A. office: World Scientific Publishing Co., Inc.
687 Hartwell Street, Teaneck NJ 07666, USA

Library of Congress Cataloging-in-Publication data is available.

**IV INTERNATIONAL WORKSHOP ON PHYSICS
OF SEMICONDUCTOR DEVICES**

ISBN 9971-50-531-2
9971-50-532-0 pbk

Printed in Singapore by General Printing and Publishing Services Pte. Ltd.

PREFACE

The Fourth International Workshop on Physics of Semiconductor Devices is being organised in the Department of Physics, Indian Institute of Technology , Madras , India during December 10-15 1987. We are grateful to all the invited speakers who have sent their manuscripts in advance to enable us to prepare the proceedings in time for the Workshop. This year the emphasis has been on microelectronics, VLSI, and special aspects related to semiconductor applications.

We would like to express our grateful appreciation of the support provided by Department of Science and Technology, Defence Research and Development Organisation and several other government departments.

We gratefully acknowledge the support provided by Messers World Scientific Publishers for having brought out the proceedings in time for the Workshop in the very short time given to them.

S C Jain
S Radhakrishna

November 15, 1987.

ACKNOWLEDGEMENTS

The International Workshop has received support from the following organisations and wishes to express its appreciation for their financial assistance.

Committee on Science & Technology
  in Developing Countries (COSTED) of
International Council of Scientific Unions (ICSU)
Defence Research & Development Organisation
Bharat Electronics Limited
Bharat Heavy Electricals Limited
Central Electronics Engineering Research Institute
Central Electronics Limited
Council of Scientific and Industrial Research
Department of Electronics
Department of Non-conventional Energy Sources
Department of Science and Technology
Indian National Science Academy
Indian Institute of Technology, Kanpur
Indian Institute of Technology, Madras
International Centre for Theoretical Physics — Trieste
National Physical Laboratory
Solid State Physics Laboratory
University Grants Commission
Deutsche Forschungsgemeinschaft — Germany

CONTENTS

MATERIALS

PHOTOVOLTAICS

GALLIUM ARSENIDE DEVICES

MICROELECTRONICS

Novel MOSFET Structures for

Integrated Magnetic Field Sensors

B.S. Gill, D. Misra and E. Heasell

Elect.Eng.Dept., Univ. of Waterloo, Ontario, Canada.

Introduction

The development of integrated, sensor devices has become a significant activity. The MOSFET structure in particular has been employed for ionic and gaseous sensing and for the detection of magnetic fields.

Magnetic field sensors, based on MOS technology, or MAGFETS, are attractive in a range of sensor applications, magnetic bar-code readers, displacement or rotation transducers etc. They lend themselves well to integration with on-chip data processing.

1, The conventional MAGFET.

A variety of magnetic field sensitive devices have been described. A recent review [1]. gives an excellent, overall picture of current work.

The devices all rely upon some variant of the Lorentz force or Hall effect. The attraction of the MOS family is based upon a naive application of the Hall effect formula :

$$V_H = R_H BI (W/t)$$

$$(1.1)$$

V_H is the Hall voltage, R_H the Hall coefficient, W the width of the conducting channel and 't' its thickness.

The available Hall voltage is governed by consideration of the permissable power dissipation in the device, and by the Hall coefficient R_H. Conventional Hall sensors exploit the larger carrier mobilities available in the III-V compounds, InAs in particular.

Conventional, flat-plate, Hall transducers, fabricated from Silicon, are less useful since the available electron mobility, even in pure material is low. The MOSFET structure allows the realisation of very large values of (W/t), it might seem ideally suited to the development of Hall effect transducers, exploiting standard n-MOS technology.

To achieve such a sensor, more-or-less conventional MOSFET devices, using two, isolated drain diffusions have been employed, [2]. Typical structures are shown in Fig.1. These devices have been called MAGFET's, [3]. Such devices exhibit a differential redistribution of the channel current, between the drains, in the presence of a magnetic field. The sensitivity is quite poor. However, when operating in saturation, the electrical output impedance is high, and standard MOS, active loads may be used to achieve larger output signals, [2].

In fact the design of these devices is at fault. The split-drain configuration is sensitive to the charge redistribution in the channel, not to the actual Hall field or voltage. The charge distribution needed to sustain even a large Hall-field, is very small. (indeed if E_H is uniform, then only a small surface charge, at the channel edges, is needed).

Charge separation can occur only over a short distance in the immediate vicinity of the drain, where the carriers are free to move laterally.

The mobility of electrons in the channel of a MOSFET is reduced both by the additional scattering resulting from the extreme narrowness of the channel, as well as by velocity saturation effects, observed also in bulk material, at high fields. The saturation of carrier velocity is especially significant in the drain region of a MOSFET, operating in saturation.

Since the chordal mobility falls, the Hall angle decreases and the deflection of the carriers, by the Lorentz force, is reduced. In addition, the carrier are only deflected in the region immediately adjacent to the drains, so that current redistribution occurs only over a very small fractin of the total channel length.

The next structure described attempts to ovecome these limitations.

2, The Graded Gate MAGFET.

To limit the longitudinal electric field in the channel, and to increase the path length over which it operates, we modify both the gate and drain structure of the device.

We use a resistive (refered to as GRADED) gate structure. Ohmic contacts to the ends of the gate permit the application of either aiding or retarding fields in the channel. The gate is made of undoped polysilicon and exhibits a resistance of around 10^9 ohms. Power dissipation, due to the gate bias is negligible, [4].

To prevent the Hall field from developing, lateral drain diffusions are placed along either side of the channel. These diffusions (refered to as lateraal-drains), are biased so that the drain potentials are more positive than the surface potential at any point in the channel. Fig.2. In some devices a third drain is placed at the end of the channel, in order to collect undeflected carriers.

The theory for the simple GGFET (without lateral drains) can be derived, assuming an Ihanolta-Moll type model (drift only). The analysis confirms that it is possible to achieve reasonably uniform, electric fields over the entire length of the channel. We can arrange operation to avoid the major effects of velocity saturation. These devices exhibit significantly higher sensitivities. The sensitivity is, as anticipated, an increasing function of the aiding, gate field. Fig.3.
Fig.3b shows the offset current versus

A first order model of the channel potential, in the presence of an external, magnetic field, predicts sensitivities in fair agreement with experiment. The device sensitivity can be determined by solving the lateral diffusion problem and integrating along the drains. Fig.4.

In the structures considered, we can regard the drain currents as a superposition of two components, the carriers deflected by the Lorentz force in the channel, and a much larger, undeflected portion, which ideally divides evenly, between the drains. In practice, due to manufacturing tolerances, the current does not divide symmetrically in the absence of an external field.

In an attempt to avoid the latter problem we consider the inclusion of an additional drain diffusion, to collect undeflected carriers.

3, Triple Drain GGFETS

In the three drain device it proves extremely difficult to prevent the entire channel current from being collected by the lateral drains. It is necessary to adjust the individual drain biases, so as to obtain a symmetrical division of the channel current, between the three drains, with B = 0. T.

It is found that the lateral drains must be biased so that their potential matches the average of the channel potential, along their length. In contrast, in the two-drain devices, the drain potentials were always much greater than the channel potential at any point between them.

The three drain devices exhibit extremely high sensitivity, but also are extremely unstable. Fig.5. A closer examination of the device behaviour discloses that a novel mechanism is present. The bias to the lateral drains pins the fermi-level in the adjacent channel. The channel potential is almost constant between the lateral-drain diffusions. At the source-end of the channel, the lateral drains COLLECT carriers, at the opposite end of the channel the lateral drains INJECT carriers. By carefull adjustment of the lateral drain potentials, these processes are made to balance.

The balance is strongly perturbed by even the small Hall voltages, developed at the source end of the channel. The imbalance produces a large redistribution in the net current between the lateral drains.

Unfortunately, it is necessary to set the potentials of the lateral drains to within $\pm$ 10 microvolts, to achieve balanced operation. Device operation, in this configuration is not practicable.

4, The Split Source MAGFET, or SSFET.

With the experience gained from the operation of the triple-drain devices, it became apparent that a more useful configuration could be achieved by operating the original GGFET's backwards ! To emphasise the effect of the Hall potential, upon injection by the two source diffusions, the gate is operated at a uniform potential. The potential gradient along the channel is governed by the channel current and the drain potential.

We have made no attempt to develop a quantitative theory for this device. Carriers injected by the sources flow toward the single drain. In the presence of an external, magnetic field, a Hall potential develops. The Hall potential modifies the barrier existing between each source and the channel. The injected current varies as the exponential of this barrier height.

It is clear that the geometry of the present devices is far from optimum. However, the SSFET's can be made with completely conventional MOS technology, and they exhibit sensitivities greater than those achieved in the GGFET.

The model suggested for the operation of these devices implies that the smaller the longitudinal field, the greater should be the effect of the Hall voltage. Experimental study confirms this behaviour, Fig.6. The source diffusions present a low, electrical impedance. It will be necessary to add current mirrors and active loads to the integrated sensor, in order achieve sensible output voltages. Such circuit additions are fully compatible with standard, MOS analogue circuitry.

FIGURE CAPTIONS

Figure 1:   The conventional MAGFET, using a split-drain configuration.

Figure 2:   Mask layout of a typical GGFET.

Figure 3:   The Sensitivity of the GGFET as a function of the voltage

difference between the ends of the gate. Fig. 3a shows the

variation of sensitivity with gate voltage difference magnetic

field for different total channel currents.

Figure 4:   Theoretical and experimental sensitivities for the two-drain,

GGFET.

Figure 5:   Table of the sensitivity for a three-drain device.

Figure 6:   The Sensitivity of the graded source MAGFET, as a function of

the drain voltage.

REFERENCES

[1]   Baltes, H. and Popovic, R.S., Integrated Magnetic Field Sensors.

Proc. IEEE, vol 74, 1107-1132, (1986).

[2]   Misra D.,Viswanathan, T.R. & Heasell, E.L.   J. Sensors and Actuators,

vol 9, 213-221, (1986)

[3]   Fry, P.W. and Hoey, S.J. A silicon MOS magnetic field transducer of
        high sensitivity. Trans. Electron Dev. ED-16, 35-39, (1969).

[4]   B.S. Gill, The Graded-Gate FET, a novel MOS structure for magnetic
        field sensing applications. MaSc. Thesis, Electrical Eng. Dept. Univ.
        Waterloo, (1987).

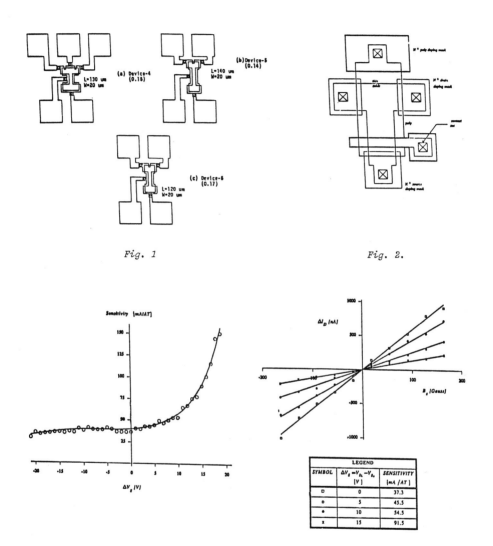

Fig. 1

Fig. 2.

Fig. 3 a.

Fig. 3 b.

LEGEND		
SYMBOL	$\Delta V_g = V_{g_L} - V_{g_s}$ [V]	SENSITIVITY [mA /AT]
□	0	37.3
○	5	45.5
●	10	54.5
x	15	91.5

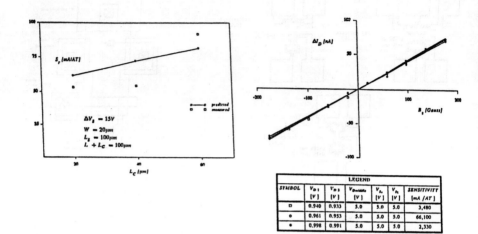

Fig. 4.

	LEGEND					
SYMBOL	V_{D1} [V]	V_{D2} [V]	$V_{Dmiddle}$ [V]	V_{S_0} [V]	V_{S_L} [V]	SENSITIVITY [mA /AT]
□	0.940	0.933	5.0	5.0	5.0	3,480
○	0.961	0.953	5.0	5.0	5.0	66,100
●	0.998	0.991	5.0	5.0	5.0	2,330

Fig. 5.

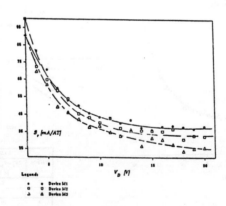

Legends
● — ● Device M1
□ — □ Device M2
△ — △ Device M3

Fig. 6.

Performance-Directed Synthesis of VLSI Systems

by

Jonathan Allen
Research Laboratory of Electronics
and
Department of Electrical Engineering and Computer Science
Massachusetts Institute of Technology
Cambridge, Massachusetts 02139

VLSI system design requires the specification and manipulation of several levels of representation. These include the function, architecture, logic, circuit, device, technology, and layout levels of representation. For a complete design, it is necessary to specify all levels of representation, although the design process may start at any of several different levels. Nevertheless, all levels must stayed aligned so that each level is a consistent projection of one single overall entity, the "complete design."

It is desirable to synthesize a complete design in a way that optimizes performance. Performance can be categorized in two areas. Conventionally, performance has centered around the circuit and layout levels of representation. This involves area, speed, and power of the circuit. In addition, there is the architectural aspect of performance, where parallelism can be introduced in order to provide increased throughput in system performance without the need to further increase circuit performance. In some application areas, such as real-time digital signal processing systems,[1] there is a need for high performance at both the circuit level and the architectural levels of representation. In light of this need to optimize performance in design, the question arises as to how performance can be considered in the overall synthesis process.

Historically, the relationship between performance and synthesis has evolved in three stages. During the first stage (which is still with us), it was necessary to synthesize the design through to the layout level, including possible fabrication, before the design could be analyzed for performance characteristics such as speed and power dissipation. This approach implies a long synthesis/analysis loop, where the entire synthesis process must be complete before it can be analyzed, usually by extensive circuit extraction from layout and simulation.

The second stage of performance-related synthesis is just beginning to emerge, and it involves more rapid feedback to the designer about the consequences of design decisions during the synthesis process. This stage involves the introduction of macromodelling techniques and macrogenerators, together with parametric performance characterizations. For example, accurate techniques for macromodelling of CMOS circuits have recently been developed[2] so that circuit delay can be estimated within 10% of comprehensive circuit simulation accuracy, but with three orders of magnitude less computational cost. In this way, it is possible to explore many different design alternatives and assess their performance based on several criteria. The result is that the synthesis/analysis loop is now much shorter, and many design alternatives can be accurately generated and characterized.

A third stage, which is just emerging, can be called "intelligent compilation." In this approach, the user submits a set of goals to the design system. The system then automatically seeks an optimum, or at least an acceptable solution, in terms of the initial design goals by internally generated exploration techniques. In this approach, it is necessary to develop representational and algorithmic techniques that permit the rapid assessment of design alternatives, as well search strategies that will lead to the best solution. Design systems with this capability are not currently available.

There are several desiderata attendant to performance-directed synthesis. These include performance, accuracy, cost, speed of design system computing, and the need for global (rather than only local) optimization. Several design issues arise from these desiderata. The first issue concerns how the optimization process is organized. In the so-called "horizontal" approach (which is usually the case), separate optimizations are performed at each level of representation. Unfortunately, this approach can lead to a

conflict between optimizations at two levels, such as the circuit and architecture levels. For example, it is possible to overly pipeline an array
multiplier using retiming transformations so that it is optimal at the
architecture level, but is slower in speed at the circuit level due to the
introduction of excessive levels of latching.[3] This conflict arises from the
granularity level of the representations, and has recently been attacked
through the "vertical" organization of representations.

In this approach, there is a three-way partitioning of the design
representations needed. First, the modules themselves (at all levels) are
considered as one unit. They may be optimized individually since all levels
of representation are present. The second component of the overall representation is the composition process itself, where the individual modules are
interconnected. This process involves placement and routing optimization
techniques which can be pursued independent of the modules. Finally, there
is a third component of the overall design process in this approach which is
concerned with module selection, and the exploration and assessment of the
overall design resulting from various compositions of the requisite modules.
This is a new approach which seems to have several advantages in terms of
rapid designs that yield a high level of performance.

From the previous discussion, it is clear that the individual representational domains overlap in terms of performance considerations. Each domain
(such as architecture, logic, circuit, and layout) has its own internal
well-formedness constraints and optimization strategies. Additionally, each
domain also has significant overlap with the other domains.

Optimizations at the circuit level overlap with those at the layout
level. For example, modern design tools allow the assessment of worst case
delay paths through a circuit, and provide corresponding device resizing
techniques which lead to the optimal speed of the circuit. On the other
hand, the consequence of these substantial device size changes can lead to
increased area and difficulty in the layout process, often with conflicting
requirements between circuit speed, layout size, and packing density.

Another example of optimization conflict in the representational domains is between the architecture and circuit level. As previously mentioned, architectural throughput is often enhanced through retiming transformations, which allow for the introduction and movement of registers within an overall design in a functionally invariant way. While these retiming transformations allow for the systematic realization of optimal throughput through constructive techniques, they do not (in themselves) allow for the characterization of circuit performance, particularly with respect to the time taken in combinational logic versus the latching of results in registers.

A final example of the optimization conflict at the different representational domains is between the logic and circuit levels. There is a variety of techniques for logic optimization, both in terms of two-level logic (e. g., program logic arrays) and in terms of multi-level logic, which is usually realized in some form of generalized inverter-type circuit. Optimization at the logic level may involve the minimization of product terms in the logic equations, or the minimization of logic gate count. But, these optimizations do not countenance circuit performance issues, and are particularly lacking in their ability to consider both delay and area costs of interconnect. In fact, despite notable achievements in optimized logic design at the circuit level, the cost of interconnect has yet to be included in any design optimization strategy.

A second major issue in optimal system design is the level of abstraction. It is desirable to introduce abstraction so that it increases focus on the desiderata, while avoiding the need to focus on irrelevant detail in the optimization of these desiderata. For example, circuit simulation provides everything in terms of knowledge of the circuit, but very little from the design optimization standpoint. In other words, at great computational cost, circuit simulation can characterize all node voltages and loop currents in a circuit. But, it does not provide help to deal with the desiderata of optimization. Instead, modelling must be introduced, which attempts to retain the accuracy of simulation while avoiding the necessity of a completely detailed characterization of the circuit at the the circuit level of representation. These macro-abstractions must provide not only enough generality to cover a significant class of circuits, but also must be

specifically focused in order to be efficient and accurate. This has recently been achieved in new techniques for delay modelling of CMOS circuits
through the appropriate characterization of waveform changes and the loading
conditions encountered by active devices.

A third issue faced by contemporary performance optimization systems is
the distinction between algorithmic optimization procedures and rule-based
systems (sometimes referred to as "expert systems"). In general, it is
desirable to utilize mathematical results such as linear programming and
graph search techniques. Yet, it is unfortunate that very few design algorithms lead to known optima. Instead, heuristics must be introduced to avoid
the complexity and computational cost of these algorithms. Nevertheless, it
has generally been found that specially designed algorithms, related to the
optimization criteria and the representational levels involved, perform
better than more generic techniques that can cover many classes of problems,
but for which there is no guarantee that a heuristically based algorithm will
not perform better. The current debate between the virtues of simulated annealing (which can be applied to many design problems) and specially focused
algorithms illustrates this conflict and controversy.[4]

From all of these considerations, it is clear that new performance-
directed synthesis techniques must: look for "reduced" abstractions which
avoid excessive detail, introduce powerful heuristic algorithms and general
mathematical tools, and focus on the performance attributes implied by the
desiderata of interest. This results in two modelling issues. On one hand,
the design procedures themselves must be developed and modelled, but the
design objects of interest must also be represented in a form that can be
readily manipulated by the optimization algorithms. For example, circuits
must be represented in a form that is physically realistic but computationally tractable within the optimization process. Insight should be
provided, but accuracy is also needed, and sometimes these are in conflict.

The remainder of this paper cites several design trade-offs at the
device and technology levels in terms of their implication on circuit performance for silicon-based CMOS design styles. It is unfortunate that currently
there is no adequate design strategy which allows for the systematic

manipulation of device and technology variables in terms of their impact on circuit design. The interactions between these variables are now beginning to become codified, and much new device-related information must be derived for submicron short-channel devices, as well as those with very thin gate dielectrics.

In light of the rapid evolution of device technology, it is important to assess its impact on circuit performance. In particular, it is important to consider the impact of new innovations at the device level in terms of their effect in realistic circuits, as opposed to test structures which are never utilized in practice. For this purpose, MOS circuits can be loosely classified into logic, memory, and analog circuits. At a more refined level, these may be cast into a seven-part framework involving densely packed static logic, dynamic logic, analog signal processing circuits, analog interface circuits, regular structures, bus-dominated logic, and special memories. The first three can be considered design intensive, since device and circuit design play a dominant role in their performance. On the other hand, bus-dominated logic, regular structures, special memories, and analog interface circuits can be considered process technology intensive, since performance generally depends on technological considerations such as interconnect pitch and multi-level metallization.

Analysis of the interaction of device characteristics with circuit performance is an important and ongoing area of research. Much has been discovered concerning the above-mentioned classes of circuits.[5] In this area, we concentrate on results corresponding to densely packed static logic. In order to characterize circuit performance, it is important to focus on device performance parameters, which in turn will be related to device design parameters. Thus, for densely packed static logic circuits (where speed is the optimization criterion), the circuit parameters include the drain saturation current, input gate loading capacitance, parasitic external loading capacitance, and supply voltage. These circuit parameters must be related to device design parameters, and then optimized to give the best overall circuit performance.

The relevant device design parameters include channel length and gate dielectric thickness. It has been found that for long channel lengths, the saturation current is less than linear with the reciprocal of channel length due to velocity saturation effects. It has also been found that saturation current increases linearly with the reduction of dielectric thickness for a given channel length. Additionally, the intrinsic gate capacitance depends on the gate area and oxide thickness; whereas the extrinsic loading capacitance is technology-dependent, and related to the interconnect technology. By conducting parametric studies, it is possible to use the saturation current and capacitive loading as circuit parameters in order to calculate propagation delay versus channel length, gate oxide thickness, and power supply voltage for different extrinsic circuit loads.

From these calculations, one finds that as the dielectric thickness is scaled down, the intrinsic gate capacitance increases at approximately the same rate as the saturation current. Thus, for a circuit dominated by intrinsic gate capacitance, oxide thickness scaling does not affect speed, but scaling the channel length does. On the other hand, if a circuit is dominated by extrinsic loading capacitance, then reduction of the dielectric thickness or channel length will increase the speed of the circuit. In any case, down-scaling a channel length is not beneficial if the oxide gate thickness is not properly scaled in a coordinated manner, since saturation current is less sensitive to the channel length for thicker gate oxide.

These recently obtained observations provide the required insight for the incorporation of device design parameters in an overall optimized system performance strategy. Design tools that permit the exploration of the device design parameter space, and how these factors influence overall system performance, are needed in order to guide new research in device innovation and enhancement. In this way, device design can take on an increasingly active role in the overall optimizations of integrated circuit systems, thus avoiding excessive effort focused on device characteristics which have minimal affect on overall system performance. The challenge for the future will be to codify these effects, and to incorporate them within the overall strategy for global optimization that involves interactions between all levels of design representation, including the fundamental importance of modern electronic devices.

References

1. Allen, J. "Computer Architecture for Digital Signal Processing," Proc. IEEE, vol. 73, no. 5, May 1985, pages 852 - 873.

2. Brocco, L. M. "Macromodeling CMOS Circuits for Timing Simulation," Research Laboratory of Electronics Technical Report 529, June 1987, Massachusetts Institute of Technology.

3. Hauck, C. E., C. S. Bamji, and J. Allen. "The Systematic Exploration of Pipelined Array Multiplier Performance." Proc. International Conference on Acoustics, Speech, and Signal Processing, 1985.

4. van Laarhoven, P. J. M., and E. H. L. Aarts. "Simulated Annealing: Theory and Applications." D. Reidel, 1987.

5. Sodini, C. G., S. S. Wong, and P. K. Ko. "Technology and Device Design Requirements for MOS Integrated Circuits," in preparation.

A STEADY-STATE MODEL FOR THE INSULATED GATE BIPOLAR TRANSISTOR

A.R. Hefner, Jr.[+], D.L. Blackburn[+], and K.F. Galloway[†]

[+]Semiconductor Electronics Division
National Bureau of Standards
Gaithersburg, MD, U.S.A., 20899

[†]Electrical and Computer Engineering Department
University of Arizona
Tucson, AZ, U.S.A., 85721

ABSTRACT

The power Insulated Gate Bipolar Transistor (IGBT) is a switching device designed to overcome the high on-state loss of the power MOSFET. The IGBT behaves as a bipolar transistor which is supplied base current by a MOSFET. The bipolar transistor of the IGBT has a wide base with the base contact at the collector edge of the base and is operated with its base in high-level injection. The usual bipolar transistor models are not adequate for the IGBT. This paper describes a model for the IGBT developed using ambipolar transport.

INTRODUCTION

A power device structure has been introduced which is designed to overcome the high on-state loss of the power MOSFET while maintaining the simple gate drive requirements of that device [1, 2]. The device is controlled at the input by a voltage, such as for a MOSFET, but the output current is characteristic of that of a bipolar transistor, hence the name Insulated Gate Bipolar Transistor (IGBT). The IGBT is a combination of both bipolar and MOSFET structures and takes advantage of the best features of both device types. This device is expected to become an important competitor for switching applications in the $100 \, kHz$, $1000 \, V$, $100 \, A$ range.

A schematic of the structure of one of the several thousand cells of an n-channel IGBT is shown in fig.1. This structure is similar to that of the VDMOSFET (Vertical Double diffused MOSFET) with the exception that a p-type, heavily doped substrate replaces the n-type drain contact of the conventional VDMOSFET. In the VDMOSFET, the current that flows through the MOSFET channels at the surface, enters the thick, low-doped epitaxial layer, and appears as drain current at the substrate. The thick, lightly doped epitaxial layer of the VDMOSFET is necessary to support high voltages in the forward blocking mode, but it contributes to a large on-state resistance. The additional p-n junction (p-substrate n-epitaxial layer) of the IGBT serves to reduce the on-state resistance by injecting minority carriers into the epitaxial layer, which modulates its conductivity.

The excess minority carriers which are injected into the epitaxial layer of the IGBT drift and diffuse through the epitaxial layer from the substrate toward the body. The excess carriers which reach the body (those which do not recombine in the base) are collected by the body-epitaxial layer junction which is reversed biased for forward conduction (anode positive). The MOSFET which is formed under the gate where the body meets the surface supplies the majority carriers for recombination in the epitaxial layer and for back injection into the substrate. Thus, the IGBT functions as a bipolar transistor which is supplied base current by a MOSFET [3]. This basic equivalent circuit of the IGBT is shown in fig. 2 and the regions of each of these components are labeled on the right half of fig. 1.

In this paper, an analytic model is derived for the IGBT. Because of the structure of the IGBT, the bipolar transistor of the IGBT cannot be treated in itional ways. The treatment necessary for describing the bipolar transistor of the IGBT is presented and the model is verified experimentally for steady-state operation.

THE BASIC IGBT MODEL

Several analytic models for the IGBT have been proposed which use the equivalent circuit of fig. 2 [4,5]. However, these models treat the bipolar transistor in traditional ways and cannot consistently describe all of the IGBT characteristics. Principally, the bipolar transistor of the IGBT consists of a low-doped, wide base region with the base contact at the collector end of the base. This bipolar transistor has a low gain ($I_n \sim I_p$) and is in high-level injection for the practical current density range of the IGBT. In general, low-gain bipolar transistors have not been important and their high-level injection characteristics have not been adequately described. In the traditional bipolar transistor models, the high-level injection characteristics are obtained by assuming that the majority carrier current is much less than the minority carrier current in the base (e.g., $I_n \ll I_p$ for a p-n-p bipolar transistor) [6,7]; but for the IGBT, this approximation is in contradiction with the low-gain condition. To describe the low-gain, high-level injection characteristics of the bipolar transistor of the IGBT, ambipolar transport must be used to describe the transport of electrons and holes in the base [8,9].

The following description is for an n-channel IGBT. The p-type substrate is the bipolar emitter and is the anode terminal of the device. The substrate-epitaxial layer (bipolar emitter-base) junction is forward biased when the IGBT is forward biased (anode positive), so holes are injected into the epitaxial layer (bipolar base) from the substrate (bipolar emitter). The injected holes drift and diffuse through the base and some recombine with electrons supplied by the MOSFET. The holes which reach the epitaxial layer-body (bipolar base-collector) junction, which is reverse biased by the drain-source voltage of the MOSFET, are swept into the collector and make up the bipolar transistor collector current.

Because of the IGBT structure, the bipolar transistor base current (electrons) supplied by the MOSFET is introduced at the collector end of the base. In the model, the region of the device at the epitaxial layer edge of the epitaxial layer-body junction, where the excess carrier concentration is zero, is designated as the contact between the bipolar transistor base and the MOSFET drain. The excess carrier concentration is zero at the epitaxial layer edge of the epitaxial layer-body junction because this junction is reverse biased when the IGBT is forward biased. The electron current that enters at this point is equal to the MOSFET current and the hole current at this point is the collector current of the bipolar transistor.

Holes injected from the emitter either travel across the entire base width and become collector current or they recombine as they travel through the base with electrons supplied by the MOSFET. The electrons enter the base at the base side of the collector-base depletion region and either recombine as they travel across the base to the emitter or are injected into the emitter, where they recombine.

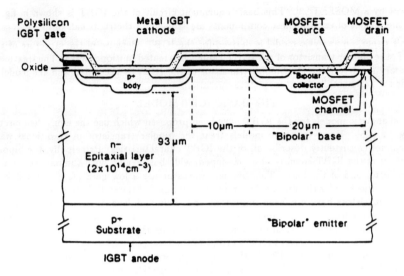

Fig. 1 A diagram of one of the diffused cells
of an n-channel IGBT.

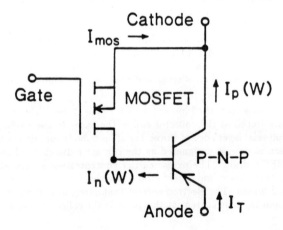

Fig. 2 The equivalent circuit model of the IGBT.

Because the excess carrier concentration is zero at the base contact, the electron and hole quasi-fermi potentials coincide with their charge-neutral, thermal-equilibrium values just as they do at an ohmic contact. The bipolar transistor emitter-contact-to-base-contact voltage difference thus includes the voltage drop through the emitter base junction and through the modulated epitaxial layer. The anode-to-cathode voltage drop of the IGBT is the sum of the bipolar transistor emitter-contact-to-base-contact voltage and the MOSFET drain-source voltage.

The bipolar transistor defined in the equivalent circuit consists of a planar emitter base junction, a wide base, and periodic base contacts at the collector junction. Wide-base bipolar transistors were not technologically important in the past due to their low gain, but the high-carrier level injection characteristics of the wide-base bipolar transistor dominates the operation of the IGBT throughout most of its operating range. Because the current gain is low and the base contact is at the collector end of the base, a proper treatment of the flow of base current through the base is essential to describe the high-level injection characteristics.

The bipolar transistor of the IGBT can be treated in one dimension because the lateral separation of the base contacts is typically an order of magnitude less than the base width and these base contacts are at the collector end of the base. Because the base contacts are at the collector, the base current flows from the collector through the base in the same direction as the collector current, and the total current (sum of the electron and hole currents) is constant throughout the base. Because the total current is constant throughout the base, the high-level injection characteristics of the bipolar transistor of the IGBT can be obtained by using the one-dimensional ambipolar transport equations to describe the transport in the base. The ambipolar transport equations are usually used to describe the transport in the base of a p-i-n diode [10]. In effect, the bipolar transistor of the IGBT is treated similarly to a p-i-n diode but with transistor boundary conditions.

Because the base is wide and the base doping concentration is low for the bipolar transistor of the IGBT, the concentration of injected carriers in most of the base becomes larger than the base background doping concentration at a small current density; e.g., approximately $0.2A/\ cm^2$ for an IGBT with the doping profile of fig. 1 and lifetime of 7.1 μs. When the excess carrier concentration is greater than the background doping concentration, the transport of electrons and the transport of holes are coupled by the electric field in the drift terms of the respective transport equations, and they cannot be treated separately.

In general, the electron and hole current densities are given by:

$$J_n = nq\mu_n E + qD_n \frac{dn}{dx} \tag{1}$$

$$J_p = pq\mu_p E - qD_p \frac{dp}{dx} \tag{2}$$

where the symbols used are defined in table 1. The first terms in eq 1 and eq 2 are due to drift and the second terms are due to diffusion. Under the high-gain conditions of the usual bipolar transistor analysis, these equations can be decoupled and the transport of minority carriers in the base is described by a simple expression, for both high- and low-level injection conditions.

However, for the low-gain, high-level injection conditions, the difference between the electron drift and diffusion currents is significant and the net electron current cannot be set to zero to approximate the electric field. In this case, the net electron current has a significant effect on the hole drift current, and the electron and hole transport equations cannot be decoupled. To describe transport in the base for this case, a simplifying transformation can be made to eqs 1 and 2 resulting in the ambipolar transport equations [11]. Assuming quasi-neutrality (i.e., $\delta n = \delta p$) and a high excess carrier level (i.e., $\delta n \gg N_B$), the current densities can be written for the ambipolar transport case by eliminating the electric field between eq 1 and eq 2:

$$J_n = \frac{b}{1+b}J_T + qD\frac{dn}{dx} \tag{3}$$

$$J_p = \frac{1}{1+b}J_T - qD\frac{dp}{dx}. \tag{4}$$

Notice that both of these expressions depend on the total current so that the transport of electrons and holes are coupled. For negligible electron current, the total current is approximately equal to the hole current and eq 4 reduces to the usual high-gain, high-level injection model [6]. But when the electron current is comparable to the hole current (e.g., in the base of the low gain bipolar transistor of the IGBT), it is apparent from this expression that the usual model is not valid.

The hole continuity equation is:

$$\frac{d\delta p}{dt} = -\frac{\delta p}{\tau_{HL}} - \frac{1}{q}\frac{dJ_p}{dx}. \tag{5}$$

From this equation and eq 4, the time-dependent ambipolar diffusion equation is obtained:

$$\frac{d^2\delta p}{dx^2} = \frac{\delta p}{L^2} + \frac{1}{D}\frac{d\delta p}{dt}. \tag{6}$$

A requirement in deriving this expression is that the total current (J_T) is independent of position in the base. This is satisfied in the IGBT because the base current flows from the collector through the base in the same direction as the collector current, whereas the base current of the usual bipolar transistor enters from the side in the middle of the base.

STEADY-STATE IGBT MODEL

In this section, a system of parametric equations is derived for the steady-state electron and hole current densities and excess carrier concentrations and for the emitter-base voltage of the bipolar transistor of the IGBT. These equations are obtained by solving the ambipolar transport equations for the boundary conditions of the bipolar transistor [8,9]. The analysis thus includes both the effects of the base transport factor and the emitter efficiency for

TABLE 1

Basic IGBT Model Symbol Definitions

J_n, J_p	electron, hole current density (A/cm^2)
n, p	electron, hole carrier concentration (cm^{-3})
δp	excess carrier concentration (cm^{-3})
ϕ_n, ϕ_p	electron, hole quasi-fermi potential (V)
E	electric field (V/cm)
ϵ_{si}	dielectric constant of silicon (F/cm)
q	electronic charge $(1.6 \times 10^{-19}\ C)$
μ_n, μ_p	electron, hole mobility (cm^2/Vs)
D_n, D_p	electron, hole diffusivity (cm^2/s)
τ_{HL}	high-level excess carrier lifetime (μs)
$J_T = J_n + J_p$	total current density (A/cm^2)
$b = \frac{\mu_n}{\mu_p}$	ambipolar mobility ratio
$D = 2\frac{D_n D_p}{D_n + D_p}$	ambipolar diffusivity (cm^2/s)
$L = \sqrt{D\tau_{HL}}$	ambipolar diffusion length (μm)
x	distance in base from emitter (μm)
W_B	metallurgical base width (μm)
W	quasi-neutral base width (μm)
N_B	base doping concentration (cm^{-3})
V_{bi}	built-in potential of the base–collector junction (V)
V_{bc}	applied base-collector potential (V)
I_{sne}	emitter electron saturation current (A)
V_{eb}	applied base-emitter potential (V)
V_A	device anode voltage (V)
R_s	device series resistance (Ω)
K_p	MOSFET channel transconductance (A/V^2)
V_T	MOSFET channel threshold voltage (V)

the low-gain, high-level injection case and describes the conductivity modulation of the epitaxial layer. The characteristics of the bipolar transistor are then combined with a MOSFET model to completely describe the steady-state current-voltage characteristics of the IGBT.

The analysis will be performed using the coordinate system defined in fig. 3. Defining $x = 0$ as the emitter edge of the base and $x = W$ as the collector edge of the quasi-neutral base, the steady-state boundary conditions for the excess carrier distribution are:

$$\delta p(W) = 0 \qquad (7a)$$

$$\delta p(0) \equiv P_0 \qquad (7b)$$

where P_0 is used as a parameter. The quasi-neutral base width is given by

$$W = W_B - \sqrt{\frac{2\epsilon_{si}(V_{bc} + V_{bi})}{qN_B}}, \qquad (8)$$

where $V_{bi} \approx 0.7\ V$. The second term on the right-hand-side of eq 8 is the base-collector depletion width (the collector doping concentration is much larger than that of the base). The effect of mobile carriers on the charge in the depletion region is ignored in eq 8.

Solving the steady-state ambipolar diffusion equation (eq 6 with $\frac{d\delta p}{dt} = 0$) in the base with these boundary conditions yields:

$$\delta p(x) = P_0 \frac{\sinh\left(\frac{W-x}{L}\right)}{\sinh\left(\frac{W}{L}\right)}. \qquad (9)$$

This equation describes the steady-state distribution of excess carriers in the base of the wide-base, high-level injection bipolar transistor. This assumes that $\delta p \gg N_B$ throughout most of the base and this is verified when the model is applied. Although for these boundary conditions the excess carrier concentration is not greater than the background doping concentration in a small region of the base near the collector, the electron and hole current densities cannot change abruptly. So when this region is small, its effect on the current densities and the carrier concentrations in the rest of the device is also small [12].

Using the quasi-equilibrium simplification [13] (i.e., the difference between the electron and hole quasi-fermi potentials is the same on both sides of the junction) and assuming high-level injection of holes into the base, the electron current at the emitter-base junction $(I_n(0))$ is related to P_0 by:

$$\frac{I_n(0)}{I_{sne}} = \exp\frac{q}{kT}(\phi_{pej} - \phi_{nej}) = \frac{P_0(N_B + P_0)}{n_i^2} \approx \frac{P_0^2}{n_i^2} \qquad (10)$$

where ϕ_{nej} and ϕ_{pej} are the electron and hole quasi-fermi potentials at the emitter base junction, respectively, and I_{sne} is the emitter electron saturation current. The approximate form follows from the previous assumption that the base is in high-level injection.

Combining eqs 3, 4, 9 and 10, the electron and hole currents in the base for an active area, A, are obtained:

$$I_n(x) = \frac{P_0^2 I_{sne}}{n_i^2} + \frac{qP_0AD}{L}\left(\coth(\frac{W}{L}) - \frac{\cosh(\frac{W-x}{L})}{\sinh(\frac{W}{L})}\right) \qquad (11)$$

$$I_p(x) = \frac{P_0^2 I_{sne}}{bn_i^2} + \frac{qP_0AD}{L}\left(\frac{\coth(\frac{W}{L})}{b} + \frac{\cosh(\frac{W-x}{L})}{\sinh(\frac{W}{L})}\right). \qquad (12)$$

These equations when evaluated at the collector edge of the base $(x = W)$ give the steady-state collector current $I_p(W)$ and base current $I_n(W)$ (fig. 2). These equations reduce to those for a narrow-base, high-level injection transistor [6] when $W/L \ll 1$; i.e., to those for the narrow base transistor obtained by setting the electron current to zero to approximate the electric field in the base.

The base contact has been defined to be at the collector edge of the quasi-neutral base where the electron quasi-fermi potential coincides with its charge-neutral, thermal-equilibrium value relative to the electrostatic potential. Because the electron quasi-fermi potential coincides with its charge-neutral, thermal-equilibrium value relative to the electrostatic potential at both the emitter and base contacts, the applied emitter-contact-to-base-contact potential is given by:

$$V_{eb} = (\phi_{pej} - \phi_{nej}) + (\phi_{nej} - \phi_{nb}), \qquad (13)$$

where $(\phi_{pej} - \phi_{nej})$ is the electron quasi-fermi potential drop in the emitter given by eq 10 (the hole quasi-fermi potential is constant in the heavily doped emitter and coincides with the electron quasi-fermi potential at the emitter contact) and $(\phi_{nej} - \phi_{nb})$ is the electron quasi-fermi potential drop across the quasi-neutral base.

The electron quasi-fermi potential gradient is related to the electron current and the electron concentration by [14]:

$$\frac{d\phi_n(x)}{dx} = -\frac{I_n(x)}{qA\mu_n n(x)}. \qquad (14)$$

The electron quasi-fermi potential drop across the quasi-neutral base is determined by integrating this equation between the emitter and the collector edges of the quasi-neutral base with $I_n(x)$ given by eq 11 and $n(x) = N_B + \delta p(x)$, where $\delta p(x)$ is given by eq 9 which assumes that most of the base is in high-level injection. The result of the integration is:

$$\phi_{nej} - \phi_{nb} = \frac{I_T W}{(1 + \frac{1}{b})\mu_n A q n_{eff}} - \frac{D}{\mu_n}\ln\frac{P_0 + N_B}{N_B} \qquad (15)$$

where

$$\frac{1}{n_{eff}} \equiv \frac{1}{W} \int_0^W \frac{dx}{N_B + \delta p(x)}$$

and using the carrier distribution of eq 9, n_{eff} has the analytical solution:

$$n_{eff} \equiv \frac{\dfrac{W}{2L}\sqrt{N_B^2 + P_0^2 \mathrm{csch}^2\left(\dfrac{W}{L}\right)}}{\mathrm{arctnh}\left[\dfrac{\sqrt{N_B^2 + P_0^2 \mathrm{csch}^2\left(\dfrac{W}{L}\right)} \cdot \tanh\left(\dfrac{W}{2L}\right)}{N_B + P_0 \mathrm{csch}\left(\dfrac{W}{L}\right) \tanh\left(\dfrac{W}{2L}\right)}\right]}. \tag{16}$$

For $P_0 \mathrm{csch}\left(\frac{W}{L}\right) \ll N_B$, the first term in eq 15 reduces to the resistance of the unmodulated epitaxial base times the electron current; and for $P_0 \mathrm{csch}\left(\frac{W}{L}\right) \gg N_B$, the quantity n_{eff} increases approximately as $P_0 / \ln\left(\frac{P_0}{N_B}\mathrm{csch}\left(\frac{W}{L}\right)\right)$. Thus, the first term in eq 15 describes the conductivity modulation of the epitaxial layer. The second term of eq 15 is due to the majority carrier concentration difference between the emitter and collector edges of the quasi-neutral base and is zero for low-level injection. The first and second terms of eq 15 cancel in a high-gain narrow base transistor, but the second term is always less than the first for the wide-base transistor.

Equations 8 through 16 are algebraic equations parameterized in terms of P_0, which can be combined to completely describe the steady-state I-V characteristics of the wide-base, high-level injection bipolar transistor. An expression for P_0 in terms of any of the terminal currents is obtained by using eq 11 or eq 12 or their sum evaluated at the collector edge of the quasi-neutral base to obtain the current of interest and by then solving the resulting quadratic equation in terms of P_0. An expression for V_{eb} in terms of P_0 is found using eqs 10, 13, 15, and 16. Using the expression found for P_0 in terms of I_T, an explicit expression for V_{eb} in terms of the total current is obtained.

The anode-cathode voltage drop of the IGBT is given by the sum of the emitter-base voltage, the MOSFET channel voltage, and series resistive drop:

$$V_A = V_{eb} + I_{mos} R_{mos} + I_T R_s. \tag{17}$$

The MOSFET current is equal to the base current, which is obtained by evaluating the expression for the electron current given by eq 11 at the collector, $I_{mos} = I_n(W)$ (fig. 2). The MOSFET is in its linear region when the IGBT is in its on-state and the MOSFET resistance is given by:

$$R_{mos} = \frac{1}{K_p(V_{gs} - V_T)}, \tag{18}$$

where the parameter K_p is equal to the product of the oxide capacitance, the surface electron mobility, and the effective width-to-length ratio of the MOSFET cells. The effect of gate voltage on the surface mobility is neglected. Thus, the IGBT on-state anode-cathode voltage drop is given explicitly in terms of anode current and gate voltage, assuming $V_{bc} \approx 0$ in eq 8.

The steady-state common emitter current gain of the bipolar transistor is defined as $\beta_{ss} \equiv I_p(W)/I_n(W)$ and is found in terms of P_0 by dividing eq 12 by eq 11, both evaluated at $x = W$. A quantity of particular interest for the IGBT is the gain enhancement factor $(1 + \beta_{ss})$, which is defined so that the total IGBT current is given by $I_T = (1 + \beta_{ss})I_{mos}$, where $I_{mos} \equiv I_n(W)$ is the MOSFET current. Using the expression found for P_0 in terms of $I_n(W)$, an explicit expression for I_T in terms of the MOSFET channel current is obtained. The IGBT is in its current saturation region when the MOSFET is in its saturation region. The saturation current of the MOSFET is given by:

$$I_{mos}^{sat} = \frac{K_p}{2}(V_{gs} - V_T)^2, \tag{19}$$

where saturation is assumed to be due to pinch-off. Using this expression and the expression for the total current in terms of the MOSFET current, the saturation current of the IGBT is described explicitly in terms of the gate voltage and anode voltage assuming $V_{bc} \approx V_A$ in eq 8.

EXPERIMENTAL VERIFICATION OF THE MODEL

In this section, the model is used to predict electrical characteristics of the IGBT. The experimental verification is performed for devices whose physical parameters are listed in tables 2 through 4 and for the different values of base lifetimes listed in table 5. The different values of lifetime were produced using reactor neutron irradiation [9].

A set of output characteristics of an IGBT is shown in fig. 4. The saturation current I_T^{sat} at a given gate voltage is defined as the current above which the anode voltage has little effect on the current. This value is noted on fig. 4 for a gate voltage of 6 V. The saturation occurs in the IGBT because the MOSFET saturates and the MOSFET saturation current, I_{mos}^{sat}, is described by eq 19. The MOSFET saturation current is multiplied by the gain of the bipolar transistor, $(1 + \beta_{ss})$, to obtain the IGBT saturation current. The value of the gain is obtained using eq 11 and eq 12 evaluated at $x = W$. The measured and predicted values of I_T^{sat} are shown in fig. 5. The calculations were made using the values in tables 2 through 5 with $V_{bc} \approx 0$ V. The value of K_p used in the predictions of fig. 5 was extracted before the lifetime was reduced by the irradiation and is not expected to change significantly for the neutron exposure levels of this experiment.

To illustrate the extraction procedure for K_p and V_T and to demonstrate the accuracy of the model, the values of the square root of I_{mos}^{sat} versus gate voltage are plotted in fig. 6 for devices before and after the lifetime was reduced by the neutron irradiation. These values are obtained by dividing the measured values of I_T^{sat} by the gain in terms of the total current calculated using the values in tables 2 through 5. This gain is shown in fig. 7 for several device lifetimes. From eq 19, it is expected that this square root of saturation current versus gate voltage should be linear, with the intercept of V_T and

TABLE 2
Physical Constants of Silicon at T=25C

n_i	$1.45 \times 10^{10}\ cm^{-3}$
μ_n	$1500\ cm^2/V\,s$
μ_p	$450\ cm^2/V\,s$
ϵ_{si}	$1.05 \times 10^{-12}\ F/cm$

TABLE 3
Measured Physical Device Parameters

N_B	$2 \times 10^{14}\ cm^{-3}$
A	$0.1\ cm^2$
W_B	$93\ \mu m$

TABLE 4
Electrically Extracted Physical Device Parameters

I_{sne}/A	$7.0 \times 10^{-13}\ A\ cm^{-2}$
K_p	$0.36\ A/V^2$
R_s	$15\ m\Omega$

TABLE 5
Different Base Lifetimes Studied

Device	$\tau_{HL}\ (\mu s)$	Neutron Dose (n/cm^2)
1	8.1	pre-rad
2	7.1	—
3	6.45	10^{11}
4	4.7	—
5	2.45	10^{12}
6	0.3	10^{13}

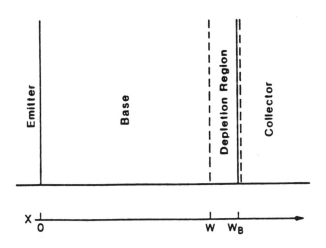

Fig. 3 Coordinate systems used in developing the basic
IGBT model.

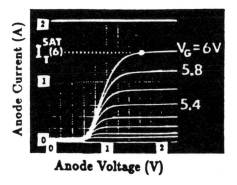

Fig. 4 An oscillogram of the steady-state characteristics
of an IGBT, defining the value of the IGBT saturation
current at a given gate voltage.

34

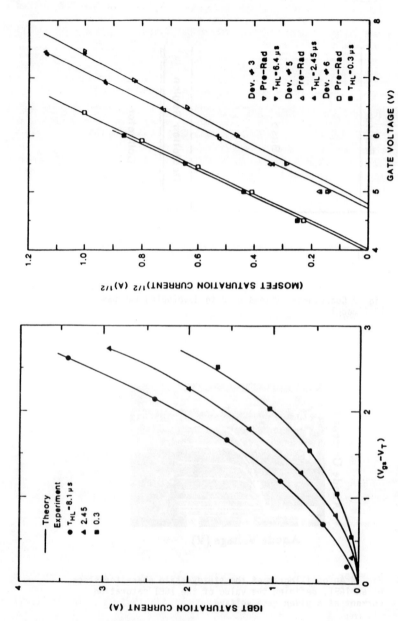

Fig. 5 The measured and predicted values of the IGBT saturation current versus gate voltage for IGBTs with different lifetimes.

Fig. 6 The calculated values of the square root of the MOSFET saturation current versus gate voltage for devices before and after the lifetime was reduced by neutron irradiation.

the slope of $\sqrt{K_p/2}$. This is observed and the value of K_p does not change for any of the neutron exposure levels of this experiment. The value of threshold voltage is only changed by 0.05 V for the highest exposure level. It follows that the values of the current gain predicted by the model are consistent with the assumption that the characteristics of the MOSFET channel do not change for the neutron exposure levels of this experiment. Previous work [15] suggests that this is a valid assumption.

The on-state voltage, neglecting the voltage drop in the MOSFET channel, is equal to the base-contact-to-emitter-contact voltage of the bipolar transistor. As an example, these ideal on-state voltage versus current characteristics are shown in fig. 8 with lifetime as a parameter. Figure 8 was generated using eqs 8 through 16 to obtain an expression for the emitter-base voltage in terms of the total current. At low currents ($I_T < 0.1$ A), the curves reflect the exponential voltage dependence of the high-level injection of holes from the emitter into the base; that is,each curve has a slope of $q\log_{10}(e)/(2kT)$ on the semi-log plot (dashed line on fig. 8).

The curves of fig. 8 roll off at higher current densities due to several different physical phenomena. In the intermediate current density range, the device with $\tau_{HL} = 0.1$ μs has an approximately ohmic on-state characteristic because of the resistance of the unmodulated base. The exponential dependence of current on voltage extends into the intermediate current range for the device with $\tau_{HL} = 8.1$ μs because the conductivity of the base is highly modulated. The remaining curves lie between these two extremes. At high currents ($I_T > 1$ A for the device with $\tau_{HL} = 8.1$ μs), the current due to injection into the emitter becomes the dominant component of the total current and the curves roll off because the total current increases much faster than n_{eff} and hence faster than the conductivity modulation. At even higher currents, the roll-off increases because of the reduction in the mobilities due to the so-called carrier-carrier scattering effect. This effect has been taken into account by using an effective reduced value of mobility calculated at each current point [16].

The on-state characteristics of the IGBT were measured for gate drive voltages from 10 V to 40 V. The on-state voltage shows a significant gate drive voltage dependence for on-state currents above 1 A due to the finite value of the MOSFET channel resistance. Therefore, to exemplify the accuracy of the wide-base bipolar transistor model, the measurements shown here are for a large gate voltage of $V_{gs} - V_T = 35$ V. The value of R_{mos} at $V_{gs} - V_T = 35$ V obtained from eq 18 using the value of K_p in table 4 is $R_{mos}(35\ V) = 71$ $m\Omega$. The theoretical and experimental values of the on-state voltage versus current at $V_{gs} - V_T = 35$ V are shown in fig. 9. The calculations were made using eqs 8 through 18. The value of the parasitic series resistance (R_s) was used to obtain the best fit. This resistance is due to the resistance of the bond wire (0.25 mm diameter), the spreading resistance at the bond, and metallization resistance. The value of $R_s = 15$ $m\Omega$ is reasonable for this device.

SUMMARY AND CONCLUSIONS

The IGBT behaves as a bipolar transistor which is supplied base current by a MOSFET. The bipolar transistor is a wide-base device operating in high-level injection with base contacts at the collector end of the base. Because the base contacts are at the collector end of the base and their separation is typically much less that the base width, the bipolar transistor can be treated in one dimension. Since the base of the bipolar transistor is in

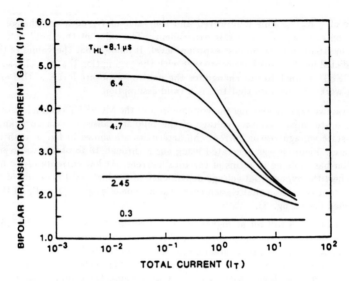

Fig. 7 The current gain versus total anode current for
devices with different lifetimes which was used to
calculate the MOSFET saturation current.

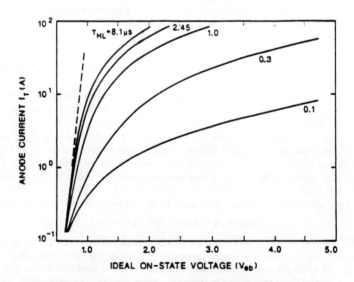

Fig. 8 The variation of the ideal (neglecting voltage drops
in the parasitics and in the MOSFET channel) on-stage
voltage versus anode current characteristic with the
epitaxial layer lifetime as a parameter.

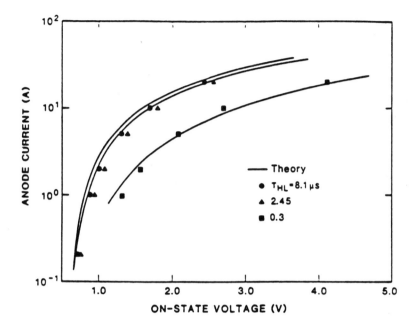

Fig. 9 Predicted and measured on-state voltage versus current for the devices with different lifetimes.

high-level injection, the transport of electrons and the transport of holes cannot be treated separately; because the base current flows from the collector, they can be described by the one-dimensional ambipolar transport equations.

An analytical model has been developed which describes the steady-state current-voltage characteristics of the IGBT. The basic element of the model is a detailed bipolar transistor analysis which uses ambipolar transport in the base. The steady-state model describes the unique features of the IGBT such as gain enhancement and conductivity modulation. A similar analytical model for the steady-state has also been presented [17]. This model [17], although formulated differently, uses the same basic principles discussed here for its development. The work described here and the work of reference 17 are the only analytical models for the IGBT to date which can consistently describe both the on-state and the saturation characteristics.

REFERENCES

1. J.P. Russell, A.M. Goodman, L.A. Goodman, and J.M. Neilson, "The COMFET–A New High Conductance MOS-Gated Device," IEEE Electron Device Lett., vol. EDL-4, 63 (1983).
2. B.J. Baliga, M.S. Adler, R.P. Love, P.V. Gray, and N.D. Zommer, "The Insulated Gate Transistor: A New Three-Terminal MOS-Controlled Bipolar Power Device," IEEE Trans. Electron Devices, vol. ED-31, 821 (1984).
3. H. Yilmaz, W.R. Van Dell, K. Owyang, and M.F. Chang, "Insulated Gate Transistor Modeling and Optimization," IEEE IEDM Tech. Dig., 274 (1984).
4. D.-S. Kuo, J.-Y. Choi, D. Giandomenico, C. Hu, S.P. Sapp, K.A. Sassaman, and R. Bregar, "Modeling the Turn-Off Characteristics of the Bipolar-MOS Transistor," IEEE Electron Dev. Lett., vol. EDL-6, 211 (1985).
5. B.J. Baliga, "Analysis of Insulated Gate Transistor Turn-Off Characteristics," IEEE Electron Dev. Lett., vol. EDL-6, 74 (1985).
6. W.M. Webster, "On the Variation of Junction-Transistor Current-Amplification Factor with Emitter Current," Proc. IRE, vol. 42, 914 (1954).
7. R.M. Warner and B.L. Grung, Transistor Fundamentals for the Intergated-Circuit Engineer. New York: Wiley, 1983, pp. 604-610.
8. A.R. Hefner and D.L. Blackburn, "Performance Trade-Off for the Insulated Gate Bipolar Transistor: Buffer Layer versus Base Lifetime Reduction," IEEE PESC Record, 27 (1986).
9. A.R. Hefner, D.L. Blackburn, and K.F. Galloway, "The Effect of Neutrons on the Characteristics of the Insulated Gate Bipolar Transistor (IGBT)," IEEE Trans. Nucl. Sci., vol. NS-33, 1428 (1986).
10. F. Berz, "A Simplified Theory of the p-i-n Diode," Solid-State Electron., vol. 20, 709 (1977).
11. W. van Roosbroeck, "The Transport of Added Current Carriers in a Homogeneous Semiconductor," Physical Review, vol. 91, 282 (1953).
12. P.E. Gray, D. DeWitt, A.R. Boothroyd, and J.F. Gibbons, Physical Electronics and Circuit Models of Transistors. New York: Wiley, 1964, pp. 12-13.
13. A.S. Grove, Physics and Technology of Semiconductor Devices. New York: Wiley, 1967, pp. 184-185.
14. W. Shockley, Electrons and Holes in Semiconductors. New York: D. Van Nostrand, 1956, pp. 302-308.
15. D.L. Blackburn, T.C. Robbins, and K.F. Galloway, "VDMOS Power Transistor Drain-Source Resistance Radiation Dependence," IEEE Trans. Nucl. Sci., vol. NS-28, 4380 (1981).
16. S. Selberherr, Analysis and Simulation of Semiconductor Devices. Wien, New York: Springer-Verlag, 1984, pp. 90-92.
17. D. Kuo and C. Hu, "An Analytical Model for the Power Bipolar-MOS Transistor," Solid State Electron., vol. 29, 1229 (1986).

RECENT DEVELOPMENTS IN PHYSICS

OF SILICON DI-OXIDE VERY THIN FILMS FOR VLSI.

NEW OXIDATION MODELS

AND ELECTRICAL CHARACTERIZATION

Georges KAMARINOS

Laboratoire de Physique des Composants à Semiconducteurs
ENSERG - INPG. UA CNRS 840
23, Rue des Martyrs
38031 GRENOBLE - CEDEX FRANCE
Tél. 76.87.69.76

INTRODUCTION

The Physics and the Technology of Silicon dioxide is the center of interest for the development of Silicon devices and Si Integrated Circuits [1].
In addition the progress of Technology and the scalling of devices need more and more thin oxide, or generally dielectric, (δ < 100 Å) layers.

This paper is divided in two main chapters :

In the first we expose the theory of the oxidation under stress which concerns rather the thin thermal grown oxide films ; In the second chapter we examine the influence of differents preparation methods on electrical parameters of very thin oxides (δ < 100 Å).

1. OXIDATION MODEL BASED ON STRESS RELAXATION

1.1. EVIDENCE OF A DECREASE OF DIFFUSIVITY IN THE Si-SiO$_2$ INTERFACE

It is well known that the initial regime of dry thermal Silicon oxidation is characterized by an anomalously rapid rate of growth. This effect has been much discussed and several models have been proposed [2...7].

In contrast to the cited approches GHIBAUDO and al in the LPCS Laboratory have developped a complete theory [8, 9, 10, 11], now validated by experiments, which is based in the assumption that, in the vinicity of the Si-SiO$_2$ interface, the rate of diffusion of O$_2$ is reduced.

The figure 1 gives typical experimental curves of the inverse of the thickness growth of the oxide.
The figure 2 gives the evidence that only a model which assumes a reduced diffusivity in the interface can explain the experimental data. The decrease in diffusivity can be attributed to the presence of high compressive stress at the interface arising from the difference in volume between Silicon and Silica.

fig. 1   fig. 2

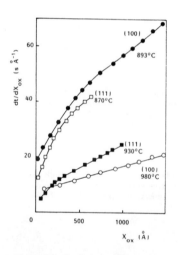

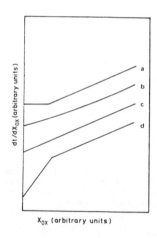

Typical experimental plots of the inverse of the oxidation rate dt/dX$_{ox}$ as a function of oxide thickness X$_{ox}$ at different oxidation temperatures for dry silicon oxidation (from Fargeix et al. (1983)).

Schematic comparison of the variation of dt/dX$_{ox}$ as obtained by different models: (a) Revesz and Evans (1969);(b) Blanc (1979) and Hu (1984); (c) Deal and Grove (1965); (d) Fargeix et al. (1983)

1.2. THE STRESS-STATE MODEL OF OXYDATION

After the observation of fig.2 G. GHIBAUDO (et al) has developed a new model that takes account of a stress-dependent diffusivity of the oxidizing species and a Maxwellian stress relaxation by viscous flow.
The basic assumptions of the model are :

(a) Deal-Grove process ; the oxygen diffuses through the oxide and reacts with Si in the interface.
So the oxidation rate in the inhomogeneous oxide film can be expressed by :

$$\frac{dt}{dX_{ox}}(X_{ox}) = \frac{1}{k_L} + \frac{X_{ox}}{vCD_{eff}(X_{ox})},$$

where C is the outer concentration of O_2 ; v the volume change and D_{eff} an effective diffusivity across the oxide of thickness X_{ox}.

(b) Coupling between the oxidant diffusivity and the amount of stress within the oxide layer :

$$D = D_0 \exp(-\sigma_{ox}\Delta V_{ox}/kT),$$

where σ_{ox} is the stress in the oxide and ΔV_{ox} the diffusion volume in the oxide.

(c) Stress relaxation mechanisms following a Maxwellian law :

$$\sigma_{ox} = \sigma_{ox_0} \exp\left(-\frac{t}{\tau}\right) \quad ; \quad \tau = \frac{\eta}{\mu} \quad ;$$

η is the viscosity and μ is the shear modulus.

It is worth noticing that this assumption is verified by the experimental in situ plot of the refractive index n with annealing time (fig. 3) [12].

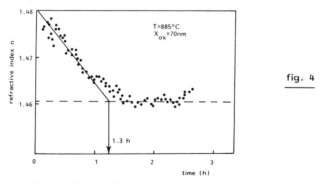

fig. 4

Experimental variation of the refractive index n with annealing time t as measured by in situ ellipsometry on low-temperature-grown oxides (from G. Ghibaudo and A. Straboni 1986, unpublished work).

The same workers in the LPCS Laboratory have prooved that a quantitative account of the reduction in diffusivity with the amount of stress can be obtained if one assumes that the activation energy of the diffusion process is shifted by the mechanical free-energy change. The explanation of the temperature dependence of the constants of Deal-Grove model is also advanced.

2. INFLUENCE OF DIFFERENT PREPARATION METHODS ON INTERFACIAL (SiO_2/Si) PARAMETERS OF VERY THIN SiO_2 LAYERS

2.1. IMPORTANCE OF VERY THIN SiO_2 LAYERS AND THEIR CHARACTERIZATION

The very thin Silicon oxide layers (thickness less than 100 Å) have a growing importance for the VLSI integrated circuits. Indeed they are used in EEPROM (memories), in a variety of bistable devices (as the MISS for example). Besides they could be used as intermediate dielectric tunnelable layers serving to enhance injection in Schottky contacts for bipolar circuit applications.

Nevertheless their fabrication is not yet standardized and different technics are in competition. In this chapter we compare the electrical properties of thin Silicon oxide films obtained by four different methods :

1. Thermal dry oxidation under low oxygen pressure (LPO_2)
2. LPCVD
3. Anodic oxidation
4. Dry thermal oxidation under chlorine ambient

The interfacial states density (and their energy distribution and cross sections) as well as the semiconductor surface potential and the voltage drop across the oxide, have been measured by a variety of experimental methods exploiting the static I-V and dynamic C-V and G-V measurements in various temperatures (77K $\leq$ T $\leq$ 300K), and levels of illumination.

Besides internal photoemission effect in the range of UV up to I-R radiations has been used, in order to obtain the metal-oxide and oxide-semiconductor barrier heights.

2.2. DESCRIPTION OF THE SAMPLES

The very thin (thickness 25 Å $\leq$ δ $\leq$ 100 Å) oxide layers have been obtained on Si by four different methods :

a) Thermal oxidation under low oxygen pressure (LPO_2). The silicon substrate (<100> oriented n-type silicon ; N_d = 10^{15} cm^{-3}) is oxidized at T = 950°C under a pressure of 1.0 Torr, during 25 minutes (δ = 25 Å).

b) Low Pressure Chemical Vapor Deposition of SiO_2 (LPCVD) [13]. The SiO_2 film is obtained by chemical reaction in vapor phase of $SiH_{2}Cl_2$ and N_2O at T = 880°C under a pressure of 0.2 Torr during 9 minutes (δ = 25 Å) ; it was afterwards deposited on <100> oriented n-type silicon substrate with N_d = 10^{15} cm^{-3}. The 25 Å tunnel devices of both technologies were fabricated after Al deposition on the oxide layer.

c) In order to obtain thin SiO_2 layers free of any contamination, anodic oxidation of silicon in ultra pure water (P = 18MΩ.cm at 20°C under a constant current density of 10μA/cm²) was used. Cr-SiO_2-Si or Al-SiO_2-Si capacitors of two different thickness (50 Å and 100 Å) were fabricated on 5Ωcm p-type <100> orien-ted silicon substrate [14].

d) Thermal oxidation under chlorine ambient[15] (0,2%HCL 10% O_2 and N_2). Cr-SiO_2-Si capacitors of several thickness (53 Å and 74 Å) were fabricated on p-type silicon <100> oriented with N_a = 10^{15} cm^{-3}.

2.3. I-V MEASUREMENTS

I-V characteristics allow us to test electrically the quality of the studied devices. In the case of very thin (30 Å) devices, the comparison between the expe-rimental and theoretical characteristics leads to the determination of interface states density using a detailed and original model of the working of MIS tunnel structures[13]. For the thicker devices (50 Å up to 100 Å) the Fowler Nordheim barrier height has been determined from these characteristics and has been found in good agreement with the barrier values obtained using optical measurements (section 2.4).

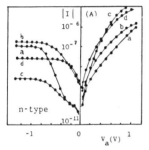

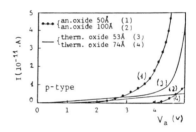

fig.4a) Experimental I-V characteristics of MIS Samples
a) LPO_2 non annealed samples
b) LPCVD non annealed samples
c) LPO_2 annealed samples
d) LPCVD annealed samples

fig.4b) Experimental I-V characteristics of anodic and thermal samples

2.4. OPTICAL MEASUREMENTS OF BARRIER HEIGHTS

The measurement of metal-oxide and oxide-semiconductor barrier height performed using internal photoemission effect, allows a complementary characteriza-tion of the quality of the insulating layers. In the case of tunnel devices, the effective metal-oxide barrier ϕ_m, as well as the Schottky barrier ϕ_{bn}, could be determined using chopped infrared radiation of photons energy hν<E_g from a mono-chromator and a silicon filter [fig.5]. ϕ_{bn} and ϕ_m values in the order of 0.1 eV and 0.8 eV have been so deduced in accordance with other results obtained using the conductance technique (section 2.5) and in connexion with the com-puter simulation of the I-V characteristics of the studied tunnel devices [16].

For the other devices, the barrier heights were measured using an UV ra-diation source with 230nm ≤ λ ≤ 300 nm.

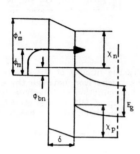

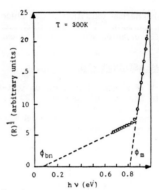

fig.5a *Energy band diagram of a MIS tunnel structure.* ϕ_m' : *Metal-oxide barrier height;* ϕ_m : *effective tunneling barrier height;* $\chi_n + E_g$: *semiconductor-oxide barrier*

fig.5b *Plot of the square root of the photoresponse* $R^{\frac{1}{2}}$ *vs photon energy* $h\nu$.

Concerning the thermal oxide devices (74 Å) the metal-oxide barrier height has been found in the range of 3eV (fig.6). The semiconductor-oxide barrier height has been determined in the range of 4eV. These value depends slowly on the applied voltage owing to the variation of the oxide layer trapezoidal barrier depth.

In the case of thinner thermal oxides (53 Å) the semiconductor-oxide barrier was found practically the same as for 74 Å thick oxides.

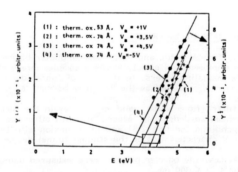

fig.6 : *Optical measurements of the metal-oxide and semiconductor-oxide barriers.*

2.5. INTERFACE STATES DENSITY MEASUREMENTS

The measurement of the interface states density N_{ss} near the silicon mid gap can be performed using the well know conductance technique [17].

In order to explore surface states located near the majority carrier energy band, we have applied these method in a range of temperature from 77 K to 300 K. The lowering of T provokes an increase of the surface states time-constant and it is possible to measure the N_{ss} using frequencies from 50Hz to 1 MHz [18].

The characterization of the other part of the gap near the minority carrier energy band can be done using the illumination technique [19]. In the presence of a sufficiently high illumination level and under reverse samples bias, it is possible to enhance exchanges between the interface states and the minority carriers energy band. By combining these two methods, a complete exploration of the silicon forbidden gap is possible[18].

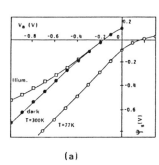 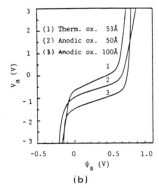

(a)   (b)

fig. 7 : Surface potential vs V_a for (a) LPCVD devices and (b) for anodic and thermal oxide devices

In order to obtain the density of states N_{ss} vs E, the surface potential $\Psi_s(V_a)$ has been determined using the Berglund integral method [20]. Fig. 7a shows typical $\Psi_s(V_a)$ plots for a very thin (25 Å) LPCVD oxide in darkness and under illumination at T = 300K. Furthermore, a $\Psi_s(V_a)$ characteristic at T = 77K is also shown. All these characteristics have been obtained by performing capacitance measurement at f = 50Hz. In the case of thermal or anodic thicker oxide devices, the $\Psi_s(V_a)$ plots have been obtained using quasistatic capacitance measurements. It is worth noticing that these devices reach the inversion regime. On the contrary, the very thin oxide (25 Å) devices do not present any inversion layer.

Fig.8a, 8b shows $N_{ss}(E)$ plots for tunnel oxide (25 Å) devices and Fig.9a, 9b for thicker oxide devices.

46

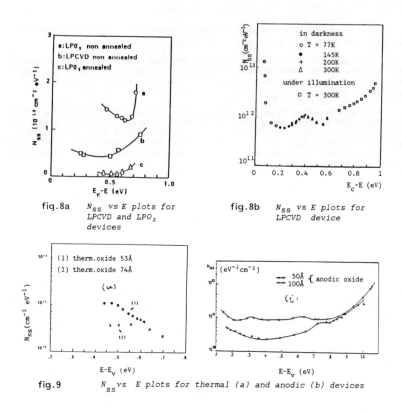

fig.8a N_{ss} vs E plots for
LPCVD and LPO$_2$
devices

fig.8b N_{ss} vs E plots for
LPCVD device

fig.9 N_{ss} vs E plots for thermal (a) and anodic (b) devices

2.6. COMPARISON OF SAMPLES

Our results show that in the case of very thin SiO$_2$ layers ($\delta \leq 30$ Å)
the deposited (LPCVD) oxides exhibit lower density of states and are more homogeneous than the thermally grown oxides (LPO$_2$).

In the case of thicker oxides ($50 \leq \delta \leq 100$ Å) we found that the interfacial parameters of anodic and thermal oxide are comparable concerning the interface states densities. However, it seems that the thermal oxides have more defects
(i.e pinholes) than the anodic insulating layers.

Agnowledgments : The author would like to thank Dr. G. GHIBAUDO and
Dr. G. PANANAKAKIS for numerous discussions and comments.

REFERENCES

[1] F.J. FEIGL, 1986, Phys. Tod. 39, 47.

[2] GROVE, A., 1969, Physics and Technology of Semiconductor Devices (New York : Wiley).

[3] TILLER, W., 1980, J. electrochem. Soc., 127, 625.

[4] REVESZ, A., and EVANS, R., 1969, J. Phys. Chem. Solids, 30, 551.

[5] BLANC, J., 1978, Appl. Phys. Lett., 33, 424.

[6] HU, S., 1984, J. appl. Phys., 55, 4055.

[7] REVESZ, A., MRSTIK, B., HUGUES, H., and Mc CARTY, D., 1986, J. electrochem. Soc., 133, 586.

[8] FARGEIX A., GHIBAUDO, G., and KAMARINOS, G., 1983, J. appl. Phys., 54, 2878.

[9] FARGEIX, A., and GHIBAUDO, G., 1983, J. appl. Phys., 54, 7153 ; 1984 a, Ibid 56, 589 ; 1984 b, J. Phys. D, 17, 2331.

[10] A. FARGEIX Thèse INPG, 1984, Grenoble.

[11] G. GHIBAUDO : Thèse INPG, 1985, Grenoble.

[12] G. GHIBAUDO : Phil. Mag B, 1987, 55, 147.

[13] G. PANANAKAKIS, G. KAMARINOS, M. EL-SAYED, Solid St. Electron, 1983, 26, 415.

[14] F. GASPARD, A. HALIMAOUI, G. SARRABAYROUSE, Revue Phys. Appl., 1987, 22, 65.

[15] J.L. PROM, F. ROSSEL, C. SOLANO, T. DOCONTO, G. SARRABAYROUSE, Journées GCIS, 18-19 Mai 1987, Toulouse.

[16] G. PANANAKAKIS, G. KAMARINOS, Surface Science, 1986, 168, 657.

[17] E.H. NICOLLIAN, A. GOETZBERGER, the Bell System J,1967, XLVI, 1055.

[18] M. EL-SAYED, G. PANANAKAKIS, G. KAMARINOS, Solid. St. Electron, 1985, 28, 345.

[19] T.C. POOM, H.C. CARD, J. Appl. Phys., 1985, 51, 6273.

[20] C.N. BERGLUND, IEEE Trans., 1966, ED-13, 701.

ELECTRICAL CHARACTERIZATION OF VERY THIN SI EPITAXIAL LAYERS USED FOR BIPOLAR VLSI

P. Spirito, S. Bellone
Electronic Dept. University of Naples, Naples, Italy
C. M. Ransom
IBM T. J. Watson Research Center, Yorktown Heights,N.Y. 10598
G. Busatto, G. Cocorullo
I.R.E.C.E, Naples, Italy

Abstract
Two new measurement techniques that allow to evaluate the recombination lifetime along epitaxial layers, and the recombination velocity at the high/low doping transition, are used to characterize the quality of very thin Si epitaxial layers used for bipolar technology. The experimental results show the capability of these techniques to give accurate and detailed informations on the quality of epi layers that could be useful in monitoring and improving the growth process. Moreover the measurements show that a simple metallization process often used in bipolar technology may introduce a large electrical damage in the whole epi layer underneath the high doped region thus worsening the electrical quality of the material itself.

1 Introduction

The growth of epitaxial layers of good quality and the realization of high/low transitions with sharp doping profile is essential in bipolar VLSI technology, because these regions will affect the performance of both the lateral PNP transistors and the vertical NPN transistor that form the basic I^2L logic cell. The "quality" of the epi layer is generally estimated by evaluating the doping profile /1,2/, and a "good" epi layer is assumed to be the one with a nearly constant doping profile along the layer and a sharp transition in doping from the buried layer or the substrate on which the epi layer is grown.

Techniques to provide a reduced out diffusion from the high doped substrate and to give rise to a "sharp" transition even in layers of submicrometer thickness are well known, but little information is reported on the recombination lifetime along the epi layer, and on the recombination at the epi/substrate transition. Only recently it has been recognized that defects at the transitions due to different process technologies could be responsible of an alteration of the quality of the epi layer; as an example it has been reported that the defects at the substrate interface due to strained lattice could have a gettering action on the recombination centers in the epi layer itself /3/.

This lack of informations is probably due to the difficulty to extract these data from diode measurements if one uses p/n junctions built into the epi layer as a test

---

* Work partially supported by grant from CNR - MEDESS

structure. From these measurements only "average" and qualitative informations on the recombination can be obtained, and the separation of the recombination at the epi/substrate interface from the one in the layer itself is very difficult to perform /4/.

In this paper we will use two recently proposed measurement techniques, respectively for the recombination lifetime along the epitaxial layer /5/, and the recombination at the epi/substrate interface/6/, to accurately evaluate the "quality" of very thin Si epi layers (< 1 μm) grown on low resistivity N+ substrates. Both measurements will use the same test structure, that is basically a p/n diode with a third (control) terminal used to control the distribution of minority carriers along the epi layer and to directly measure the relevant quantities related to the parameter we want to evaluate. The use of both techniques allow one to have two different parameters, namely the recombination lifetime profile along the epi layer and the recombination velocity at the substrate boundary, both related to the "quality" of the epi layer when used for making bipolar structures. As a case study, these measurements will be applied to the evaluation of two different metallization technologies, used in VLSI process.

2 Measurement principle

The test structure utilized for both measurements is basically formed of alternate P and N high doped regions built on the top surface of the epitaxial layer, with a third N+ region that could be either the substrate or the buried layer (see Fig. 1). The P+NN+ diode could be either the top surface

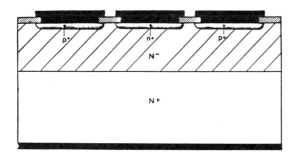

Fig. 1 - Schematic of the test structure

diode between the two following N and P doped regions, or the vertical structure formed between the top P region and the substrate. This diode structure is used to inject minority carriers into the epi layer, while the remaining N+ region (respectively the substrate or the top N region) acts as a control electrode, and it is used to carry on the two measurements.

In the first measurement technique, one can obtain the

recombination velocity at the epi/substrate interface if the structure is biased as in fig. 2a. Here the surface P+NN+ diode

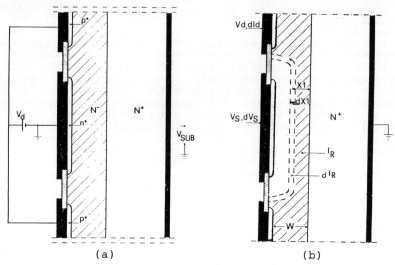

(a)   (b)

Fig 2 - a) Recombination velocity measurement; b) Lifetime
         profile measurement

is forward biased to inject holes in the epi layer, and the substrate acts as the control terminal, which is left open. By measuring the voltage V_{sub} between the substrate and the top N+ region one can obtain the hole gradient along the epi layer and hence   the   recombination   velocity   at   the   epi/substrate interface, as demonstrated in /6/.

     In the second measurement technique, the p/n diode is now the vertical P+NN+ structure created between the top region and the substrate (see fig. 2b). Here the top N+ region acts as the control electrode, and in this case it is positively biased with respect to the substrate so that minority carriers are repelled from it and confined to a distance X from the substrate, that shrinks as the control voltage is increased. By measuring the a.c. current dI_d induced in the diode mesh by a modulation dV_s of the control voltage one can obtain the lifetime at the abscissa X. Then, by varying the d.c. voltage V_s the entire layer can be scanned and the lifetime profile is obtained as demonstrated in /5/. The control terminal could in principle be the substrate as well, and the P+NN+ diode in this case is the surface structure; however the configuration of fig. 2b is preferred in our case because, as shown in /7/, the profiling accuracy is increased when the control voltage is increased and the recombination region width X is made thinner. So if one uses the top N+ region as the control electrode, the lifetime profile is more accurate near the substrate interface, where we espect a variation in recombination.

     In both measurements the voltage across the injecting

diode is merely used to inject minority carriers into the epi layer, and the measurements are made by detecting an electrical quantity that is more closely connected with the parameter under test. In the recombination velocity measurement it is the voltage V_{sub} detected at the substrate that is directly related to the voltage drop at the high/low transition and to the recombination velocity, while in the lifetime measurement the a.c. component dI_d is related to the recombination in the layer dX depleted by the a.c. voltage dV_s. Both quantities become small if the material under test is of good quality, namely if the recombination at the epi/substrate transition is negligible and if the lifetime in the epi layer is very high. Moreover these quantities are dependent on the thickness of the epi layer, and they become smaller if the thickness is reduced.

To asses the sensitivity of both measurements in the evaluation of epi layers intended for VLSI technology, a theoretical evaluation of the dependance of the measurable quantities V_{sub} and dI_d as a function of the parameter to be extracted has been done. The results are plotted in fig. 3 for the measurement of recombination velocity at the high/low transition, and in fig. 4 for the lifetime profile measurement. In both cases a thickness of 1 μm and a nominal doping of 1 10^{16} cm^{-3} was assumed for the epitaxial layer.

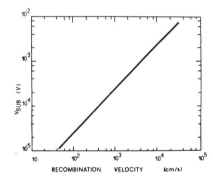

Fig. 3 - Sensitivity of the recombination velocity measurement

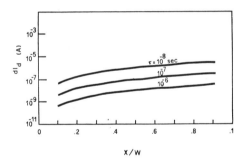

Fig. 4 - Sensitivity of the lifetime profile measurement

For the case of fig. 3, an injection level $P_0/N_d = 2$ was assumed (the measurement technique requires high injection condition to develop voltage drop across the high/low transition). From the plot it comes out that a detection of recombination velocities S in the range of 10 - 100 cm/sec is feasible in that it requires the measurement of voltages V_{sub} in the range of tens of microvolts, on the reach of sensitive voltmeters.

For the second technique, a low injection condition (P_0/N_d = .1) was assumed as the heaviest one, and the detectable quantity dI_d has been plotted versus the normalized abscissa

X/W of the epi layer, being W the thickness of the layer itself, for different lifetime values. It can be seen from fig. 4 that the lowest values of dI_d correspond to relatively high lifetimes and to sampled abscissas X near the substrate (X/W<< 1). However for a relatively high lifetime of 1 μs and at a distance less than .2 μm from the substrate is still possible to measure dI_d values in the range of nanoamps using a lock-in amplifier with a current imput.

It can be concluded that the above techniques are well suited even for the evaluation of very thin epi layers, with a sensitivity much greater than the one of alternative techniques that use simple p/n diodes for the same measurements.

3 Experimental

The wafers under test have been obtained by growing Si epitaxial layers 1.2 μm thick arsenic doped on low resistivity substrates <100> oriented and antimony doped with a resistivity of about 10 mohm-cm. Before growth all substrates were given a BHF dip, and a subsequent oxide layer 100 nm thick was grown by dry oxidation mixed with hydrocloridric acid. The entire oxide layer was then removed by BHF dip. After that all wafers received a high temperature prebake at 1150° for 15 min., and an undoped epitaxy was carrieed out for 2 min as a capping layer before doped layer deposition. The doped layer was finally deposited at 1050° for 7 min.

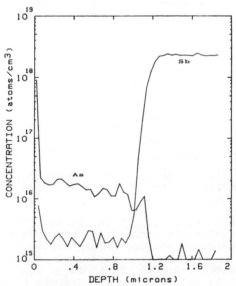

Fig. 5 - Doping profile of the As-grown epi layer, obtained by SIMS analysis.

In fig. 5 the SIMS profiles of the As and Sb concentration along the as-grown epi layer are reported. A doping concentration of $2 \cdot 10^{16}$ cm^{-3} was obtained near the surface, with an abrupt transition at the epitaxial-substrate interface, showing a two decades resistivity change in .2 micrometers.

The test structures were built into these wafers with standard planar technology, by implanting both the N+ and the P+ surface regions. The test pattern was made of many devices of different areas, composed of one or more basic cells as reported in fig. 1, with a stripe geometry for the top regions.

Two runs were carried out to study the effect of different metallization layers on the epitaxial recombination, as shown in fig. 6. For the first run the N+ top region was obtained by phosphorous implant, and a simple metallization with Al-Cu 5%/Si was used to contact both the N+ and P+ regions. After annealing the depth of the implanted regions, measured by SIMS analysis, was found to be .35 μm for the P+ implant and .3 μm for the N+ implant. In the second run an arsenic implant was used for the N+ top region, and a Ti/Al-Cu/Si metallization was given. In this case the depth of arsenic implant after annealing was found to be .18 μm.

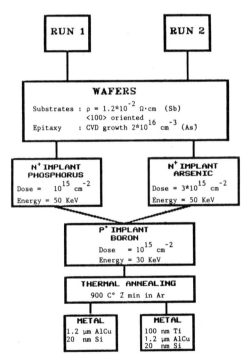

Fig. 6 - Process steps for runs 1 and 2.

The use of a Ti barrier in the metal contacts is known to reduce the contact resistance and avoid Si dissolution into the Al film /8/. In our experiment we wanted to see if the two metallization processes give comparable effects in the underneath epi layer, or if the former would degradate also the quality of the layer under the high doped region, and to what extent.

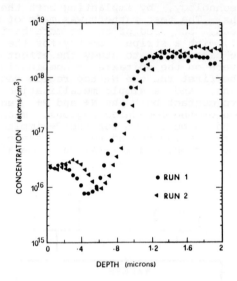

Fig. 7 - Doping profile of the two runs, after technological processes.

The doping profiles of the two runs after device fabrication, obtained by spreading resistance analysis, were almost identical, except for the two N+ implant regions, showing a diffusion of the substrate dopant into the epi due to the annealing step, and a widening of the transition as espected (see fig. 7). In spite of the very similar doping profiles, the P+NN+ structures between the top layer and the substrate showed very different behaviours. In fact the I-V curves in forward bias for samples of run 1 presented a poor behaviour and large recombination currents, while a good diode behaviour was obtained for samples of run 2.

4 Results

The recombination measurements with the techniques of sec.2 were carried out for both runs by using an automated measurement set-up controlled by computer through IEEE 488 interface.

The measurement of recombination velocity at the epitaxy/substrate interface allows one not only to evaluate the effective recombination velocity S at the transition, but also to separate it in the two components, namely the component S_r due to the recombination at the transition region and the component S_n due to the recombination in the high doped substrate, as shown in /6/. In our case the evaluation of the two components separately is important because the quality of the transition region is related only to the component S_r and not to the substrate component S_n. In Table I the values of S_r for both runs are reported, measured on 20 test structures for each run.

TABLE I

Recombination velocity S_r at the substrate transition

RUN 1		RUN 2	
S_r	Stand. dev. σ	S_r	Stand. dev. σ
$3.2 \cdot 10^4$	$5.4 \cdot 10^3$	< 100	--

The transition region of run 1 is much more defective than the one of run 2, the former being more than two orders of magnitude larger than the latter one. According to /9/, the recombination belocity in the transition region can be related to an "average" lifetime in this region as:

$$S_r = \frac{W_r}{\tau_{eff} \ln (N^+eff/N_d)} \tag{1}$$

where W_r is the width of the transition region, N^+eff is the effective doping of the high doped substrate, taking into account the bandgap narrowing correction, and N_d is the epi doping. In our case, assuming an average doping of epi layer 10^{16} cm^{-3} and a transition region width of .4 µm, from eq. (1)

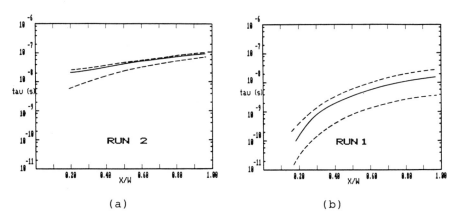

(a)   (b)

Fig. 8 - Lifetime profiles respectively (a) for run 2 and (b) for run 1. In each case the transition region is at X/W = 0.

it is seen that very low lifetime values (< 1 ns) have to be
assumed to give rise to the high S_r values shown by run 1.

The second measurement technique gives the lifetime
profile along the epi layer, and the results are plotted in
fig. 8a and b respectively for run 2 and 1. Here the normalized
distance X/W is the distance between the transition region
(located at X/W=0) and the bottom of the N^+ doped surface
region (located at X/W=1). The solid line curve is the profile
for the "mean" device, while the dashed line curves correspond
to the best and the worst case measured, over 15 devices.

For the case of run 2 (fig. 8a), the lifetime shows
acceptable values, in the range of 10^{-7} s, and a relatively
flat profile. The profiles do not show a strict correlation
with the doping profile because the drop toward the substrate
interface (X/W << 1) should suggest an increase in doping along
the epi layer while the plots of fig. 7 show a decrease in
doping. This drop is due to an increased defect density near
the transition region that probably came out from the substrate
during the epi growth.

For the case of run 1 (fig. 8b), the lifetime profiles
show a dramatic decrease in values respect to the plots of fig.
8a, demonstrating that the metallization layer of Al-Cu has
degraded significatively the whole epi layer. The large drop
of lifetime toward the substrate seems to be attributed to a
some "gettering" action of the substrate interface on the
defects that diffuse from the surface metal layer. It is
believed that this large increase of defect density should be
mainly attributed to Cu atoms, that have a very fast diffusion
coefficient even at the relatively low temperatures ($\approx 400°$)
used for sintering the metal layer. The lifetime values
measured near the substrate transition are in agreement with
the lifetime values given by eq. (1) for the corresponding S_r

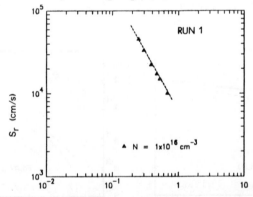

minority carrier lifetime (nsec) at x/w = 0.20

Fig. 9 - Correlation between recombination in the transition
region and lifetime in the epi layer for run 1.

measured. To proof further if the lifetime values near the
transition region are dependent on the quality of the

transition region itself, the values of the recombination velocity S_r measured on single devices of run 1 versus the corresponding lifetime measured at a distance of .2 μm from the transition have been reported in fig. 9. A strict correlation can be found between these two quantities suggesting that the quality of the interface is largely responsible of the recombination of the whole epi layer, due to the very small thickness of the layer itself.

5 Conclusion
The quality of thin epitaxial layers intended for bipolar applications is evaluated using two new measurement techniques that allow to extract separately the lifetime profile along the epi layer and the recombination velocity at the substrate transition. Results show that the lifetime profile is not strictly correlated with the doping profile and it can be largery altered by processes that do not change significantly the doping profile. Moreover the influence of the recombination at the substrate interface on the recombination in the epi layer is demonstrated, and a strict correlation between the two quantities has been found in the case of a large degradation.

BIBLIOGRAPHY

/1/ V.J. Silvestri, G.R. Srinivasan, B. Ginsberg, J. Electrochem. Soc., vol. 131 n. 4, p. 877, 1984

/2/ S.M Fisher, M.L. Hammond, N.P. Sandler, Solid State Technology, vol. 29 n. 1, pp. 107-112, 1986

/3/ H. Beneking, P. Narozny, P. Roentgen, M. Yoshida, IEEE Electron Device Letters, vol. EDL-7 n. 2, p. 101, 1986

/4/ F.N. Gonzales, A. Neugroschel, IEEE Trans. Electron Devices, vol. ED-31,pp. 413-416, 1984

/5/ P .Spirito, G.Cocorullo, IEEE Trans. Electron Devices, vol.ED-32, n.9, pp.1708-1713, 1985.

/6/ S. Bellone, A. Caruso, G.Vitale, IEEE Trans. Electron Devices, vol.ED-32, n.9, pp.1771-1775, 1985.

/7/ P.Spirito, G.Cocorullo, Proceedings of the Third International Workshop on Physics of Semiconductor Devices, Madras (India), Nov.27-Dec.2, 1985.

/8/ Y. Pauleau, Solid State Technology, vol. 30 n. 4, pp. 155-162, 1987

/9/ G.V. Ram, M.S. Tyagi, Solid State Electron., vol. 24, pp. 753-761, 1981

HEMT LSI: STATUS AND TRENDS

T. Mimura, M. Abe, J. Komeno, and K. Kondo

Fujitsu Laboratories Ltd., Fujitsu Limited

Atsugi 243-01, Japan

I. INTRODUCTION

Since the introduction of high electron mobility transistor (HEMT) 1 Kb SRAMs in 1984 [1], HEMT LSI technology has progressed rapidly. HEMT 4 Kb SRAMs were introduced later in 1984 [2], 1.5 K gate arrays in 1986 [3], and 4.1 K gate arrays, in 1987 [4], forming a new generation of high-speed low-power integrated circuits.

The main technological factors that have made such rapid progress possible are as follows: First, because HEMTs have a simple device structure, only a simple fabrication process is needed. Second, the short-channel effect in HEMTs is very small, even when the gate length is reduced to submicron dimensions. Third, because of the high low-field mobility and high Shottky-clamping voltage, noise margins large enough for fabricating LSI are easily obtained.

In this paper, we describe the current developments and trends in HEMT LSI technology, based on the main points above.

II. STATUS OF HEMT LSI

Fabrication technology

Direct-coupled FET logic (DCFL) circuitry is used in the basic gates of HEMT LSI, because it is simple and has a low power consumption. A vital issue when using DCFL circuitry is how to achieve uniform and controllable threshold voltages, V_{TH}, for the enchancement (E) and depletion (D) HEMTs. To do this, we introduced a unique epi-structure and fabrication process, as shown in Figure 1.

The basic epilayer structure consists of a 600 nm undoped GaAs layer, a 35 nm $Al_{0.3}GA_{0.7}As$ layer doped to 1.4×10^{18} cm^{-3} with Si, and a 70 nm GaAs cap layer sequentially grown on a semi-insulating substrate by molecular beam epitaxy (MBE). The electron mobility was found from Hall measurements to be 8000 $cm^2V^{-1}s^{-1}$ at 300 K, and 38000 $cm^2V^{-1}s^{-1}$ at 77 K. The concentration of two-dimensional electron gas (2DEG) was 1.1×10^{12} cm^{-2} at 300 K and 9.0×10^{11} cm^{-2} at 77 K.

A thin $Al_{0.3}Ga_{0.7}As$ layer is embedded in the cap GaAs layer if E- and D-HEMTs are to be fabricated on the same wafer to act as a stopper against selective dry etching. The fabrication for E- and D-FET circuits starts with ion implantation of O^+ down to the undoped GaAs layer, to localize the active region (Figure 1 (a)). The area next to the source and drain of the E- and D-HEMTs is metallized with a Au-Ge eutectic alloy, with Au overlay alloying, to form a good ohmic contact with the 2DEG (Figure 1 (b)). The gate pattern for E-HEMTs is defined by the photoresist. The thin $Al_{0.3}Ga_{0.7}As$ stopper is removed by wet chemical etching (Figure 1 (c)). After the gate pattern has been defined, selective dry etching using an etching gas composed of $CCl_2 F_2$ and He is carried out to remove the GaAs layer, exposing the top surfaces of the AlGaAs layers (Figure 1 (d)). Next, Al is deposited and lifted off to form the gate (Figure 1 (e)). Finally, a cross over insulator made of SiON is deposited by plasma-enhanced CVD. Electrical connections, composed of Ti-Pt-Au are provided through windows etched in the SiON film, from the interconnections to the device terminals (Figure 1 (f)). This unique epi-structure, using the selective dry etching technique, is essential for a simple fabrication process. If needs only six photomasks, and LSI circuits with precisely-controlled characteristics can be fabricated.

LSI implementation

The present state-of-the-art in LSI implementation using the fabrication technology described above, is a 4.1 K gate array (Figure 2). The basic gate is a 3-input NOR gate. The chip contains 4096 NOR gates, and measures 6.3 mm x 4.8 mm. A basic gate delay of 40 ps has been achieved. To test the gate array's performance, a 16 x 16-bit parallel multiplier was implemented. The multiplier, which uses 93% of the gate array, consists of registers for 16 x 16-bit multiplier products, 15 half-adders, 210 full-adders, and a carry-look-ahead circuit. The multiplication time for the critical path of the multiplier was 4.1 ns; the supply voltage was 1.1 V, and the power consumption was 6.2 W. The 4.1 ns multiplication time is the fastest one ever reported for a 16 x 16-bit multiplier.

To show clearly the improvements in performance that HEMT LSI circuits offer, Table 1 lists the main device parameters for the 1.5 K and 4.1 K gate arrays. The gate delay for the 4.1 K gate array with a standard load (FI = FO = 3, wiring length = 2 mm) was improved by a factor of about 1.7 over that of the 1.5 K array. This is because the 4.1 K gate array has a shorter gate length. The power dissipation of each gate of the 4.1 K gate array was reduced to one quarter of that of the 1.5 K gate array by using a smaller cell size.

Therefore, reducing device dimensions is essential for producing high performance HEMT LSI. Next, we discuss the effects of scaling-down on device performance.

III. PERFORMANCE OF SCALED-DOWN HEMTS

Short-channel effect

In FETs, short-channel effects, such as threshold voltage shifts, normally occur when the gate length is reduced. To suppress the short-channel effect in conventional FETs such as GaAs MESFETs, the doping concentration of the channel must be increased to raise the gate-to-channel capacitance. In this case, the electron mobility eventually decreases with increasing doping concentration in the channel, degrading the high-speed performance.

However the HEMT structure has the inherent advantage of reducing the short-channel effect. This is because the gate-to-channel capacitance can be increased by raising the doping concentration in the n-AlGaAs layer using the modulation doping technique. This shields the drain fields, without degrading the semiconductor mobility. This would give easily-designable and stable current-voltage characteristics for gates in the submicron range.

Figure 3 shows how the threshold voltage varies with gate length for both HEMTs and self-aligned gate GaAs MESFETs. It can be seen that the threshold voltage variation of a HEMT whose gate is in the submicron range is much smaller than that of a GaAs MESFET. For a HEMT, the variation was less than 30 mV when the gate length varied from 1.4 μm to 0.28 μm. Therefore, exising HEMTs can potentially allow 0.25 μm LSI to be produced.

Propagation delay time

It is very important that large enough noise margins are maintained to allow the LSI to operate stably. Figure 4 shows the basic propagation delay time, t_{pd}, versus the gate length, L_G, for both high- and low-level noise margins (larger than 200 mV). The supply voltage, V_{DD}, is 1 V. It can be seen for Figure 4 that t_{pd} increases proportionally with L_G. The average value of t_{pd} for L_G = 0.5 μm, over a 2-inch wafer, was 23 ps, with a standard deviation of 1 ps. The standard deviation at the high level was 12 mV, and that at the low level was 15 mV.

If we consider that Si dRAM will be the technology leader, we can expect a gate dimension of 0.5 μm at production level by 1992. We can anticipate that by then, HEMT 10 K gate arrays with a 100 ps propagation delay per gate, and HEMT 16 Kb SRAMs with 0.5 ns access time, will be feasible.

IV. MOCVD FOR HEMT LSI

The quality and quantity of the epitaxial wafers is a prerequisite for more advanced HEMT LSI development. At present, serious efforts are being made to produce commercial MBE machines, and research on metal organic chemical vapor deposition (MOCVD) technology is being carried out.

In this section we discuss successful 3 multi-wafer growth for HEMT LSI using MOCVD. Figure 5 shows the uniformity of the threshold voltage for HEMTs grown by MOCVD. The standard deviations over a 2-inch wafer were 23 mV for E-HEMTs and 35 mV for D-HEMTs. These values are comparable with those for state-of-the-art, MBE-grown wafers.

Table 2 shows the device characteristics of a MOCVD-grown HEMT. The transconductance was 250 mS/mm, and the current-gain cutoff frequency was 23 GHz for a HEMT with a gate length of 0.8 μm. These values compare favorably with those for MBE-grown HEMTs.

V. SUMMARY

In summary, we have described the current state of HEMT technology and its present trends. Factors affecting technology, such as simplifying LSI processing, achieving able characteristics in the scaling-down process, and developing flexible device structures will play a vital role in the future development of more advanced HEMT LSI. The success of multi-wafer growth of HEMT wafers by MOCVD indicates that large improvements in the throughput of HEMT LSI wafers can be expected in the near future. Considering the results to date, we believe that HEMT technology will be one of the most important semiconductor technologies that has emerged so for, and efforts are being made worldwide to develop it.

Acknowledgment

The authors wish to thank their colleagues, whose many contributions have made possible the results described here.

The present research effort is part of the National Research and Development Program on "Scientific Computing Systems", conducted under a program established by the Agency of Industrial Science and Technology, Ministry of International Trade and Industry.

References

[1] K. Nishiuchi et al.: ISSCC Dig. Tech. Papers, 48 (1984).

[2] S. Kuroda et al.: IEEE GaAs IC Symp. Dig. Tech. Papers, 125 (1984).

[3] Y. Watanabe et al.: ISSCC Dig. Tech. Papers, 80 (1986).

[4] K. Kajii et al.: CICC Dig. Tech. Papers, 199 (1987).

[5] J. Komeno et al.: EMC Tech. Program Abstract, E-8 (1987).

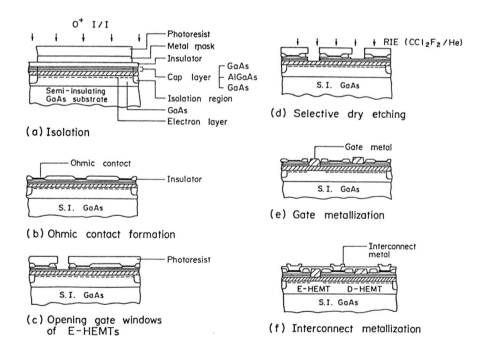

Figure 1 Basic processing steps for HEMT LSI

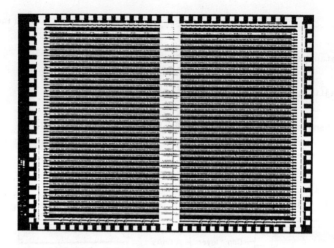

Figure 2   HEMT 4.1 K gate array

Table 1   Comparison of HEMT gate arrays

	1.5K	4.1K
Gate Length	1.2 μm	0.8 μm
Cell Size	132x54 μm	37.5x45 μm
Line/Space	2/2 μm	2/2 μm
t_{pd}/gate (FI/FO=3/3, l=2 mm)	178 ps	108 ps
P_d/gate	6.7 mW	1.6 mW

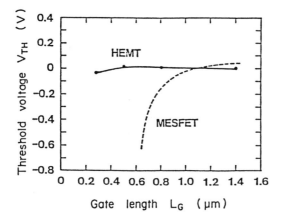

Figure 3   Threshold voltage, V_{TH}, versus gate length, L_G, for HEMTs and GaAs MESFETs.

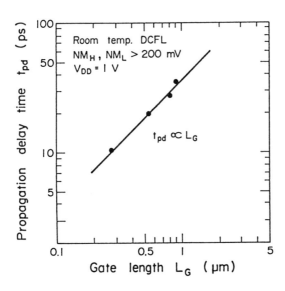

Figure 4   Propagation delay time, t_{pd}, versus gate length, L_G

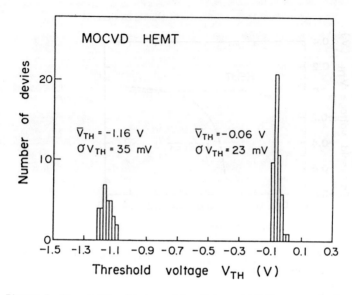

Figure 5   Threshold voltage uniformities of MOCVD-grown HEMTs over
a 2-inch wafer

Table 2   Device parameters of MOCVD-grown HEMT

Gate Length (L_G)	0.8 μm
Gate Width (W_G)	50 μm
Transconductance (g_m)	250 mS/mm
Current Gain Cut-off Frequency (f_T)	23 GHz

ISIT: Ultra High Speed Ballistic Devices

Jun-ichi Nishizawa

Research Institute of Electrical Communication
Tohoku University
2-1-1 Katahira, Sendai, 980, Japan

Abstract

This paper describes the ideal static induction transistor (ISIT) in which carriers travel in the channel without lattice scattering. The novel fabricated technology in which the processing temperature is below $450°C$ using GaAs and $Al_xGa_{1-x}As$ photo-stimulated molecular layer epitaxy (PMLE) is also introduced.

I. Introduction

The ideal static induction transistor (ISIT) in which the carriers travel without scattering by the lattice of a semiconductor crystal will be suitable device for the amplification up to 1000 GHz (1 THz) (1).

This device was proposed as the theoretical consideration to improve the SIT frequency performances prior to the ballistic transistor by Eastman et al. in 1979.

The SIT itself was invented by Y. Watanabe and J. Nishizawa in 1951 (3) and the SIT made of Si was realized in 1969 in Japan (1) by using the Si perfect crystal technology. Then Si power SIT, the integrated circuits such as I^2L (4), CML (5), CMOS (6), the photo SIT and its integrated image converter (7), the static induction thyristor (SIThy) and photo triggering and quenching SIThy have been developed as the SIT family devices. Their device performances have confirmed to be superior to those of the bipolar (BPT) and the field effect transistors (FET).

The SIT shows the unsaturated I-V characteristics never obtained by the former BPT and FET. This comes from the operating principle of the SIT that the carrier injection from the source region is controlled not to occur the current saturation by the potential barrier formed in front of source through the gate and the drain bias voltage by virtue of the small channel resistance. Now the X band SIT has been developed (1, 9).

The carriers of the ISIT moves very fast with the thermionic emission mode. The mean free path of an electron is very short, so the dimension of ISIT is very small. This leads to the small bias voltage, the high g_m and the high f_T.

The atomic layer epitaxy of II-VI compounds was proposed by Suntola et al. using the alternative evaporation of II and VI element and the alternative injection of two gases which contain the II and VI column element (10).

The GaAs single crystal combining the alternative injection of TMG and AsH_3 and the ultra violet illumination to the ultra high vacuum chamber during the crystal growth has successfully realized in 1984 by our group (11-13). The accurate growth layer control to atomic accuracy and the low growth temperature below 270°C has been obtained.

The GaAs, $Al_{1-x}Ga_xAs$ and Si PMLE and their doping techniques have been developed for the device application purpose (14-17). In this paper the heterojunction gate ISIT fabricated by using GaAs and $Al_{1-x}Ga_xAs$ PMLE under 450°C device process temperature will be presented.

II. ISIT

The cross sectional view and its potential diagram of the center of the channel from the source to drain of the ISIT is shown in Fig. 1. The channel length is set to be shorter than the mean free path of an electron. The injected electrons from the source to the channel are not scattered by the crystal lattice and drift very fast in the channel, so the f_T of ISIT is increased much higher than that of the conventional SIT. The transit time in the channel of ISIT is so small that the f_T of ISIT is determined by the input capacitance which includes the stray capacitance and g_m.

The f_T of the ISIT is given by (1);

$$f_T = \frac{g_m}{2\pi\ Cin} \simeq \frac{1}{2\pi\ Wg'} \sqrt{\frac{kT}{2\pi\ m^*}} \tag{1}$$

The f_T has been estimated up to 718 GHz with the gate length of 0.1 μm (1). The f_T over 1000 GHz (1 THz) will be attained with the decreasing of the gate length.

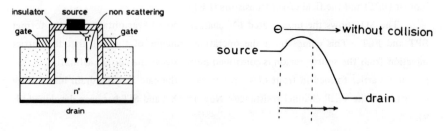

Fig. 1 (a) Schematic cross sectional view of the ideal SIT (ISIT) and
(b) potential distribution from source to drain along the center
of channel of the ISIT.

The MIS structure gate which eliminates the storage effect occurred in the pn junction will be suitable for the GaAs ISIT. However there is no good insulation material to GaAs. Then the heterojunction gate ISIT has been proposed by the author. The high resistive $Al_{1-x}Ga_xAs$ gate to GaAs in which the lattice constant is much fitted to the insulator and is matched to GaAs by the introduction of P as $Al_xGa_{1-x}As_{1-y}P_y$ will be performed as like MIS gate.

The ideal static induction thyristor (ISIThy) will be also realized by the p^+ anode region in place of n^+ drain region of the ISIT. The low loss and the ultra high speed thyristor will be obtained. The ISIT is useful for the integrated circuit for both the analogue and the digital applications. The proposed inverter construction is shown in Fig. 2 with its equivalent circuit. The drain, source and recessed heterojunction gate region is formed on the surface of the wafer. The ultra high speed and low power consumption ISIT IC in a room temperature operation, which will be competitive to the Josephson junction device, will be realized.

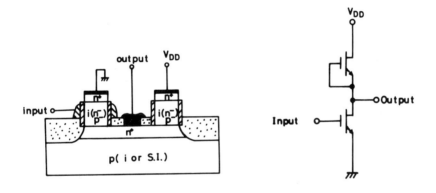

Fig. 2 (a) Cross section of ISIT inverter and (b) its equivalent circuit.

III. SITT and ISITT

The storage effect between the source and the potential barrier maximum point to an electron in the channel will degrade the frequency characteristics of ISIT. This will be solved by the introduction of the tunnel injection source (p^+ - n^+) in place of n^+ source as shown in Fig. 3. The device was proposed in 1980 and was named as the static induction tunnel transistor (SITT) (18).

The reverse biased p^+ - n^+ - i (v or p^-, n^-) - n^+ channel structure is as same as that of the Tunnett diode which oscillates up to 338 GHz (19). The amount of tunnel injection from the p^+ source to the channel is controlled by the gate region. The ideal SITT, in which channel length is shorter than the mean free path of an electron, will be operated much higher than that of ISIT. f_T of the ISITT will be as high as 100 THz by virtue of the ultra fast quantum mechanical tunneling injection. The minimum voltage to establish the high elective field intensity at the p^+ - n^+ junction is calculated to be smaller than the diffusion potential with very thin δ - function like n^+ layer. The drain voltage can be made to a small value, so this device will operate with the ultra high speed, the low power consumption and the low noise.

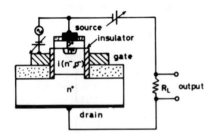

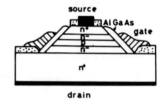

Fig. 3 Static Induction Tunnel Transistor (SITT).

Fig. 4 Cross section of heterojunction gate GaAs ISIT.

IV. Fabrication of the $Al_xGa_{1-x}As$ heterojunction gate GaAs ISIT

The GaAs ISIT in which the channel length is less than 840 Å has been fabricated using two times GaAs and Al_xGa_{1-x} As PMLE under 450°C. The other processing temperature are also less than 450°C. Therefore the process temperature using PMLE is surprisingly lower compared to those of other GaAs technology such as using ion implantation, MOCVD and MBE etc. The cross sectional view of fabricated device structure which operates as the surface conduction mode is depicted in Fig. 4 (20).

The apparatus of GaAs and $Al_xGa_{1-x}As$ PMLE is shown in Fig. 5. The GaAs substrate is placed in the ultra high vacuum chamber and is heated by the infrared lamp through the quartz plate. Also the ultraviolet ray such as the Hg lamp and eximer laser is also illuminated to the substrate through the quartz window. The gases which contain the Al, Ga, As and the dopant are introduced alternatively onto the GaAs substarte. The detail of this process is described elsewhere (13-17).

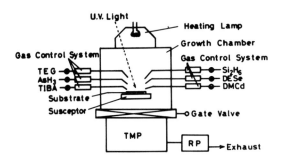

Fig. 5 PMLE apparatus for GaAs and $Al_xGa_{1-x}As$.

The gas injection sequence of GaAs, $Al_xGa_{1-x}As$ and the doping process of GaAs is illustrated in Fig. 6. The DESe and Si_2H_6 are used for the doping gas of n type dopant and the DMCd is used for the p type doping, respectively. The TIBA is used to obtain the high resistive $Al_xGa_{1-x}As$ layer.

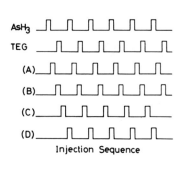

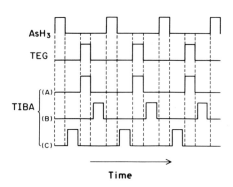

Fig. 6 (a) Gas injection sequences for doping in GaAs

(b) Gas injection sequences for $Al_xGa_{1-x}As$.

The fabrication processes of the heterojunction gate ISIT are as follows;

[1] $n^+ - n^- - p - n^- - n^+$ GaAs PMLE at 450°C,

[2] Si_3N_4 plasma CVD on the epitaxial layer, the source region definition by the photolithography and the plasma etching of Si_3N_4 film,

[3] mesa etch process to form the recessed portion in the channel, Si_3N_4 plasma CVD deposition and opening of the Si_3N_4 layer in the recessed gate region,

[4] $Al_xGa_{1-x}As$ (x = 0.3 ~ 0.4) PMLE to exposed GaAs recessed portion to form the high resistive heterojunction gate at 450°C, and

[5] metalization of gate region and the source and drain ohmic contacts, respectively.

The top view of the fabricated GaAs ISIT is shown in Fig. 7. The I-V characteristics of the GaAs ISIT have been obtained. The resistor load ring oscillator circuit is fabricated now to measure the expected high speed and the low power consumption.

The initial success to fabricate the GaAs ISIT below 450°C processsing temperature using PMLE will be promising to fabricate the ISIT, SITT, ISITT and their integrated circuits in the near future.

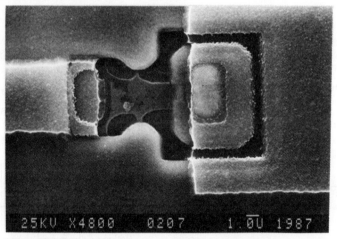

Fig. 7 Top view of fabricated GaAs ISIT.

V. Conclusion

The ISIT, SITT, ISITT and their integrated circuit as the ultra high speed ballistic devices are introduced. The GaAs ISIT with the channel length less than 840 Å has been fabricate at the whole process temperature less than 450°C by using GaAs and $Al_xGa_{1-x}As$ PMLE. These kind of ballistic SIT and IC will expand the f_T up to 100 THz region where there is no active devices.

References

(1)   J. Nishizawa, T. Terasaki and J. Shibata, RIEC Technical Report, TR-36 (Oct. 1973).

J. Nishizawa, T. Terasaki and J. Shibata, Int. Electron Device Meeting, 1972.

J. Nishizawa, T. Terasaki and J. Shibata, IEEE Trans. Electron Devices, Vol. ED-22, p. 185 (1975).

J. Nishizawa and K. Yamamoto, IEEE Trans. on Electron Devices, Vol. ED-25, No. 3, pp. 314-322 (1978).

J. Nishizawa, Digest of Tech. Papers of 11th Conf. on Solid-State Devices, Tokyo p. 1 (1979).

J. Nishizawa, Jpn. J. Appl. Phys, Vol. 19, 19-1, p. 3-11 (1980).

J. Nishizawa, Conf. on Microwave Solid State Electron, Vol. 3, Gdansk, Poland, pp. 79-99, Oct. 27-31 (1980).

(2)   M.S. Shur and L.F. Eastman, IEEE Trans. Electron Devices, Vol. ED-26, No. 11, pp. 1677-1683 (1979).

(3)   Y. Watanabe and J. Nishizawa, Japanese Patent No. 205068 (1950).

(4)   J. Nishizawa and B.M. Wilamowski, Proc. 1976 Int. Conf. Solid State Devices (Tokyo), Jap. J. of Appl. Phys. suppl., Vol. 16-1, pp. 151-154 (1977).

J. Nishizawa, Denshi-Gijutsu, p. 6 (1980).

J. Nishizawa, T. Nonaka, Y. Mochida and T. Ohmi, J. of Solid-State Circuits, Vol. SC-14, No. 5, p. 873 (1979).

Y. Mochida and T. Nonaka, Semiconductor Technology, Ohm North-Holland (1979).

(5)   J. Nishizawa, N. Takeda and T. Ohmi, Proc. of first Speciality Conf. on Giga Bit Logic for Microwave Systems, May (1979).

(6)   T. Nakamura, M. Yamamoto, H. Ishida and M. Shinoda, IEEE J. of Solid State Circuits, Vol. SC-13, No. 5, pp. 572-176, Oct. (1978).

S. Suzuki, Report of JRDC, p. 3-8 (1983).

N. Takeda, Symposium Proc. of Nishizawa Perfect Crystal Project, J. Nishizawa ed., pp. 33-41 (1986).

(7)   J. Nishizawa, T. Tamamushi, and T. Ohmi, IEEE Trans. Electron Devices, Vol. ED-26, pp. 1970-1977, Dec. 1979.

J. Nishizawa, T. Tamamushi and S. Suzuki, in JARECT in Semiconductor Technologies, Vol. 8, J. Nishizawa, Ed., OHM & North-Holland, pp. 219-242 (1983).

J. Nishizawa, T. Tamamushi, K. Nonaka and S. Suzuki, IEEE Electron Device

Lett., Vol. EDL-6, No. 1, pp. 17-19 (1985).

A. Yusa, J. Nishizawa, M. Imai, H. Yamada, J. Nakamura, T. Mizoguchi, Y. Ohta and M. Takayama, IEEE Trans. Electron Devices, Vol. ED-33, No. 6, pp. 735-742 (1986).

(8) J. Nishizawa, K. Muraoka, T. Tamamushi and Y. Kawamura, IEEE Trans. Electron Devices, Vol. ED-32, p. 822 (1985).

T. Terasawa, A. Mimura, and K. Miyama, IEEE Trans. Electron Devices, Vol. ED-33, p. 91 (1985).

J. Nishizawa, K. Muraoka, Y. Kawamura and T. Tamamushi, IEEE Trans. Electron Devices, Vol. ED-33, p. 507 (1986).

K. Muraoka, Y. Kawamura, Y. Ohtsubo, S. Sugawara, T. Tamamushi and J. Nishizawa, PESC '86 Record, p. 94 (1986).

J. Nishizawa, IEEE Denshi Tokyo, No. 25, pp. 102-106 (1986).

(9) Y. Kajiwara, M. Aiga, Y. Higaki, M. Kato, Y. Yukimoto and K. Shirahata, Proc. 1979 Int. Conf. Solid State Devices (Tokyo), Jap. J. Appl. Phys., Suppl., Vol. 19-1, pp. 305-308 (1980).

T. Shino, H. Kano, K. Aoki and S. Ono, ibid., pp. 283-287 (1979, 1980).

A. Cogan, R. Regan, I. Bencuua, S. Butler, and F. Rock, IEEE IEDM, Tech. Dig. 9.5 (1983).

R. Regan, I. Bencuya, and S. Butler, Microwave & RF, pp. 63-68 (1985).

(10) M. Ahonen, M. Pessa and T. Suntola, Thin Solid Films, 65, 301 (1980).

T. Suntola, Extended Abstracts of 16th Conf. on Solid State Devices and Materials, Kobe, Japan, p. 647 (1984).

(11) J. Nishizawa, J. of Metal Society of Japan, Vol. 25, No. 5, A 149 (1961).

J. Nishizawa, ibid., Vol. 25, No. 6, A 177 (1961).

M. Kumagawa, H. Sunami, T. Terasaki and J. Nishizawa, Japan J. Appl. Phys. Vol. 7, No. 11, 1332 (1968).

(12) J. Nishizawa, Y. Kokubun, Extended Abstracts of 16th Int. Conf. Solid State Devices and Materials, Kobe, Japan, p. 1 (1984).

(13) J. Nishizawa, H. Abe, and T. Kurabayashi, Group meeting of semiconductor and transistor of IECE of Japan, SSD 84-55, pp. 73-77 (1984).

J. Nishizawa, H. Abe and T. Kurabayashi, J. Electrochem. Soc., Vol. 132, p. 1197 (1985).

(14) J. Nishizawa, H. Abe, T. Kurabayashi and N. Sakurai, J. Vac. Sci. Technol. A (4), p. 706 (1986).

(15) J. Nishizawa, T. Kurabayashi, H. Abe and N. Sakurai, J. Electrochem. Soc., Vol. 134, No. 4, pp. 945-951 (1987).

(16) J. Nishizawa, T. Kurabayashi, H. Abe and A. Nozoe, Surface Science, Vol. 185, pp. 249-268 (1987).

(17) T. Kurabayashi, H. Abe, N. Matsumoto, K. Aoki and Y. Kokubun, Symposium Proc. of Nishizawa Perfect Crystal Project, J. Nishizawa ed., pp. 1-24 (1986).

(18) J. Nishizawa, "Microwave SIT", Conf. on Microwave Solid State Electronics, Vol. 3, Gdansk, Poland, pp. 79-99, Oct. 27-31 (1980).

(19) J. Nishizawa, K. Motoya and Y. Okuno, Proc. of the 9th European Microwave Conference, pp. 463-467 (1979).

(20) S. Suzuki, Symposium Proc. of Nishizawa Perfect Crystal Project, J. Nishizawa ed., pp. 25-31 (1986).

76

Andrzej Jakubowski and Alfred Świt
Institute of Microelectronics and Optoelectronics
Warsaw University of Technology
Warsaw, Poland

PHYSICAL LIMITS OF ULSI

INTRODUCTION - GENERAL DISCUSSION OF LIMITS

The fact, that microelectronics is developing much quicker than any other field of technical sciences and that this development brings great economical and social changes, leads to our fascination with it. One of the most common questions put here is how far it can become developed, what are the physical limits of its development and what technological barriers we have to over-come.

To describe its development we use logistic curve or similar "s" shapes curves. Graphic illustration of such development is presented in Fig.1 .

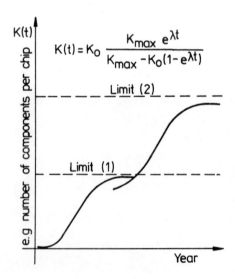

$$K(t) = K_0 \frac{K_{max}\, e^{\lambda t}}{K_{max} - K_0(1 - e^{\lambda t})}$$

Fig.1. Logistic curve

Development of semiconductor technology indicates that many of its

parameters increase (or decrease) exponentially in time function. The

most common case is the so called "Moore's law" [1]. The decrease in

progress rate and beginning of stabilization may be the result of one

of the following factors acting as limiting mechanism :

(i)   Gaps in the knowledge,

(ii)   Change in interests,

(iii) Technical or economical limits.

In the discussed case of silicon technology the first two are surely

not dangerous at the moment. Therefore, the only reason may be the

latter factor. Let us try to name in brief the technical limits. They

are :

- technology barriers, nonideal materials and technological

  processes, problem with yields, limits due to the used

  equipment, ... etc.

- physical limits resulting from the known physical laws or from

  the device physics,

- limits resulting from the circuits complexity, also problems

  with design and testing of very complex IC's,

- limits resulting from the difficulty in defining fields of

  applications for very complex logic circuits (which of course

  does not concern memories).

After Meindl [2] ultra large scale integration is governed by a

hierarchy of limits. The levels of this hierarchy can be codifield as:

1) Fundamental

2) Material

3) Device

4) Circuit

5) System

The knowledge of limitations, however unprecise it may be, could

constitute an important induction as how to treat them most

effectively. Therefore so many publications are the subject of the

above mentioned questions [e.g. 3-10].

GROWTH FACTORS

Before we start discussing limitations let's consider what factors cause the increase of components number in integrated circuits. They are : (i) increase of chip size (S), (ii) reduced feature size (F), (iii) cleverness or packaging efficiency (U).

The example of dynamic RAM MOS shows that this can be proved by approximate calculations. In Fig.2 the increase of chip size and decrease of the feature size in succesive years is shown. The evolution of three - transistor cell (1 kb) in dynamic RAM's to one-transistor type and trench type capacitor or stacked capacitor cell (1Mb) leads to increase of packaging efficiency. Since 1971 systematically every three years a new generation DRAM has appared on the market with four times higher number of gates as compared to the previous one. This can be described by such a formula :

$$N(bit) \simeq 10^3 * \exp [0.46 * (Y - 1971)]$$

where Y represents the year under consideration.

The area of individual component is proportional to feature size F to the second power. The relation between the minimum feature size and time can be expressed as follows :

$$F(um) \simeq S * \exp [-0.13 * (Y - 1070)]$$

The value of F decreased about 1.4 to 1.5 times per RAM generation, which makes up the resulting decrease of F by the factor of 2 per 6 years. The chip area increased according to the formula :

$$S(mm^2) \simeq 10 * \exp [0.13 * (Y - 1970)]$$

The increment was about 1.5 per generation.

Finally, using "Moore's law" we can determine the packaging efficiency factor U in dynamic RAM's. In the 15 years it is equal aproximately to 2.9, as it ts shown by the following calculation:

$$U = \frac{N}{\frac{1}{F^2} * S} = \frac{1000}{49*7} \simeq 2.9$$

As a result of all just mentioned trends, the price of a single bit in RAM's systematically decreases. It decreases even faster than it is expressed by a formula :

$$K(\$/bit) \simeq 7 * 10^{-3} \exp [-0.4 \; (Y - 1972)]$$

At the same time, the rapidn rise in memory bits production is observed. This trend is described by the next formula :

$$P(bit/year) \simeq 4.5 * 10^7 * \exp [0.85 \; (Y - 1972)]$$

where Y as before represents succesive year. It is expected, that in the year 2000, one year production of memory bits will become 10^{20}, which gives 10^{12} of 64 Mb memory circuits. If we, as mankind succeed in this it would mean about 100 of these memory circuits per person.

ENERGY MEASURE OF LOGIC CIRCUIT QUALITY

The measure of logic integrated circuits performance can be expressed as a number of logic switching which can be performed in a time unit. This performance is limited in two ways :

- by the total number of logic units, which may be formed on one chip, divided by the time of single logic switching,

- by the maximum power that can be dissipated from the circuit.

To judge performance of MOS IC's let us consider the elementary ideal MOS inverter whose operation we will approcimate as charging of the capacitor C through the nonlinear resistance. Energy used to perform elementary logic switching is equal to :

$$E = p\mathcal{T} = 1/2 \; CV^2_{DD}$$

where V_{DD} is applied voltage, p is the electric power dissipated in the gate, $\mathcal{T}$ - is the switching time. If we remind ourselves that maximum power that can be dissipated in the chip is P_T, clock frequency is $f_c = 1/4\mathcal{T}$ and number of logic operations is equal to the number of gates - N, we have :

$$N < \frac{P_T}{2 \; CV^2_{DD} \; f_c}$$

A convenient way of measuring the performance of IC's is the following product (functional throughput rate)

$$FTR = \frac{number\ of\ gates\ *\ clock\ frequency}{per\ chip} = \frac{P_T}{2\ CV^2_{DD}}$$

Fig.3 [10] presents the value of the FTR parameters for integrated circuits at the beginning and at the end of the eighties.To obtain 10^{13} gates * Hz, the submicron technology is needed.

Concluding, we would like to say that the main measure of the quality of logic circuits is the power-delay product which is dependent on the design and technology of the IC's.

MINIMUM DIMENSIONS OF THE MOS TRANSISTOR

Minimum dimensions of MOS devices are influences by both deterministic and stochastic processes. However, litography is the factor mainly discussed in the literature, which influency the minimum dimensions of the cell. It has already been shown how the feature size reduction during recent years can be expressed. If this trend will remain constant then in the beginning of the nineties we will reach the limits of optical litography. Therefore, further progress can only rely on other litographic techniques, e.g. x-ray, electron or ion beam. Their limits has already been determined many times in the literature. The fundamental limits of litographic techniques are the results of Heisenberg's principle. The other limit is of quantum character and determined by the probablity of tunneling through the thin layers. In case of MOS structures this determines the minimum gate dielectric layer thickness, which for SiO_2 is assumed to be about 5 nm. Gate dielectric thickness for last fifteen years has changed approximately as follows

$$t_{ox}\ (nm) \cong 120\ *\ exp\ [-0.12\ *\ (Y\ -\ 1970)]$$

If this trend will remain constant oxide layer thickness limit will be reached in the mid nineties. We shduld also consider source-drain punch through effect which leads to the determination of the minimum channel lenght as being between 0.07 μm and 0.3 μm. In several works maximum number of gates per square centimeter was estimated as between 10^7 and 10^8. Still another limit results from charge carrier flictuation. When the components become extremaly small, there is a considerable probability that there will be devices with dopant concentration value considerably smaller and considerably higher than the mean value. It gives limitations similar to the ones resulting from punch through effect.

MINIMUM BIAS VOLTAGE, MINIMUM ENERGY OF LOGIC OPERATION

It is obvious that minimum bias voltage should be higher than the electrokinetical potential. It is therefore easy to express minimum voltage as follows : V_{MIN} = M kT/q, where M is between 10 and 30 depending on the complexity of the IC's. Assuming V_{MIN} = 0.7 V and C = 1 fJ we get E_{MIN} = 0.25 fJ. It seems reasonable to estimate 1 fJ as an energy per gate limit for silicon IC's of great complexity. Above mentioned expression for minimum voltage shows also how effective reduction of power ✱ delay product may be obtained by the decrease of the operating temperature (pT is proportional to the second power of temperature).

FUNDAMENTAL LIMITS

The progress in logic IC's production demands the decrease in energy used during swithing. For the IC's which are the most representative the years considered the exponential relation between energy and time holds for the last 20 years. Nowadays MOS IC's have the power delay product of about 0.05 pJ.

Let's compare this value to the values of the critical energy which can be determined from the fundamental limits. Analysis of thermodynamic principle gives the value E_{MIN} = kT ln 2, which is about 7 orders of magnitude smaller than in the presently produced IC's.

Second fundamental limit is the result of Heisenberg"s principle (quantum mechanical uncertainty principle) and can be expressed as E > ℏ/T, where ℏ is the planck's constant. If we assume f_c = 1 THz then we get the result shown in the Fig.4 [11]. Quantum − mechanical limits are dominating for T < 11 K. Concluding, we can say that we are still a from the fundamental limits. However if we introduce the problem of probability of an error in logic circuit due to the thermal noise − the situation will change considerably. There are several ways of treating of this problem nevertheleas all approaches give the similar results. It is obvious that the energy corresponding to difference between "0" and "1" state must be much higher than kT, to eliminate accidental errors due to thermal fluctuations [12]. If we would also consider the distribution of threshold voltage across the chip - the limit may become of the order of 1000 times kT [13]. Our discussion also shows, that decreasing of the operating temperature bring us closev to the fundamental thermodynamic limits since we have

$$CV^2_{MIN}/ \ kT \div T$$

OVERVIEW OF LIMITS

A representative overview of the hierarchy of limits governing VLSI's is obtained by a power - delay plot as illustrated in the Fig.5 [2]. The fundamental limits are due to thermal noise and uncertainty principle. Fundamental limits due to thermal noise are different for simle and complex circuits. The material limit on carrier transit time in silicon is imposed by scattering limited carrier velocity and critical field strength. The MOS device limit is set by velocity saturation and punch-through effects.The circuit limit applies to a CMOS gate with minimum dimensions. The dotted line represents the state-of-art power - delay product around mid. eighties.

SUMMARY

The whole discussion proves that we have a lot of space left before we reach the limits. However this does not mean that we will reach it easily. Many barriers are met, e.g. technological, material, etc. In fact the semiconductor industry men know that at the moment these barriers are the number one problem. IC's manufacturing has come to the point where well known and until now effective method are no more efficient. We belive that CMOS technology will play more and more important role, among others due to the smallest critical power limits. The progress rate will depend not only upon the technological barriers, but also upon the new solutions in the design of very complex IC's. Some hopes are also connected with SOI technology and formation of two (or even more) levels of the devices on chip. Also different concepts of WSI have been appearing which are very promising. Full utilization of silicon MOS technology gives us a chance of formation at the end of this century chips including about 10^8 or 10^9 cells. We believe that this optimistic forecast is real.

REFERENCES

[1] G.E.Moore; IEDM Techn.Dig., p.11, Dec.1975.

[2] J.D.Meindl; IEDM Techn.Dig., p.8, Dec.1983.

[3] "Transition to One Micron Technology", Semiconductor International, Special Issue, 1984.

[4] R.W.Keyes; Proc.IEEE, v.69, p.267, 1981.

[5] Ch.Svensson; Integration, the VLSI J., v.1, p.3, 1983.

[6] O.G.Folbreht, J.H.Bleher; Microelectronics J., v.10, No 3/4, p.33, 1979.

[7] A.Reisman; Proc.IEEE, v.71, p.550, 1983.

[8] A.Jakubowski, A.Swit; Elektronika, v.25, No 6, p.14, 1984.

[9] K.Goser; in: H.Painke (ed.) "Digital Technology, Status and Trends", Oldenbourg, Munchen, 1982.

[10] J.L.Prince; in: D.F.Barbe (ed.) "Very Large Scale Integration VLSI", Springer, Berlin 1980.

[11] R.T.Bate; in: N.G.Einspruch (ed.), "VLSI Electronics, v.5, Academic Press, New York, 1982.

[12] K.U.Stein; IEEE J.Solid-St.Circ., v.SC-17, p.527, 1977.

[13] A.Jakubowski, A.Cetner; IEEE J.Solid-St.Circ., v.SC-21, No 4, p.366, 1986.

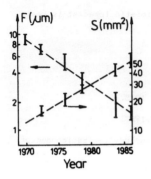

Fig.2. Chip area and feature size
(MOS DRAM)

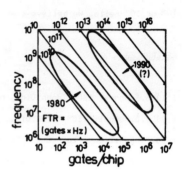

Fig.3. Values of the FTR
parameters [10]

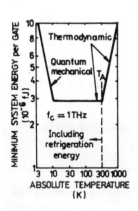

Fig.4. Fundamental limits [11]

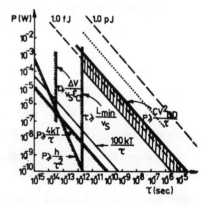

Fig.5. Overwiev of limits [2]

K.INIEWSKI, A.JAKUBOWSKI, B.MAJKUSIAK

Institute of Microelectronics and Optoelectronics,
Warsaw University of Technology,
Koszykowa 75, 00-662 Warsaw, Poland

PROBLEMS OF EXTRACTION OF MOS TRANSISTOR

PARAMETERS

1. Introduction

Extraction of MOS transistor parameters is very important problem in modelling and device optimization for VLSI. It works also as a diagnostic tool on technological line for control purposes.

Two approaches are commonly used. The first is a global extraction in which the set of MOS transistor parameters is calculated using optimization algorithms [1-6]. The second one is a detailed extraction in which parameters are determined from specific range of measurements under specific conditions [7-12]. Using global extraction very good agreement between theoretical I-V characteristics and measured ones can be obtained. However the parameters loose their physical interpretation and have rather formal meaning. On the other hand using detailed extraction the physical meaning of the parameters is more or less retained although agreement mentioned above might not be so good.
The leading idea of this paper is to discuss an agreement between values of parameters obtained from detailed extraction methods and values resulting from physical definitions. The problem follows from the fact that procedures of detailed extraction are usually based on simplified models of MOS transistor. The clear

character of relations between particular elements of simple models enables easy interpretation of transistor characteristics and indicates the way of analysis of measurement data. However, one often forgets that magnitudes obtained in such a way become only the actual parameters of the model describing better or worse the characteristics of transistor measured. These parameters can differ considerably from magnitudes of concrete physical meaning in regard to the values as well as to the character of dependence on other physical magnitudes.

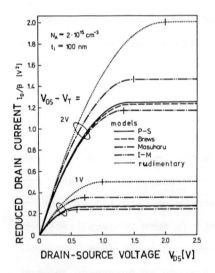

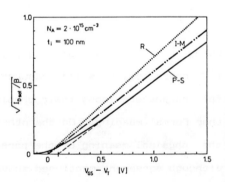

Fig.1 Family of output characteristics.

Fig.2 Transverse characteristics in saturation mode.

2. Considerations

Fig.1 shows output characteristics resulting from various models of long-channel MOS transistor. The Pao-Sah model [13] is the most accurate one, serving usually as a source of reference characteristics. Other models employ various physical assumptions and simplifications. In the Brews' model [14] the inversion layer charge is expressed by difference between a total semiconductor charge and a depletion layer charge. Neglection of the diffusion of channel carriers reducing the current leads to the model described by Masuharu et.al [15]. If the increment of surface potential in the strong inversion state above its threshold value i.e. the doubled Fermi potential is neglected additionally as in the Ihantola-Moll model [16], the minority carrier charge and in consequence, the channel current are overestimated. The rudimentary model, neglects additionally the increase of deple-tion layer charge along the channel giving additional overestima-tion of the current value. Therefore, as it is shown in Fig.1, deviation of results of various models for the same values of transistor parameters is very large. The opposite statement must also be valid: the analysis of the same experimental data basing on various transistor models must give various values of transis-tor parameters. The more inaccurate the model is, the greater errors in extraction of parameters could be expected. The most accurate models are too complicated, they require the use of numerical methods. Therefore the procedures usually used for the detailed extraction follow from the rudimentary or Inhantola-Moll model. Both models are very inaccurate.

Let us consider for example the procedures of extraction of the threshold voltage V_T. The rudimentary model predicts that the square root of drain current in saturation should be linear function of gate voltage and in consequence, the threshold voltage can be extracted from an intercept of such a straight line with the gate voltage axis.

Fig.2 shows a comparison of so-constructed transverse characteristics predicted by various models. The Ihantolla-Moll model gives nearly linear dependence whereas more accurate models including changes of surface potential in the strong inversion state do not. Therefore, if we treat the characteristic obtained from the Pao-Sah model as a result of experiment, we can see that the extrapolation of the linear part of it to the voltage axis gives the intercept voltage above 100 mV greater than the true threshold voltage.

In order to avoid phenomena affecting the current in the saturation region as for example the modulation of effective channel length, the threshold voltage is often determined from extrapolation of linear part of transverse characteristics $I_D(V_G)$ for very small drain-source voltage V_{DS} i.e. in the non-saturation range. Fig.3 shows such a characteristic resulting from the Pao-Sah model and it can be seen that it is also nonlinear even in the non-saturation region. This non-linearity which is the source of erroneous determination of threshold voltage is also caused by the changes of the surface potential in the strong inversion state with the gate voltage.

The fact that procedures of detailed extraction give as a result parameters differing considerably from their true values does not mean that the models, on which the extraction procedures are based, can not be used for simulating the transistor characteristics. For example, Fig.4 shows the family of output characteristics resulting from the Pao-Sah model and the Ihantolla-Moll model. After increasing the threshold voltage by $_\Delta U_T$ = 0.089 V and reducing the parameter β = $\mu_{eff} C_i$ (W/L) to 0.937 its real value Ihantolla-Moll model fits the Pao-Sah characteristics very well. The differences are practically invisible provided the gate voltage is greater above 0.5 V than threshold voltage when the changes of the surface potential at the source end of the channel with gate voltage are very small.

The error of extraction of the threshold voltage from analysis of transverse characteristics of MOS transistor cannot be determined in practice. It is dependent on other parameters of transistor as the oxide thickness and the substrate impurity concentration. Additionally, the real characteristics are affected also by transverse field dependence of mobility, by charging of interface traps, drain and source resistance influence and small geometry effects. Therefore, if we want to interprete the dependence of threshold voltage on other parameters or if we use its value in further physical consideration we must take care.

As a good example of appearance of relations obtained from inattentive analysis of experimental data the relation between the effective mobility of carriers in the channel and the transverse effective field can serve. The effective mobility is usually determined from measurement of output conductance in linear region of output characteristics g_{d0} :

$$\mu_{eff} = \frac{g_{do}}{W/L \ Q_{inv}} \qquad (1)$$

where W/L is the channel width to lenght ratio and Q_{inv} is the inversion layer charge. A traditional way of expressing Q_{inv} charge is to multiply the insulator layer capacitance C_i and the difference between the gate U_G and threshold U_T voltages :

$$Q_{inv} = - C_i (U_G - U_T) \qquad (2)$$

The first problem occuring in practice is how to determine correctly the threshold voltage, this problem has been discussed above. The second problem follows from the fact that the relation (2) is only an approximation neglecting the changes of the surface potential in the strong inversion state as in the Ihantola-Moll and the rudimentary models. The inversion layer charge should be expressed more exactly by taking into consideration the increments of surface potential ϕ_s and depletion layer charge Q_B above their values at the threshold voltage. Therefore the effective mobility determined from the experimentally measured conductance in the traditional way is smaller than the true value of mobility μ_{eff} even if the threshold voltage U_T is determined exactly.

As an example, Fig.5 shows the results of an analysis of experimental data performed in such a simple way [17]. The hole mobility was found to depend strongly on oxide thickness. After recalculation that data with taking into consideration the surface potential increment, the hole mobility appeared to be independent of oxide thickness [18], as it can be seen in Fig.6.

A comparison of two latter figures can serve as an excelent example and also warning, how the use of too simple model at extraction of transistor parameters may deform an image of physical phenomena in transistor. In order to determine the inversion layer charge and in turn, the effective mobility without the need of knowledge of the threshold voltage a "split" C-V measurement technique can be used [19,20]. This technique of mobility determination is not convenient due to aparature requirements and the need of integration of data. It also could not be used for samples with significant interface trap density such as transistors after hot-carrier injection. As an alternative new method recently proposed could be employed [21]. It is based on measurements of output conductance at small drain voltage versus gate voltage for transistors with various oxide thicknesses. One of the advantages of the method is that threshold voltage need not be extracted and that simultanously effective metal-semiconductor work function defference is determined.

The last parametr, extraction of which will be considered here is a doping concentration in the substrate of MOS transistor. Stricly speaking the doping under the gate is never in practice uniform as a result of ion implantation and/or high-temperature processes. However for a device modelling purposes the doping profile is quite often considered as an uniform. This assumption allows simplification of the expressions for drain current to an analytical from suitable for parameter extraction. As a result of this simplification the questions arise : how doping concentration should be determined and how close to the real doping profile this value is.

To extract doping concentration from I-V measurements of MOS transistor two approaches are generally used. The first approach uses body effect. The doping is determined as a slope of V_T vs $\sqrt{V_{SB} + 0.7\, V}$ curve. In the second approach doping concentration is calculated to fit theoretical value of threshold voltage V_{TO} (V_T for $V_{SB} = 0$) to a measured value. Using the first method value of doping concentration close to bulk doping is obtained. As a result threshold voltage dependence on source-bulk voltage is modelled properly. However V_{TO} calculated for this concentration is far from measured value. This is due to higher doping concentration in the surface region of semiconductor (we assume here that shallow ion implantation is used during device fabrication). As a result flat-band voltage V_{FB} must be changed from its real value to fit theoretical and measured values of V_{TO}. In this way V_{FB} looses its physical meaning and could not be used for evaluation of effective oxide charge Q_{eff} which is a measure of quality of technological process. Moreover inaccurate value of V_{FB} could also lead to inproper modelling of subthreshold current. Using the second approach the value of doping concentration close to the average doping concentration in the surface region of the transistor is obtained. In this way above mentioned problems are partially avoided. However threshold voltage dependence on source-bulk voltage is now modelled inaccurratively. This is due to the fact that body factor is higher than its real value. Some new models of MOSFET with nonuniform doping have been proposed recently [22-27], however their I-V equations are much more complicated which make the extraction of parameters difficult.

3. Conclusions

Extraction of MOSFET parameters basing on simple, analytical models is widely employed in MOS device modelling. As shown above the parameters determined this way are the actual parameters of the model which fit theoretical characteristics to measured ones. It does not always mean that they have exact physical values. This difference is caused by neglecting many second-order effects during extraction. As a result one should be very carefull about physical meaning of obtained results.

REFERENCES

1. K.Doganis, D.L.Scharfetter, IEEE Trans. Electr. Dev., vol.ED-30,pp.1219-1228,1983.
2. W.Maes, K.De Meyer, L.Dupas, Electron. Lett.,vol.21,p.23,1985.
3. W.Maes, K.De Meyer, L.Dupas, IEEE Trans. Comp. Aided Des., vol.CAD-5,pp.320-325,1986.
4. O.Melstrand et al, IEEE Trans. Comp. Aided Des., vol.CAD-3, pp.47-51,1984.
5. R.J.Sokel, D.B.Macmillan, IEEE Trans. Electr. Dev., vol.ED-32, pp.2110-2116,1985.
6. S.J.Wang, J.Y.Lee, C.Y.Chang, IEEE Trans. Comp. Aided Des., vol.CAD-5,pp.170-179,1986.
7. A.B.Bhattacharyya et al, IEEE Trans. Electr. Dev., vol.ED-32, pp.545-550,1985.
8. M.F.Hammer, Proc. IEEE, Pt.I, vol.133, pp.49-54, 1986.
9. C.Hao, Solid-St. Electr., vol.28, pp.1025-1030, 1985.
10.W.Marciniak, H.Madura, Electron Technology, vol.17, pp.69-96, 1985.
11.M.J.Thoma, C.R.Westgate, IEEE Trans. Electr. Dev., vol.ED-31, pp.1113-1116, 1984.
12.M.J.Thoma, C.R.Westgate, IEEE Trans. Electr. Dev., vol.ED-33, pp.312-313, 1986.
13.H.C.Pao, C.T.Sah, Solid-St. Electr., vol.9, p.927, 1966.
14.J.R.Brews, Solid-St. Electr., vol.21, p.345, 1978.
15.M.Masuharu et al,IEEE Trans. Electr. Dev.,vol.ED-21,p.363,1974.
16.H.K.Ihantola, J.L.Moll, Solid-St. Electr.,vol.7,p.423,1964.
17.H.Q.Su et al,IEEE Trans. Electr. Dev.,vol.ED-32,pp.559-561, 1985.
18.B.Majkusiak, A.Jakubowski, IEEE Trans. Electr. Dev., vol.ED-33,pp.1717-1721,1986.
19.C.G.Sodini,T.W.Ekstedt,J.Moll,Solid-St. Electr.,vol.25,pp.833-841,1982.
20.M.S.Liang et al,IEDM Techn. Dig.,pp.152-156,1984.
21.B.Majkusiak, A.Jakubowski, to be published.
22.C.Y.Wu, Y.W.Daih, Solid-St. Electr., vol.28, pp.1271-1278,1985.
23.C.Y.Wu et al, Solid-St. Electr., vol.28, pp.1263-1269, 1985.
24.C.Y.Wu, G.S.Huang, H.H.Chen, IEEE Trans. Electr. Dev., vol.29, pp.387-394, 1986.
25.P.K.Chatterjee, J.E.Leiss, G.W.Taylor, IEEE Trans. Electr. Dev., vol.ED-28, pp.606-607, 1981.
26.D.M.Rogers, J.D.Hayden, D.R.Rinerson, IEEE Trans. Electr. Dev., vol.ED-33, pp.955-964, 1986.
27.G.T.Wright, IEEE Trans. Electr. Dev., vol.ED-34, pp.823-833, 1987.

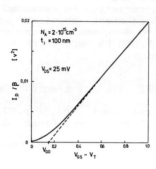

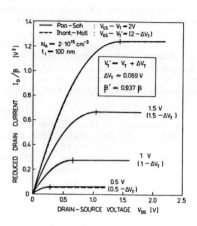

Fig.3 Transverse characteristics in
linear mode (Pao-Sah model).

Fig.4 Output
characteristics.

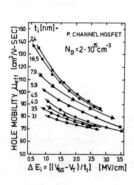

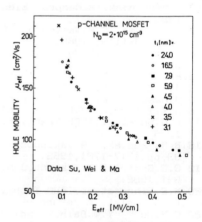

Fig.5 Hole mobility for different
oxide thicknesses [17].

Fig.6 The same as Fig.5,
data recalculated [18].

AN ANALYTICAL MODEL AND COMPUTER SIMULATION
OF THE J-V CHARACTERISTICS OF N^+PN^+ (P^+NP^+) TWO-TERMINAL DEVICES

Z. J. Staszak[*], S. C. Jain[**] and R. H. Mattson

Department of Electrical and Computer Engineering,
University of Arizona, Tucson, AZ 85721, USA

Abstract: Computer simulations show that the analytical model of an n^+-p-n^+ or bulk-barrier diode is valid only for small and very large applied voltages. For a considerable range of intermediate voltages, accurate results can be obtained only by computer simulations. Reasons for the limitations of the analytical model are discussed.

I. Introduction

Early theoretical work on the transport of injected carriers over a barrier in a semiconductor with relatively few mobile carriers was done by Watanabe and Nishizawa [1], Shockley et al. [2,3], and Dacey [4]. A good summary of this and other experimental and theoretical investigations related to the development of the Bipolar Static Induction Transistor is given in the papers of Nishizawa and coworkers [5,6,7]. A similar carrier transport mechanism is the dominant process in short-channel MOSFETs [8,9], and ultrasmall punch-through MOSFETs [10].

Simple n^+-p-n^+ (or p^+-n-p^+) devices having only two contacts at the n^+ (or p^+) layers (known as bulk-barrier diodes) exhibit similar transport mechanisms and are of considerable interest [8-15]. Carriers flow from a heavily doped part of the device,

---

[*] Present address: Gdansk Polytechnics, 80-952 Gdansk, Poland.

[**]Former address: Solid State Physics Laboratory, Delhi 110007, India. Permanent address: 39 New Campus, IIT Hauz Khas, New Delhi 110016, India.

i.e., from the n^+ (or p^+) source into the depleted p (or n) region. In most cases the carriers have to pass over a potential barrier moving from one n^+ (or p^+) region to the other.

These devices have proved very useful in many applications such as microwave applications (BARITT diodes [8]) and high speed photon controlled switches [11]. More recently [12,13], the use of n^+-p-n^+ structures operating in the punch-through mode has been suggested as load resistors for applications in silicon integrated circuits. They can be fabricated using standard bipolar process technology; and, high resistor values can be obtained in very small dimensions.

For many applications bulk-barrier devices are fabricated by epilayer technology. Then the thickness of the lightly doped layer is small compared to the other two dimensions. Their charac-teristics can be predicted by a one-dimensional (1-D) model developed by Sze and coworkers [8,14,15].

There are two weaknesses in this model; however, both occur when the applied voltage is large, close to the flat-band voltage. In this case the junction voltage across the forward biased junction is so large (the barrier is almost zero) that the depletion layer approximation at the junction is not valid [16]. In addition the effect of mobile carriers in the p (or n) region becomes important. Sze et al [14,15] have integrated Poisson's equation taking into account mobile carriers. However, they assumed that the electric field or potential were equal to zero at the same point for all values of applied potentials [15]. This is not a good approximation as shown by our computer simulations.

In view of the importance of this structure we have made a more exact computer calculation to solve the Poisson's, continuity and transport equations, and to predict the J-V characteristics. We have also investigated whether the current flow is due to thermionic emission or is diffusion controlled [8]. The results of these investigations are reported in this paper.

II. Computer Simulation

Computer simulation of small geometry high-speed devices for VLSI applications requires a general approach that does not employ frequently used simplifying assumptions (e.g., space charge neutrality, depletion approximation, neglecting recombination, dominance of one type of carrier, neglecting fast transitions that are shorter than relaxation times). A GEneral semiconductor device performance SIMulation program (1-D), GESIM1, developed by Wilamowski et al. [17] allows the user to simulate the static and dynamic performance of a device without such simplifying assumptions. The program solves five basic differential equations: transport, continuity and Poisson's in one dimension. The simulation includes all effects and provides detailed information about the static and transient behavior of semiconductor devices such as the potential, electric field, charge, hole and electron concentrations, hole and electron currents, etc. The required input data consists of the impurity concentration distribution as a function of distance, and the applied terminal voltages. The steady-state solution is used in this paper. A transient analysis can be performed if desired.

III. Previous Results

Figure 1 shows an n^+-p-n^+ device, its electric field (at punch-through) and potential distribution (before and after punch-through) based on a depletion layer approximation[1]. In this case no appreciable current flows until the applied voltage

---

[1] Because of the similarity of these two-terminal devices to MOS devices, the forward biased junction (1) contact will be called the source and the other (2) the drain.

V exceeds the punch-through voltage V_{PT}. When $V > V_{PT}$, the potential barrier is reduced by an amount ϕ_B.

At low values of ϕ_B, the depletion layer approximation is quite good. Using this approximation and assuming that the flow of carriers is controlled by thermionic emission, the current density J is given by [8]

$$J = A_1 \ T^2 \ \exp(-\frac{qV_{bi}}{kT})[\exp(\frac{q\phi_B}{kT}) - 1] \tag{1}$$

where A_1 is the effective "Richardson constant" and is independent of the channel length or doping concentration, T is the temperature, V_{bi} is the built-in potential barrier, ϕ_B is the reduction in the potential barrier (barrier lowering) of the forward biased junction, and kT/q (= V_T) is the electrothermal potential.

If the transport of carriers is assumed to be diffusion controlled, then [5-7]

$$J = A_2 \ \exp(-\frac{qV_{bi}}{kT})[\exp(\frac{q\phi_B}{kT}) - 1] \tag{2}$$

where A_2 is given by

$$A_2 = \frac{q \ D \ (N_S - N_{BC})}{d} \tag{3}$$

and q is the electron charge, D is the appropriate diffusion constant, N_S is the density of carriers in the highly doped source region, N_{BC} is the background impurity concentration, and d is the diffusion length in the channel region or the effective width of the potential barrier, whichever is smaller.

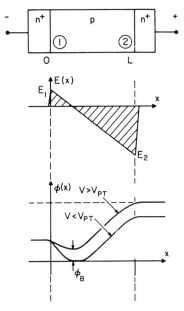

Figure 1. n^+-p-n^+ structure, electric field distribution at punch-through, and potential distribution before $(V<V_{PT})$ and after $(V>V_{PT})$ punch-through.

The relation between the applied voltage and the forward-biased barrier height ϕ_B is given by [8,14,15]

$$V_{bi} - \phi_B = (V_{FB} - V)^2/4V_{FB} \qquad (4)$$

where V_{FB}, the flat band voltage of such devices, is the applied voltage at which the potential barrier is reduced to zero, and based on the depletion layer approximation is given by [8]

$$V_{FB} = \frac{q\ N_{BC}\ L^2}{2\ \epsilon_o\ \epsilon_s} \qquad (5)$$

with L being the spacing between highly doped regions (or device channel), ϵ_o - permittivity ($8.85418 * 10^{-14}$ F/cm), and ϵ_s the silicon dielectric constant (11.9).

Thus the current density J - voltage V relationship is given as

$$J \quad \alpha \quad \exp[- \frac{q}{kT} \frac{(V_{FB} - V)^2}{4 \ V_{FB}}] \tag{6}$$

in both the thermionic and the diffusion controlled model. The preexponential factor (constant of proportionality in Eqn. 6) depends whether the diffusion or thermionic emission mechanism is a dominant mechanism for the flow of carriers.

When the applied voltage is so large that ϕ_B is close to V_{bi}, the mobile carrier density becomes large, the depletion layer model breaks down, and the above treatment is not valid. If the mobile carriers dominate, the current density is roughly given by the space charge formula [8]

$$J = \frac{2 \ v_{sat} \ \epsilon_s \ \epsilon_0}{L^2} V \tag{7}$$

where v_{sat} is the saturated carrier velocity.

IV. Results of computer simulation and comparison with analytical solutions

A number of n^+-p-n^+ devices were simulated using the GESIM1 program [17]. Three parameters were varied: the doping concentrations of the p intervening region $N_{BC} = N_A$, the doping concentration of the highly doped regions N_S, and the spacing L between the n^+ regions. The inserts to the figures that follow show the assumed doping profiles.

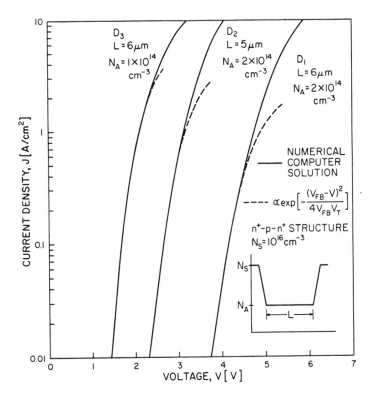

Figure 2. Numerically computed J-V characteristics of various n^+-p-n^+ structures compared with those calculated analytically using Eqn. (8).

Figure 2 shows the resulting J-V relationships for 3 devices, D_1 through D_3. The analytical calculations were made using the form of Eqns. (1) or (2)

$$J = J_0 \exp[-\frac{q}{kT} \frac{(V_{FB} - V)^2}{4 V_{FB}}]$$
(8)

where J_0 is chosen to give the best fit of the analytical results with the simulated results at low current levels. The values of J_0 that meet this criteria are shown in Table 1.

Table 1. Values of J_O for computed devices.

DEVICE	$V_{FB}[V]$	L [μm]	N_A [cm^{-3}]	N_S [cm^{-3}]	J_O [A/cm^2]
D_1	5.47	6	$2*10^{14}$	10^{16}	1.7
D_2	3.8	5	$2*10^{14}$	10^{16}	2.5
D_3	2.74	6	$1*10^{14}$	10^{16}	3.5

The analytical results compare well with the simulated predictions for values of $V<V_{FB}$, but deviate substantially from the computer simulated predictions as V approaches V_{FB}. As mentioned before, this deviation is presumably due to the effect of a large number of mobile carriers present in the p-region and the breakdown of the depletion layer approximation. The actual point at which the analytical curve deviates from the computer simulated curve also depends upon the fact that A_2 or J_O is a function of the applied voltage. However, this does not change the discussion and conclusions reached in this paper.

We now discuss whether thermionic emission or diffusion is more important in determining the J-V relation. It is known from Eqns. (1) and (2) that in the case of thermionic emission, J_O is independent of the doping concentrations and the dimensions of the device. In the case of the diffusion controlled mechanism, J_O depends on the doping concentration as well as the width of the barrier (and thus the spacing between the highly doped regions). The barrier width also changes with the applied potential and this introduces a small variation in J_O with applied potential. Tab. 1 shows that J_O is indeed different for the three devices, supporting the diffusion mechanism theory. It is possible that both thermionic emission and diffusion are important. However, it has been argued recently [18], in the case of transport over a barrier due to a grain boundary, that the transport must be diffusion controlled because unlike a metal-semiconductor contact, the carriers are transported and accepted much more rapidly when there is silicon on both sides of the barrier. The arguments given for the grain boundary generated barriers also apply to n^+-p-n^+ or p^+-n-p^+ structures. Temperature effects reported in reference [19] support the diffusion theory and cannot be explained by the thermionic emission theory. We therefore conclude that simple thermionic emission theory is not adequate to describe (in general) characteristics of n^+-p-n^+ diodes.

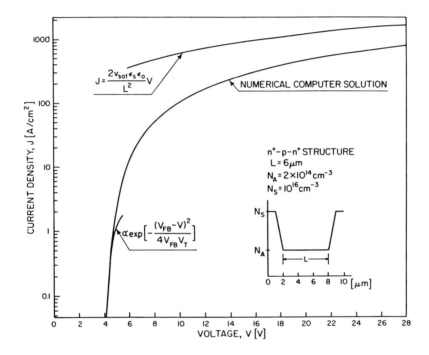

Figure 3. *Numerically computed J-V characteristics of a 6-μm n^+-p-n^+ device compared with those obtained analytically using Eqns. (7) and (8).*

Figure 3 shows a plot of Eqns. (7) and (8) along with the computer simulations for device D_1. It is seen that there is a significant discrepancy between the computer calculations and the plot of Eqn. (7). This discrepancy can be explained if one considers the plots of electric field E calculated by the computer program **GESIM1** versus distance for various applied voltages (Figure 4). Note that **E** is substantially smaller than $5*10^4$ V/cm near the forward biased junction, even at the highest voltages studied. Eqn. (7) is therefore not rigorously applicable to such problems because of the electron velocity, **v**< v_{sat} in a part of the channel. This is an important result since Eqn. (7) has been used extensively to interpret results in a variety of structures [8,14,15].

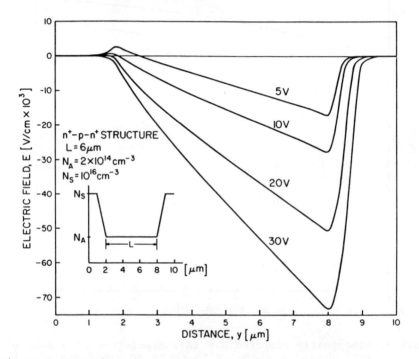

Figure 4. Numerically computed electric field distribution for the 6-μm n^+-p-n^+ device (as in Figure 3) as a function of the applied potential.

Figure 5 shows results of numerically computed **J-V** characteristics for various background concentration and source doping levels. The computer plots show a dependance on the doping levels, but the plots of J approach the analytical curves of Eqn. (7) asymptotically as V increases and becomes independent of the doping levels. Eqn. (7) is a good approximation only in the limit of very large applied voltages.

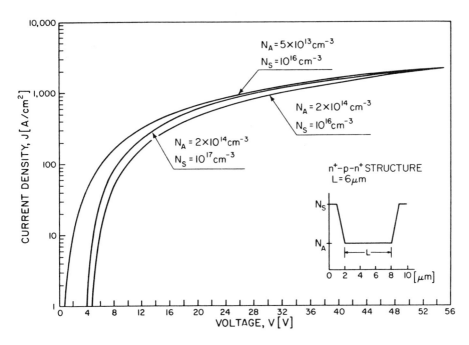

Figure 5. Numerically computed J-V characteristics of 6-μm n^+-p-n^+ devices (N_A and N_S varied).

V. Summary and conclusions

To summarize, we can say that at very low applied voltages the analytical model agrees with the simulations. The thermionic emission theory to predict current flow is not adequate to interpret simulated (and experimental [20]) results for n^+-p-n^+ or p^+-n-p^+ structures. When the applied voltage is such that ϕ_B is close to V_{bi}, none of the equations (6) or (8) holds. Eqns. (6) or (8) become invalid because the depletion region approximation fails. At large values of applied voltage the analytical model is inadequate since it predicts the current assuming saturated carrier velocities throughout the channel. Eqn. (7) fails because there is always some portion in the vicinity of the source junction where the electric field is not sufficiently large for the carrier velocity to be saturated. The computer simulation provides an improved estimate of the J-V relationship.

REFERENCES

[1] Watanabe, Y. and J. Nishizawa, Japanese Patent 205068: published No. 28-6077 (Fig. 15): Application Date Dec.1950.
[2] Shockley, W. and R. C. Prim, Phys. Rev., vol. 90, p. 753, 1953.
[3] Shockley, W., U.S. Patent 279-0037, Mar. 14, 1952.
[4] Dacay, G. C., Phys. Rev., vol. 90, p. 749, 1953.
[5] Nishizawa, J., Terasaki, T. and J. Shibata, IEEE Trans. Electron Devices, vol. ED-22, pp. 185-197, April 1975.
[6] Nishizawa, J., Ohmi, T. and H. Chen, IEEE Trans. Electron Devices, vol. ED-29, pp. 1233-1243, August 1982.
[7] Nishizawa, J. in Semiconductor Technologies, 1982, Nishizawa, J. ed., pp.201-219, Ohmsha, Ltd., Tokyo, Japan.
[8] Sze, S. M., Physics of Semiconductor Devices, Wiley, New York, 1981.
[9] Jain, S. C and P. Balk, Solid-State Electronics, vol. 30, pp. 503-511, 1987.
[10] Grossman, B. M., Hwang, W. and F. F. Fang, Solid-State Electronics, vol. 27, pp. 1083-1090, December 1984.
[11] Auston, D. H. in Picosecond Optoelectronic Devices, C. H. Lee, ed., Academic Press, New York, 1984.
[12] Mattson, R. H. and B. M. Wilamowski, Proc. of the Second International Workshop on the Physics of Semiconductor Devices, pp. 45-52, Dec. 5-10, 1983, Delhi, India.
[13] Wilamowski, B. M., Mattson, R. H., Staszak, Z. J. and A. Musallam, 36th Electronic Components Conference, pp. 399-404, May 5-7, 1986, Seattle, WA.
[14] Sze, S. M., Coleman, D. J. and A. Loya, Solid-State Electronics, vol. 14, pp. 1209-18, 1971.
[15] Chu, J. L., G. Persky, and S. M. Sze, J. App. Phys., vol. 43, pp. 3510-3515, August 1972.
[16] Gray, P. E. et al, Physical Electronics and Circuit Models of Transistors, SEEC 2, chapter 10, Wiley, N.Y. 1964.
[17] Wilamowski, B. M., Staszak, Z. J. and R. H. Mattson, University of Arizona, ECE Dept., Tucson, Az, unpublished draft.
[18] Plotka, P. and B. M. Wilamowski, Solid-State Electronics, vol. 24, pp. 105-107, 1981.
[19] McGonigal, G. C. et al., Physical Review B, vol. 28, p. 5908, 1983.
[20] Jain, S. C., Staszak, Z. J. Musallam, A. and R. H. Mattson, to be published in J. Appl. Physics.

A Quasi One-dimensional Model for Short-Channel MOST

C.R.Viswanathan
Electrical Engineering Department
University of California
Los Angeles, CA 90024

ABSTRACT

This paper describes a one-dimensional model that accurately predicts the characteristics of a short-channel MOS transistor. The procedure is based on a current-driven model and successfully includes the field-dependent mobility. The saturation characteristics are accurately predicted. The model can predict the behavior of the device at low temperatures as well as at higher temperatures. The model in addition to being useful in anlalog circuit application is valuable for gaining an insight into the physics of short-channel devices.

Introduction

Short-channel effects arise due to the proximity of the source and the drain. Due to the perturbation of the electric field in the space-charge region under the gate the field lines diverge laterally instead of proceeding vertically towards the gate. This means that the gradual channel approximation cannot be used to determine the inversion-charge density in the channel. Due to the two-dimensional effects present in the device, modeling short-channel devices becomes complicated. Numerical analysis programs such as Pisces [1] or Minimos [2] are typically used to model and simulate short-channel devices. Circuit simulation programs such as Spice [3] incorporate expressions involving a number of empirical parameters to include short-channel effects. A model called BSIM [4] is another example of models using empirical parameters to take into account short-channel effects. However, the physical significance of these parameters is obscure. These models do not give an insight into the physics of the device. In this paper, a quasi one-dimensional model is presented which has several attractive features such as correct inclusion of the field-dependent mobilty, a smooth transition from linear to saturation regions of operation, and an accurate prediction of output resistance in saturation region. The model provides an understanding of the physics of the device. The model assumes that the divergence of field lines in the oxide layer between the inversion layer and the gate can be neglected for the purpose of calculating the current upto the onset of saturation. In the following sections the model is described and the results of the application of the model are compared with experimental results. The model enables the determination of the variation of the mobilty as a function of position in the channel.

Current-Driven Model

In an earlier paper [5], a procedure was outlined by which the drain voltage was calculated for a given current passing through the device whereas normally in MOS models, the current through the device is calculated for a given value of drain voltage applied across the device. To distinguish the former from the latter, the former is called the current-driven model and the latter is called the voltage-driven model.

The general expression for the current at some distance y in the channel is

$$I_D = W\mu Q_i \frac{dV}{dy} \tag{1}$$

where W is the width of the device, μ is the effective mobility, Q_i is the inversion charge density and V is the Fermi potential (Voltage) at y, measured with respect to the source at y=0.

In the voltage-driven model, the drain current[6] is obtained as

$$I_D = \frac{1}{L} \int_0^{V_D} \mu W Q_i dV \tag{2}$$

where L is the length of the channel. In simple cases such as uniform impurity concentration in the substrate and constant mobility, the integration is readily performed to obtain closed form solutions. However, the mobility in general is a function of both perpendicular (x direction) and parallel (y direction) components of the electric field in the channel.

$$\mu = \mu(E_x, E_y) \tag{3}$$

$$E_y = \frac{d\psi_s}{dy} \tag{4A}$$

$$E_x = \frac{1}{\varepsilon_s}(\frac{Q_i}{2} + Q_{dep}) \tag{4B}$$

where Q_{dep} = depletion charge density at y, Q_{inv} = inversion charge density at y, and ε_s = permittivity of silicon.

It is not possible to include the field-dependence of the mobility in equation (2), since there is no explicit space dependence of the voltage V in the integral. What is usually done[7] is to average the mobility over the entire channel length and to take it outside the integral. A second disadvantage of the voltage-driven model is that no information can be gained as to how the electric field or the voltage drop varies as a function of position in the channel.

In the current driven model, equation (1) is rewritten as

$$dV = \frac{I_D}{\mu W Q_i} dy \tag{5}$$

Integrating between 0 and y, the voltage at any value of y is obtained as

$$V(y) = I_D \int_0^y \frac{dy}{\mu W Q_i} \tag{6}$$

Thus $V(y)$ and therefore the electric field are obtained as a function of position in the channel. Hence the mobility can be corrected for field dependence at every point in the channel. The drain voltage is obtained as

$$V_D = \int_0^{V_D} dV = I_D \int_0^L \frac{dy}{\mu W Q_i} \tag{7}$$

We obtain the drain voltage V_D for a specific value of I_D and the electric field dependence of the mobility at every point in the channel is properly included when carrying out the integral in equations (6) and (7).

Q_i is a function of the surface potential $\psi_s(y)$ and the voltage $V(y)$. Considering only the magnitudes of charge and voltage

$$V_G = \frac{Q_{sc}(\psi_s(y), V(y))}{C_{ox}} + \psi_s(y) \tag{8}$$

and

$$Q_i = Q_{sc}(\psi_s(y), V(y)) - Q_{dep}(\psi_s(y)) \tag{9}$$

Any two of these four quantities V_G, $Q_i(y)$, $\psi_s(y)$ or $V(y)$ must be known, to calculate the other two. Thus the integration cannot be carried out in a closed form. However at the source $V(y) = 0$ and V_G is given. Hence Q_i and ψ_s can be determined at y = 0. A numerical integration can then be carried out to obtain $V(y)$ and V_D. The numerical integration is accomplished by dividing the channel into a number of elementary sections of length dy. The voltage drop dV across each elementary section is given by equation (5). A summation of the voltage drop across all elementary sections yields the drain-to-source voltage. The summation is carried out as follows.

From measurements on a long channel device at a low drain voltage, the mobility has been obtained as a function of perpendicular electric field. The mobility dependence followed the universal curve published by others [8,9]. At the source the perpendicular electric field is calculated for the given values of the terminal voltage and the mobility is corrected. The mobility is not corrected for tangential electric field dependence at the source. For an assumed drain current, the voltage across the first elementary section dy is calculated using equation (5). The value of V(y) in the second elementary section is taken as the voltage drop across the first section. In the second section Q_i is calculated and mobility is corrected for both perpendicular and longitudinal electric fields. The longitudinal field in any section between y and $y - dy$ is approximated as

$$E_y(y) = \frac{\Psi_s(y) - \Psi_s(y-dy)}{dy}$$
(10)

while the perpendicular field E_x is obtained using equation (4B). Thus in every elemental section dy the mobility is corrected for field dependence. This stepping procedure is repeated until the drain is reached. At each step a new value for $V(y)$ is obtained by adding the incremental voltage given by (5) to the channel voltage at the previous step. The summation of the voltage drops across all elementary sections of the channel corresponds to a numerical integration of equation (5) along the channel.

The expression of Cooper and Nelson [10], which was obtained using time-of-flight technique, was used for longitudinal field dependence and is given by

$$\mu = \frac{\mu_0}{\left[1 + \left[\frac{\mu_0 E_y}{v_s}\right]^\alpha\right]^{1/\alpha}}$$
(11)

They obtained a good fit to measured data with $\alpha=1.92$ and $v_s=9.23\times10^6$ cm/s. Here, μ is the effective mobility in the inversion layer, μ_0 is the mobility degraded only by the perpendicular field, E_y is the tangential field and v_s is the saturation velocity. We obtained a good agreement between measured curves and simulated data using equation (11) with the parameter values of Cooper and Nelson.

For the calculation of $Q_i(y)$ a MOS analysis program, based on a one-dimensional Poisson solver [11] is used. Input to the program are temperature, oxide thickness, doping profile and the distribution of interface traps in the bandgap. Using a finite difference technique, the program solves the Poisson equation iteratively for the boundary conditions of bias voltages specified by the user. In the physical formulation, dopant freeze-out effects and degenerate statistics are accounted for. The total charge density in the space-charge region, the inversion charge density and the depletion charge density as well as the low and high frequency capacitance data are obtained from this program as a function of device bias. The advantage of using the program is that it can take into account any arbitrary non-uniform impurity distrubution. It can also be used at low temperatures.

Under conditions of saturation, the integration can be carried out only up to the pinch-off point ($y = L_p$) and the pinch-off voltage V_p is obtained.

The current-driven model can thus be used to obtain the drain voltage V_D for each value of I_D in the pre-saturation regime and the pinch-off voltage V_P and the pinch-off point L_p in the saturation regime. The I_D - V_D characteristic in the saturation regime can be generated using the technique described in a later section. Mobility is properly accounted for electric field dependence at every point in the channel since, $\psi_s(Y), V(Y)$ and their gradients are known in the channel.

Short-Channel Modeling

As stated earlier, the determination of the inversion charge-density as a function of position in the channel can not be accomplished by assuming gradual channel approximation. The band-bending (the surface potential) is a minimum in the middle of the channel and increases symmetrically towards the source and the drain for low drain voltages [12]. The inversion charge-density, hence increases from the mid-channel value to larger values as the source is approached (and also as the drain is approached at low values of drain voltage.) This gives rise to increased conductance near the source (and the drain) region. The inversion charge- density in the channel at a distance y from the source is the sum of $Q_{iu}(y)$, the unperturbed inversion charge density and $\Delta Q_i(y)$, the increase in inversion charge density due to the proximity of the source. $\Delta Q_i(y)$ is obtained as follows: From the basic charge balance equation for a MOS structure one obtains

$$V_{GS} = \Psi_s(y) + \frac{Q_{dg}(y)}{C_{ox}} + \frac{Q_i(y)}{C_{ox}} \tag{12}$$

where Q_{dg} denotes the charge density in the depletion region that is induced by the gate. Equation (12) assumes a negligible divergence of induced by the gate. position measured from field lines in the oxide. In the middle of the channel

$$V_{GS} = \Psi_s(\tfrac{1}{2}L) + \frac{Q_{dg}(\tfrac{1}{2}L)}{C_{ox}} + \frac{Q_i(\tfrac{1}{2}L)}{C_{ox}} \tag{13}$$

As y approaches zero, Ψ_s increases due to charge sharing effects, Q_{dg} decreases and Q_i increases.

$$V_{GS} = \Psi_s(\tfrac{1}{2}L) + \Delta\Psi_s(y) + \frac{Q_{dg}(\tfrac{1}{2}L) - \Delta Q_{dg}(y)}{C_{ox}} + \frac{Q_i(\tfrac{1}{2}L) + \Delta Q_i(y)}{C_{ox}} \tag{14}$$

However, $Q_i(y)$ is exponentially increasing with Ψ_s and hence $\Delta\Psi_s$ is small compared with the terms $-\Delta Q_{dg}/C_{ox}$ and $+\Delta Q_i/C_{ox}$. Therefore neglecting $\Delta\Psi_s$,

$$V_{GS} = \Psi_s(\tfrac{1}{2}L) + \frac{Q_{dg}(\tfrac{1}{2}L) - \Delta Q_{dg}(y)}{C_{ox}} + \frac{Q_i(\tfrac{1}{2}L) + \Delta Q_i(y)}{C_{ox}} \tag{15}$$

Thus it is reasonable to assume that the increase in inversion charge density as the source (or drain) is approached is equal to the decrease in gate induced depletion charge density. To determine ΔQ_i as a function of distance from the drain/source, threshold-voltage measurements are carried out on a set of MOS transistors with varying length. The threshold voltage is measured in strong inversion from the intercept with the V_{GS}-axis of the extrapolated maximum slope of the I_D-V_{GS} curve obtained at low V_{DS}. ΔV_T, the difference in threshold voltage between a long-channel device and a short-channel device of length L is approximated as

$$\Delta V_T = \frac{\Delta Q_i\ (\tfrac{1}{2}L)}{C_{ox}} \tag{16}$$

where $\Delta Q_i\ (\tfrac{1}{2}L)$ symbolizes ΔQ_i in the middle of the short-channel device. Since everything is symmetric (small V_{DS}) the contribution to ΔV_T from the drain should be equal in magnitude to the contibution from the source. The contribution from drain or source at a distance $\tfrac{1}{2}L$ is then

$$\Delta Q_i\ (\tfrac{1}{2}L) \approx C_{ox}\ \tfrac{1}{2}\ \Delta V_T \tag{17}$$

The threshold measurements on different length devices yields the inversion charge-density in the middle of the channel and this can be used to obtain the extra inversion charge-density as a function of distance $\Delta Q_i(y)$, employing equation (17). An example of this is shown in Figure (1).

Saturation and Output Impedance

At large values of the drain voltage when the channel is pinched off, the current-driven model can only yield the pinch off voltage and the location of the pinch-off point. The voltage-drop between the drain and the pinch-off point is difficult to predict because of two dimensional effects. The simple one dimensional p-n junction model used by Reddi and Sah [13] or the refinements[14] of that model are not adequate. In our procedure we make the assumption that this voltage drop between the pinch-off point and the drain is determined only by the doping profile in this region, by the oxide thickness and by the gate voltage and is independent of the distance between the source and the pinch-off point. This means that the voltage drop ΔV as a function of ΔL, the distance of the pinch-off point from the drain, is assumed to be independent of the channel length. A single $\Delta V(\Delta L)$ function can therefore be used to calculate the drain voltage in saturation for a transistor of any length for a given gate voltage and a given fabrication process.

Ideally the function $\Delta V(\Delta L)$ can be determined using a 2d numerical analysis program such as Pisces or Minimos. However, in practice, this is found to be more difficult than a simpler method we have adopted. Using the experimental data on a device of one channel length and results of the current-driven model on this device, the $\Delta V(\Delta L)$ function is generated for each gate voltage as shown in Figure (2).

This function $\Delta V(\Delta L)$ is used to determine the I_D-V_D characteristic in the saturation regime of devices with different channel lengths. The results of application of this procedure are shown in Figures (3) and (4) for devices with three different channel lengths (1.2, 1.7 and 2.7 μm) for gate voltages of 2V and 5V respectively[15]. The current-voltage characteristics were obtained using the current-driven model up to the onset of saturation and the saturation characteristics were obtained using the above procedure. As can be seen, the agreement between measured and simulated data is very good. In particular, the transition between the linear and saturation regimes is well modeled. This is usually a weak point in other models.

Discussion

The results of applying the current-driven model to obtain the mobility, the channel voltage and the electric field as a function of position in the channel in a long-channel device $(L=49.7\mu m, V_G=5V \text{and} V_D=3.6V)$ and in a short-channel device $(L=0.7\mu m, V_G=5V \text{and} V_D=1.75V)$ are shown in Figures (5) through (8). The bias voltages were so chosen[15] as to keep the device operating below saturation in both cases. Figure (5) yields the channel voltage as a function of position in the channel. Figure (6) shows the tangential field (along y) for the two cases. In the long-channel cases the electric field is small except near the drain where it reaches a value slightly in excess of $2KV/cm$. In the short-channel device, the electric field is very high through out the channel ($>8KV/cm$). Figure (7) shows the mobility variation in the long channel device as a function of position in the channel. Both μ_0 (the part that depends only on the perpendicular field) and μ, the total mobility are plotted. Due to the small tangential field, μ is very close to μ_0. A slight deviation is obtained near the drain where the electric field is higher. We observe the mobility to increase towards the drain because the normal electric field decreases as the drain is approached.

On the other hand, in short- channel devices, the mobility variation is quite different as shown in Figure (8). μ_0 still increases towards the drain. However, the total mobility μ is reduced by a large tangential field even at the source and decreases still further as the drain is approached.

Conclusion

A quasi one-dimensional model that provides an insight into the channnel conductance process has been described. The model predicts the experimentally observed behavior of the device very accurately at all regions of operation of the device and at low temperatures. The model brings out explicitly the mobility dependence on longitudinal field in short-channel devices. The model provides an experimental vehicle for testing different expressions for field-dependent mobility. The extra inversion charge near the source can be accurately predicted by a two-dimensional numerical analysis program although in this work experimental data on actual devices were used. The model also presribes a procedure for predicting the output resistance in saturation. Again here, experimental data on one short-channel device was used to predict the output resistance of all short-channel devices fabricated in the same process, a numerical two-dimensional Poisson solver can provide the $\Delta V(\Delta L)$ data. The model demonstrates that the prediction of short-channel current-voltage characteristics can be made

112

accurately for all channel lengths provided a knowledge of two parameters both dependent on the fabrication process is available. One is the extra inversion charge density variation near the source and the other is the function $\Delta V(\Delta L)$ for the pinch-off region.

References

[1]   M.R.Pinto et al., "PISCES II:Poisson and Continuity Equation Solver", Stanford Electronics Laboratories, Technical Report, Sept. 1984

[2]   S.Selberherr et al., "MINIMOS-A Two-Dimensional Transistor Analyzer" *IEEE Trans.Electron Devices*,Vol.ED-27, August 1980, pp1540-1550

[3]   A.Vladimirescu and S. Liu, " The Simulation of MOS Integrated circuits using SPICE2", *UCB/ERL M80/7* Feb. 1980

[4]   B.J.Sheu et al., "BSIM: Berkeley Short-Channel IGFET Model for MOS Transistors" *IEEE Journal of Solid State Circuits,PVol.SC-22, August 1987, pp558-566*

[5]   C. R. Viswanathan, "MOS MOdeling," Proceedings of the Custom Integrated Circuits Conference, pp 199-204, Rochester, N. Y., 1982.

[6]   S.M.Sze, *Physics of Semiconductor Devices*,2nd Edition, Wiley, New York, 1977

[7]   M.H.White et al., "High Accuracy MOS models for computer aided design" *IEEE Trans.Electron Devices*,Vol.ED-27, pp899-906, May 1980

[8]   A. G. Sabnis and J. T. Clemens, "Characterization of the Electron Mobility in the inverted <100> Si Surface," in *IEDM Tech. Dig.*, vol. 18, 1979.

[9]   S. C. Sun and J. D. Plummer, "Electron Mobility in Inversion and Accumulation Layers on Thermally Oxidized Silicon Surfaces," *IEEE Transact. Electron Devices*, vol. ED-27, p. 1497, 1980.

[10]   J. A. Cooper and D. F. Nelson, "High-Field Drift Velocity of Electrons at the $Si-SiO_2$ Interface as Determined by a Time-of-Flight Technique," *J. Appl. Phys.*, vol. 54, pp. 1445-1456, 1983.

[11]   R. C. Jaeger, F. H. Gaensslen, and S. E Diehl, "An Efficient Numerical Algorithm for Simulation of MOS Capacitance," *IEEE Trans. CAD*, vol. CAD-2, pp. 111-116, 1983. 914 (1980).

[12]   C. R. Viswanathan, B. C. Burkey, G. Lubberts, and T. J. Tredwell, "Threshold Voltage in Short-Channel MOS Devices," *IEEE Trans. Electron Dev.*, vol. ED-32, pp. 932-940, 1985.

[13]   V. G. K. Reddi and C. T. Sah, "Source to Drain Resistance Beyond Pinch-Off in Metal-Oxide-Semiconductor Transistors (MOST)," *IEEE Trans. Electron Dev.*, vol. ED-12, pp. 139-141, 1965.

[14]   D. Frohman-Bentchkowsky and A. S. Grove, "Conductance of MOS Transistors in Saturation," *IEEE Trans. Electron Dev.*, vol. ED-16, pp.108-113, 1969.

[15]   Jon Wikstrom, *UCLA Ph.D Dissertation 1987*

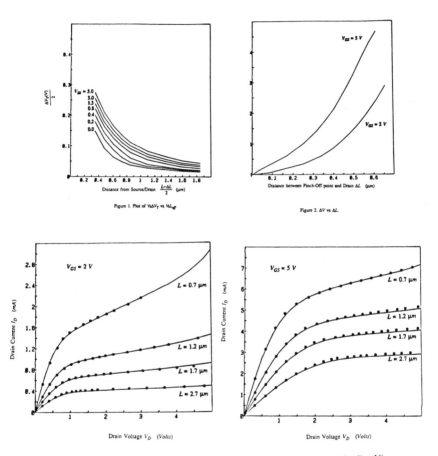

Figure 1. Plot of ½ΔV_T vs ½L_eff

Figure 2. ΔV vs ΔL.

Figure 3. Drain current vs drain voltage V_{GS} = 2 V.

Figure 4. Drain current vs drain voltage V_{GS} = 5 V.

114

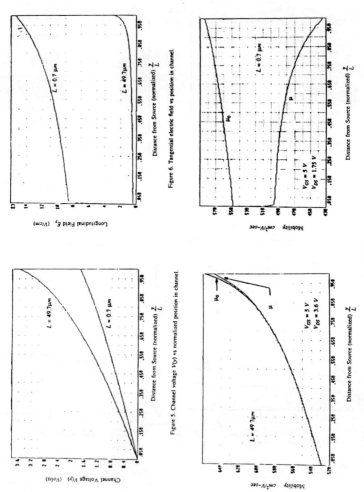

Figure 5. Channel voltage V(y) vs normalized position in channel.

Figure 6. Tangential electric field vs position in channel.

Figure 7. Long-channel effective mobility vs position in channel.

Figure 8. Short-channel effective mobility vs position in channel.

VLSI PACKAGING DESIGN METHODOLOGY CONSIDERATIONS*

J. L. Prince, O. A. Palusinski, Z. J. Staszak and M. R. Scheinfein

Department of Electrical and Computer Engineering
University of Arizona, Tucson, AZ   85721, USA.

I.   Introduction

The development of models and simulation tools for design of electronic packaging structures has lagged the development of tools for chip design. This is especially true in the electrical performance area. Trends in chip and system characteristics [1,2] and the current state of the art indicate that the ability to verify packaging designs, and optimize materials and geometries before hardware is available, is as necessary for packaging structures as it is for chips. This paper discusses a research program which addresses the problem of electrical and thermal performance of Level 1 and Level 2 integrated circuit packaging structures by developing models and simulation tools, and characterization procedures. The program has as a long term goal the integration of simulation tools into an interactive design/analysis system such as is shown schematically in Figure 1 [1]. In such a system electrical and thermal/mechanical modeling and simulation tools are integrated into modules which interact with each other and with a CAD data base which contains the geometrical information about the design. This design system should allow the prediction of electrical, thermal and mechanical (thermally induced stress) characteristics of Level 1 and Level 2 packages. It should also have built-in some degree of expertness, so as to aid the designer and speed up the simulation process.

Following sections will discuss electrical performance modeling and simulation tools, and thermal/mechanical performance modeling and simulation tools developed under this research program. Although research work is now underway in development and integration of the design system itself, it is too early to discuss those developments.

II.   Electrical Modeling and Simulation Tools

TEM-Wave Parameter Modeling Tools

The first step in approaching the modeling of the multi-conductor, multiple-dielectric problem was to make the following simplifications:

a)   assume the validity of the TEM-wave approximation; thus Maxwell's equations are reduced to Poisson's (or Laplace's) equation in the low frequency limit (all charges and currents on the surfaces of infinite-conductivity conductors).

*Research supported in part by Semiconductor Research Corporation under Contracts 84-06-050, 86-07-086, and 87-MP-086.

116

b) assume two-dimensional structures so that the cross-section of the structure is independent of distance along the conductor.

Under these simplifications two modeling tools were developed. One, called UAC1.1, depends on the semi-analytic Greens' function method. This particular technique of determining the short circuit capacitances for a system of conductors is based on Week's [4] method. It is assumed that the conductors are ideal and that all of the charges and currents lie on the surface. This method has been developed for the three special cases shown in Figure 2. These are case (1) N conductors between two infinite ground planes with one dielectric present, case (2) N conductors, one infinite ground plane and one dielectric interface, and case (3) N conductors, one finite ground conductor, and one dielectric interface. Presently, these are the only configurations treated by this method due to the availability of applicable Green's functions. Some discussion of numerical techniques embodied in UAC1.1 has been published [5]. Note that development of the UAC1.1 program is based upon the knowledge of the real space Green's function for a specified geometry. Although the Green's function is known for several configurations important in the modeling of electronic packages, in general not all configurations can be handled. Further, for multiple dielectric\ground plane layers, the Green's functions contain infinite series, which must be numerically truncated resulting in extended computation times.

To overcome the difficulties associated with the above method, we have implemented a very general two-dimensional quasi-static TEM transmission line parameter extractor. The integral equations for the boundary value problem are cast by using the free space Green's function and the polarization charge along dielectric/dielectric interfaces. These equations are solved by the method of moments using pulses for expansion and point matching. This tool, called UAMOMENT1.0, may be used to evaluate the capacitance [C], inductance [L], and conductance [G] matrices for an arbitrary number of conductors and an arbitrary number of dielectric layers [6]. The conductors may have arbitrary shape given that they must be discretized by straight line sub-intervals and the dielectric layers are constrained to lie in parallel layers (a constraint which can be removed). An arbitrary number of finite sized ground conductors may be included, as well as one or two infinite ground planes. The conductors may be of finite cross-section or they may be infinitely thin. A composite of the geometric cases comprehended by this model is shown in Figure 3.

The UAMOMENT1.0 program has been implemented and internally tested. Convergence studies and detailed examples are included in a user's manual. The program runs in two environments. The PC version will handle up to 6 conductors and 6 dielectric layers and a total of 60 sub-intervals. The PC version has graphics capabilities to analyze input configurations and equipotential surfaces. The VAX/VMS, UNIX SYSTEM V, and BSD #4.2 UNIX versions will handle up to 10 conductors and 10 dielectric layers and a total of 200 sub-intervals. At the present time, this VAX version has no graphics capabilities. No further work on static two dimensional models is foreseen.

Some examples of the utility and flexibility of these tools are shown in Figures 4 and 5. Figure 4 shows the geometry of buried microstrip line, and calculated results of characteristic impedance Z_0 versus geometric parameters [7]. Figure 5 shows a four-conductor system, with one dielectric interface and one grounded conductor, and the calculated capacitance matrix.

A three-dimensional quasi-static TEM parameter extractor which uses the method of moments has been implemented [8]. It will calculate the three dimensional charge density distribution, subject to the normal surface charge approximation, and other conditions discussed below. The algorithm is identical to the algorithm used to solve the 2-dimensional problem in UAMOMENT1.0. The program will accept either one or two infinite ground planes and/or an arbitrary number of finite sized ground conductors. The conductors may be of finite volume or infinitely thin. The dielectric layers are constrained to lie in horizontal planes. This program, as a preliminary version, constrains the bounding planes of the conductors to lie in one of the three cartesian planes. This allows the matrix elements to be evaluated analytically, thereby saving a factor of 100 in run time.

This program has been implemented and internally tested for a number of simple packaging configurations. As well as outputting total capacitance values, the program will discretize the conductors lengthwise (in z) and output a per-unit-length capacitance $[C(z)]$, inductance $[L(z)]$, and conductance $[G(z)]$ matrix for each interval along the conductor length. This will provide the data input into a non-uniform line transient response TEM line simulator.

TEM-Mode Transient Waveform Simulators

The general N-coupled-line problem which must be simulated, in performing interconnect design, is shown in Figure 6 [9]. Simulation of TEM-mode transmission line electrical characteristics (crosstalk, pulse distortion, etc.) requires both knowledge of the line parameters and application of a simulation program or tool. SPICE is one such simulation tool. Extensive experimentation with SPICE [6] has shown that it is flexible, versatile, and extremely profligate of CPU time when used for coupled transmission line simulations. Two alternate tools were developed to perform TEM transmission line simulations more efficiently. Both tools presently suffer from the inability to incorporate line non-uniformities and non-linear driving-point impedances. However, they provide simulation results which are very close to SPICE results, and in good agreement with available experimental data, for "long" coupled lines. CPU times are two to three orders of magnitude faster for the alternate tools, compared to SPICE.

One lossless coupled line simulation tool is named UATL. It is based on analysis exploiting the fact that the solutions to the wave equation have components of the form $v^m(X,t) = V_o^m g^m(t \pm X/c_m)$ for $m = 1,2. \ldots, M$, where V_o^m is an eigenvector and the function g^m represents the incident (- sign) or the reflected (+ sign) wave. The constant c_m represents the modal velocities calculated from $det(1/c_m^2 - LC) = 0$. It was assumed in developing the first version of UATL that the modal solutions are determined by the resistive boundary conditions (near and far end voltage generators with internal resistances). For the second lossless coupled line simulation tool (PCTEM) a matrix method was developed whereby the lines are decoupled via a convergence transformer transformation [9]. Once the lines have been effectively decoupled, the voltage node equations are solved in the time domain. Results of applying PCTEM and SPICE to simulation of two long coupled lines is shown in Figure 9 [9]. Also shown in this figure are experimental results for the two-line system [10].

A lossy (low-loss case) TEM-mode transmission line simulator has been developed, based on modal analysis in the frequency domain. Termination networks must be linear and time-invariant. The time domain excitations in these networks are transformed into the frequency domain with an FFT. Modal analysis of the line system is performed at each frequency to calculate the response. Finally, the signals are inverse transformed to determine the time domain response.

III. Thermal Modeling and Simulation Tools

For thermal modeling of the complex-geometry packaging and interconnect systems, multi-dimensional modeling is necessary. Conduction, convection and radiation must be addressed, as must temperature dependent material parameters, with possible inhomogenity and anisotropy included. Also, necessary are interfaces with mechanical (structural) and electrical simulation modules, and real time input/output for interactive design. To address these problems, a Thermal and Thermally Induced Stress Modeling/Simula-tion System for VLSI-based packaging and assemblies was developed [11,12]. It is organized into four basic modules: PREPROCESSOR with an EDITOR, INTERNAL DATABANK, "NUMBER-CRUNCHER(S)" and POSTPROCESSOR. These modules have the following functions:

PREPROCESSOR: generates and edits the geometrical and material thermal and thermal-mechanical systems. It takes geometry descriptions, material definitions, boundary and initial conditions, and loads (thermal/mechanical). From this data, the preprocessor generates the input file for the specific numerical solution code ("number-cruncher") implemented in the system. A built-in EDITOR allows the user to make changes in the inputted file.

INTERNAL DATABANK: contains the list of physical, thermal and mechanical parameters such as density, thermal conductivity, specific heat, Young's modulus of elasticity, Poisson's ratio, coefficient of thermal expansion for materials which are or may be used in packaging applications. It also lists heat exchange modes applicable to packaging problems, and contains a library of models/structures.

"NUMBER-CRUNCHER": takes the combined output of the preprocessor and the databank, and provides numerical codes to solve the appropriate equations.

POSTPROCESSOR: generates and displays results. It provides the spatial distribution of results (temperatures, displacements/stresses) in arbitrary selected planes for a particular type of analysis (thermal/mechanical). Basic statistics and critical parameters of the analyzed structure are included. It also generates temperature data (thermal loads) for thermally induced stress analysis.

In this modular structure each module functions as a "stand-alone" program that interfaces with the rest of the system through a pre-defined and well-established structural I/O. The "stand-alone" concept allows the system to be developed with independent modules of varying degrees of sophistication and complexity.

A preliminary version of the system, called the Packaging Thermal Mechanical Calculator (UAPTMC1.0) comprises a 3-D thermal analysis capability (steady-state and transient conditions), a 2-D thermally induced stress analysis capability (elastic, steady-state and quasi-transient conditions), and an internal databank of physical parameters (thermal and mechanical material properties) and thermal and thermal-mechanical models. Presently in the system thermal modeling is based on the finite difference electro-thermal analogy, and thermal-mechanical analysis makes use of the finite element approach. UAPTMC includes/supports the following "number-crunchers": SPICE2G (widely available) for 3-D steady-state and transient thermal analysis, and the "in-house" developed programs NODAL(T) for 3-D steady-state and transient thermal analysis, and NODAL(M) and TIS for 2-D steady-state and quasi-transient thermally induced stress (elastic) analysis [11]. UAPTMC has been written in FORTRAN 77 and is currently implemented on a VAX 11/780 computer.

In a parallel effort, a prototype of an interface program for a general-purpose non-linear code ANSYS TM has been developed [12]. The program leads the user through the process of creating appropriate thermal and thermally induced stress analysis models and other operations necessary to run ANSYS. The tool, ANSYS Packaging Thermal Mechanical Calculator (APTMC), may be included in future versions of the UAPTMC system.

UAPTMC is being methodically tested on a variety of packages and under various heat transfer and load conditions. Examples range from (a) 16- and 40-pin plastic packages on PCB with various silicon-to-plastic area ratios, to (b) pin grid array packages, (c) multisource hybrid packages, and (d) packaged thermal test chips. A typical example of results obtained using the thermal modeling approach is shown in Figure 8. The figure shows a model of a packaged chip for thermal analysis, a unit volume used in thermal modeling and typical modeling results. Extensive thermal characterization of VLSI packaging is underway using both commercially available and in-house developed test equipment and test structures/chips. The displayed experimental data in Figure 9 is the response of the sensor cells that lie on the diagonal cross section of the 4x4 cell UA TTC03 test chip due to thermally pulsing the edge of the chip.

References

1) J. L. Prince, D. J. Hamilton, E. M. Matz and Z. J. Staszak, "The Role of Universities in Electronic Packaging Engineering" (Invited Paper), Proc. IEEE 73, pp. 1416-1423, Sept. 1985.

2) C. M. Val, "Trends in Packaging", Int'l J. Hybrid Micro-electronics 7, p. 21, June 1984.

3) L. W. Schaper and D. I. Amey, "Improved Electrical Perfor-mance Required for Future MOS Packaging", IEEE Trans. Compo-nents, Hybrids, and Manufacturing Technology, CHMT-6, p. 283, September 1983.

4) W. Weeks, "Calculation of Coefficients of Capacitance of Multiconductor Transmission Lines in the Presence of a Dielectric Interface", IEEE Trans. on Microwave Theory and Technique MTT-18, p. 35, 1970.

5) O. A. Palusinski, J. C. Liao, P. E. Teschan, J. L. Prince and F. Quintero, "Electrical Modeling of Interconnections in Multilayer Packaging Structures", IEEE Trans. Components, Hybrids, and Manuf. Technology CHMT-10, pp. 217-223, June 1987.

6) M. Scheinfein, J. C. Liao, O. A. Palusinski and J. L. Prince, "Electrical Performance of High Speed Interconnect Systems", accepted for publication in IEEE Trans. Components, Hybrids, and Manuf. Technology CHMT-10 (Sept. 1987).

7) J. Prince, M. Scheinfein, R. Senthinathan and O. Palusinski, "Electrical Characteristics of Buried Microstrip in the TEM Approximation", Proc. of the Technical Conference, Sixth International Electronics Packaging Conference, pp. 424-30, November 1986.

8) Michael Scheinfein and John Prince, "Electrical Performance
    of Integrated Circuit Packages: Three Dimensional
    Structures", Proc. of the 37th IEEE Electronic Components
    Conference, pp. 377-83, May 1987.

9) R. Senthinathan, J. Prince and M. Scheinfein, "Character-
    istics of Coupled, Buried Microstrip Lines by Modeling and
    Simulation", accepted for publication in IEEE Trans. Compo-
    nents, Hybrids, and Manuf. Technology CHMT-10 (Dec. 1987).

10) D. K. Lynn, C. S. Meyer and D. J. Hamilton, Analysis and
    Design of Integrated Circuits, New York: McGraw-Hill Book
    Company, 1967, p. 352.

11) Z. J. Staszak, J. L. Prince, B. Cooke and D. Shope, "Design
    and Performance of a System for VLSI Packaging Thermal
    Modeling and Characterization", accepted for publication in
    IEEE Trans. Components, Hybrids, and Manuf. Technology CHMT-
    10 (Dec. 1987).

12) J-J. Shiang, Z. J. Staszak and J. L. Prince, "ATPMC: An
    Interface Program for Use with ANSYS for Thermal and
    Thermally Induced Stress Modeling/Simulation of VLSI
    Packaging", 1987 ANSYS Conference Proc., pp. 11.55-11.62,
    Mar. 1987.

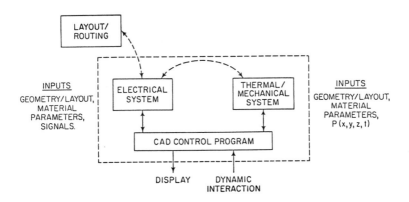

Figure 1.   Design and Simulation System for VLSI Packaging [1]

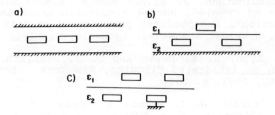

Figure 2. Geometric Cases Comprehended by Semi-Analytic Real-Space Green's Function Tool UAC1.1

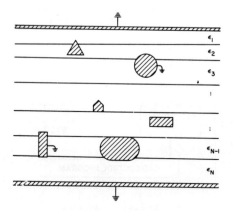

Figure 3. Composite of Geometric Cases Comprehended by Semi-Analytic Free-Space Green's Function "Method of Moments" Tool UAMOMENT1.0

SINGLE-CONDUCTOR GEOMETRY
NO OVERCOAT

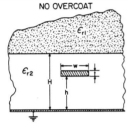

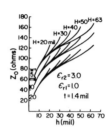

Figure 4.   a) Buried Microstrip Geometry
            b) TEM-Mode Z_0 for Buried Microstrip [7]

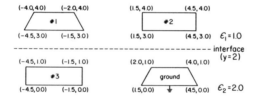

Figure 5.   Capacitance Example Showing Coordinate Inputs and
            Maxwell Capacitance Matrix Output in pF/cm

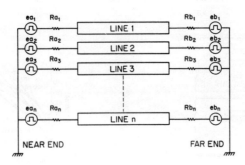

Figure 6. General N-Coupled Line Circuit for Electrical
Simulation

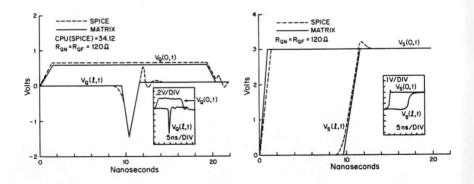

**Figure 7. Experimental and Simulation Results for Coupled
Transmission Lines; SPICE and PCTEM Tools [9]**

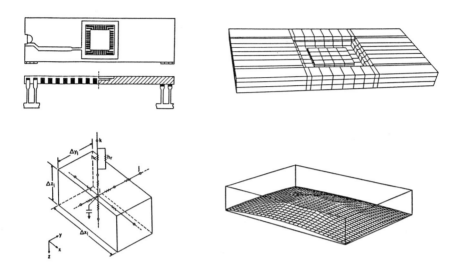

Figure 8. Typical Aspects of Packaged VLSI Thermal Modeling: Packaged Chip, Unit Volume, Model and Typical Results

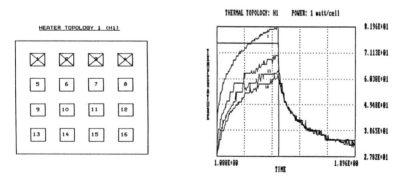

Figure 9. The Experimental Response of Sensors 1,6,11,16 to a .4 Second Pulsing of Heaters 1-4 with a 1.0 Watt/Heater Power Level

126

MOSFET MODELING FOR VLSI SIMULATION

Narain D. Arora

Digital Equipment Corporation
77 Reed Road, Hudson, MA 01545 USA

1 Introduction

In today's VLSI circuit design, circuit simulators
have become indispensable tools. However, their
utility depends on the accuracy of the models used
to represent the devices. The purpose of a model
in the simulator is to calculate the device currents,
and the capacitances between various nodes for a
given bias condition. Since these calculations are
carried out thousands of time during the transient
analysis, it is imperative that the model be computationally efficient as well as accurate, although
there is always a tradeoff between the accuracy and
speed of simulation.

With the ever decreasing device dimensions, a
fully physical model cannot be incorporated into
circuit simulators because of the 3-D nature of the
small-geometry effects. The desire to achieve more
accurate modeling and to improve the computational efficiency of the model created the need of
adding empirically based parameters to the existing physical parameters. These semi-empirical analytical models retain the basic functional form of
fully physical models, while replacing the complicated equations by empirical equations to represent small-geometry effects and process variations.
Practically all the models used in today's circuit
simulators falls in this category, and range from
simple to more complex models [1-5 and references
therein].

The simulation of the transient and ac response of MOS circuits is critically dependent on
the device capacitances. Recently [5-7] the use
of charge equations derived from the DC current
model has been shown to guarantee charge conservation, and at the same time represent more accurate capacitance model, and as such are more appropriate to be used for circuit simulators. At micron and submicron geometries device 'parasitics'
like source-drain (S-D) resistance, S-D overlap and
fringing capacitances become important in determining device behavior and thus must be modeled
properly. In the next section we will discuss semi-
empirical model for the enhancement type devices
in the different region of operation of the device.
Successively we will discuss the charge and capacitance model and finally extraction of the model
parameters.

2 Current-Voltage Characteristics

Present state-of-the-art VLSI technology uses
CMOS process for device fabrication. The use of
ion-implantation in adjusting the threshold voltage
of MOSFET is now universal. Both the p- and n-
channel CMOS devices, are enhancement type devices and, as such, are normally off. In an n-channel
device the threshold adjust implant is of the same
type as that of the substrate (see figure 1a) while
a p-channel device normally has an implant opposite in type to that of the substrate (figure 1b).
Since the latter implant is very shallow, the surface
layer in p-channel devices is normally depleted at
zero gate bias and the current flows at the interface. Thus p-channel devices could be modeled in
the same way as the usual n-type enhancement devices. Throughout this chapter we will assume that
we are dealing with n-channel devices, unless otherwise stated. It is further assumed that the device
is operating in quasistatic mode which is normally
true. Figure 2 shows cross-section of a n-channel
MOSFET, showing symbols for the voltages, currents and charges.

The analysis of long channel MOSFET is
based on the *gradual channel approximation* (GCA)
which assumes that the lateral electric field E_x
along the channel is small fraction of the normal
electric field E_y from gate to substrate. This approximation is still assumed for small geometry devices while the major perturbations in the results
are then modeled from the first principle.

2.1 Threshold Voltage

The MOSFET threshold voltage (V_{th}) is very important device parameter. For effective chip design its value must be controlled within certain
prescribed limits. It is therefore essential that V_{th}
be modeled accurately. The threshold voltage is
determined from the requirement that the charges
outside the semiconductor (at the gate and in the
gate insulator), and inside the semiconductor area
be equal. For implanted profiles, the widely used
expression for the threshold voltage V_{th} is [5]

$$V_{th} = V_{fb} + 2\phi_f + \gamma\sqrt{2\phi_f + V_{sh}} + \gamma_0\left(2\phi_f + V_{sh}\right) \tag{1}$$

where V_{fb} is the flat-band voltage, ϕ_f is the Fermi
potential with respect to the the intrinsic level (
substrate), γ is the body factor coefficient and V_{sh}
is the substrate bias. The fitting parameter γ_0 ad-
justs the body-effect relationship for any nonuni-

form doping. In most cases, a negative value of γ_0 will be needed to offset the high value of γ resulting from higher concentration at the surface. Strictly speaking, the concept of flat-band is only applicable for uniformly doped substrate, but it is still used because of it being an important reference potential. In fact it has been shown that for nonuniform substrate, a modified definition of V_{fb} could be used [2,8].

2.1.1 Threshold Voltage: Short-channel Effect

In the derivation of V_{th} we have assumed that the gate controls the entire bulk charge Q_b under the gate. This is approximately true for long channel devices, however, as the channel length becomes shorter, the charge ΔQ_l controlled by the source and drain can no longer be neglected in comparison to the charge that is controlled by the gate. Thus Q_b reduces resulting in decrease in threshold voltage. For the sake of simplicity, in CAD models the decrease in V_{th} with channel length is invariably calculated using a charge sharing scheme based on geometrical approximations. Various assumptions are made in calculating ΔQ_l using this method, some of which are discussed by Nguyen et al [9] and Viswanthan et al [10], nonetheless different charge sharing schemes reported in the literature are a variation of the one used by Yau [11] which is shown in Figure 3. The key in the derivation is the introduction of a trapezoidal region representing the gate controlled charge. The charge beyond that region (shaded area) is controlled by source and drain.

Yau's simple geometrical approximation has been modified by many others. Some of these are reviewed by Fichtner et al [12] and Akers et al [13]. Very often increased complexity in these equations is not warranted for CAD models. Since ΔQ_l as calculated by geometrical approximations is only approximate, in CAD models it is often approximated by [2,5,8]

$$\Delta Q_l = (2\phi_f + V_{sb})\gamma_l/L_{eff} \qquad (2)$$

where γ_l is a fitting parameter, and L_{eff} is the effective or electrical channel length which is different from the drawn channel length (mask dimensions) [8].

2.1.2 Threshold Voltage:Narrow-width Effect

In a modern MOSFET isoplanar process, there is always some variation in the substrate doping due

to the field implant and gate-induced electric field occurring at the device periphery where the transition from the thin oxide (active) area to thick oxide (field) isolation area takes place. This is shown in Fig. 4. Because the gate induces a fringing field around its edges, there is an extra depletion charge ΔQ_w under the gate at each side (see Fig. 4). Since this additional charge must also be supported by the gate, it causes the threshold voltage to increase by an amount $2\Delta Q_w/C_{ox}$ where C_{ox} is the gate oxide capacitance per unit area. Thus, decreasing the device width increases the threshold voltage. Similar to the short channel effect, the effect of ΔQ_w is included empirically in the expression for V_{th}. Different empirical approaches have been followed [8,12-13 and references therein] to calculate the value of ΔQ_w. One which has been found to work well for different technologies down to W_{eff} of 1 μm is based on charge control analysis. Assuming the field implant concentration N_{fld} is much larger than the channel surface concentration N_s, and assuming there is no short channel effect, it has been shown [8] that

$$\Delta Q_w = \epsilon_{si}\left(\phi_{fld} + V_{sb}\right)/W_{eff}, \qquad (3)$$

where ϕ_{fld} is the surface field potential. The over simplification made in arriving at equation (3) can easily be corrected by multiplying right hand side of (3) by a factor γ_w. The value of γ_w is always less than unity and is technology dependent. It should be pointed out that taking ϕ_{fld} as a parameter rather than calculating it from N_{fld} gives better fit to the experimental data. For CAD models N_{fld} is often assumed to be equal to N_b so that ϕ_{fld} equals $2\phi_f$.

Assuming the short and narrow channel threshold effects are additive, the threshold voltage at low V_{ds} for short geometry devices becomes

$$V_{th} = V_{fb} + 2\phi_f + \gamma'\sqrt{(2\phi_f + V_{sb})} - \frac{\Delta Q_l}{C_{ox}} + \frac{\Delta Q_w}{C_{ox}} \qquad (4)$$

where γ' is determined by substrate bias and doping concentration. For uniformaly doped substrate $\gamma' = \gamma$.

2.1.3 DIBL Effect

The above analysis of short and narrow channel effects on the threshold voltage assumed that the source to drain voltage V_{ds} is very small ($< 0.1V$). However, as the channel length is reduced and V_{ds} is also increased, the drain depletion region moves closer to the source depletion region resulting in a significant field penetration from drain to source. Due to this field penetration, the potential barrier at the source is lowered resulting in an increased injection of electrons by the source giving rise to increased subthreshold current. This process is called *drain induced barrier lowering* or simply DIBL [14-15]. The degree of the lateral field penetration or the DIBL depends on the channel length, the gate oxide thickness t_{ox}, the source/drain junction depth X_j, the channel doping profile concentration, and the back gate bias V_{sb}.

In short channel devices, the DIBL effect is reflected in a threshold voltage reduction using a simple empirical equation, which assumes a linear dependence of V_{th} on V_{ds}. Thus

$$V_{th} = V_{th0} - \sigma V_{ds} \tag{5}$$

where V_{th0} is threshold voltage at low V_{ds} ($< 0.1V$) given by (4) and σ is the DIBL parameter given by

$$\sigma = \sigma_0 \epsilon_{si} / C_{ox} L_{eff}^n \tag{6}$$

where σ_0 is a fitting parameter used to allow for geometry dependence of DIBL effect over a given range of X_j and N_b. The exponent of L varies in the range 1-3. In the above equation, the substrate bias dependence of σ has not been included. This effect can be significant depending on the particular process conditions.

It is interesting to note that since the drain modulates the potential barrier in the channel, it is sometime called the second gate and σ is called the coefficient of static feedback. DIBL is a strong effect for short channel devices operating near threshold. For such devices operating in saturation, DIBL is the principal factor determining the output conductance.

2.2 Effective Mobility

Our understanding of the effective mobility is largely based on experimental studies [16-19]. Recent experimental measurements of effective mobility [16-17] at low V_{ds} show a "universal curve" when plotted against the effective normal field E_{eff} as shown in Figure 5. It can be seen that mobility is almost independent of substrate-bias, and doping concentration over a wide range. The experimental data appear to fit the following relation

$$\mu_s = \frac{\mu_0}{1 + \theta E_{eff}} \tag{7}$$

where μ_0 is the low field surface mobility whose value is 550-650 cm^2/(V.s) for the electron, and

200-300 cm^2/(V.s) for holes, and θ is experimentally determined mobility degradation coefficient with values between 0.03 to 0.09 $cm.V^{-1}$.

The effective normal field, when averaged over the electron distribution in the inversion layer, is given by

$$E_{eff} = (Q_b + \eta Q_c) / \epsilon_{si}, \tag{8}$$

where Q_c is the channel charge density. For n-channel devices, using $\eta = 0.5$ gives a good fit to experimental data. However for p-channel devices $\eta = 0.25 - 0.30$ has been found to be more appropriate [19].

The mobility degradation due to the lateral electric field has more significant effect on the device current equations than does the normal field. This is because an increase in the lateral field eventually causes velocity saturation of the carriers. For a given normal field $E_y (= E_{eff})$, the velocity is proportional to E_x at low lateral fields, and the proportionality constant is the mobility μ_s discussed above and plotted in figure 5. However, as E_x increases the velocity tends to saturate. Combining the effects of mobility degradation due to the normal and lateral fields in the channel, we can write to a first approximation the effective mobility μ_{eff} as

$$\mu_{eff} = \frac{\mu_0}{(1 + \theta E_{eff})(1 + E_x/E_c)} \tag{9}$$

where E_c is the critical field where carriers gets velocity saturated and is related to saturated carrier velocity v_{sat} by

$$E_c = v_{sat}/\mu_s \tag{10}$$

2.3 Drain current calculation

Under the assumptions that (1) recombination and generation can be neglected, (2) hole current can be neglected, and (3) μ_{eff} and the electron quasi-Fermi potential V is constant along the channel, the drain current I_{ds} can be obtained

$$I_{ds} = W_{eff} \mu_{eff} Q_c(x) \frac{dV}{dx} \tag{11}$$

where V is the electron quasi-Fermi potential at a distance x along the length of the channel from the source toward the drain.

2.3.1 Drain current in the linear region

The linear region implies that $V_{gs} > V_{th}$. Under the assumption that diffusion current can be neglected compared to drift current, (11) can be integrated from $V = 0$ (at source) to $V = V_{ds}$ (at drain end) to obtain I_{ds} as

$$I_{ds} = \frac{\beta}{(1 + \theta E_{eff})\left(1 + \frac{V_{ds}}{L_{eff}E_c}\right)}[V_{gs} - V_{th} - 0.5\alpha V_{ds}]\,V_{ds}$$

(12)

where

$$\beta = \mu_0 C_{ox} W_{eff}/L_{eff} \quad and \quad \alpha = 1 + \delta\gamma$$

Here δ arises because of replacing the square root term in Q_b by some linear approximation [2-5].

For long channel devices the velocity saturation term in equation (12) is omitted, and often $\alpha = 1$ is assumed i.e., the bulk charge Q_b is constant along the length of the channel.

Classically that point in the channel at which channel charge $Q_c = 0$ is called the "pinch-off" point. The pinch-off condition is then described by

$$V_{dsat} = (V_{gs} - V_{th})/\alpha \qquad (13)$$

and is called the saturation voltage. The corresponding current is called the saturation current I_{dsat} and is obtained by replacing V_{ds} in (12) by V_{dsat}. Assuming no carrier velocity saturation, the condition for pinch-off is equivalent to the condition that $dI_{ds}/dV_{ds}=0$. Physically, at pinch-off, the normal field E_y is inverted which pushes the mobile carriers away from the surface, flowing into the drain area.

For modern short channel devices, the classical pinch-off condition ($Q_c = 0$) is not realistic because the carriers reach velocity saturation before the pinch-off condition are fulfilled. When the carrier velocity saturates, the current can be approximated as

$$I_{dsat} = W_{eff}Q_c v_{sat} = W_{eff}C_{ox}v_{sat}(V_{gs} - V_{th} - \alpha V_{dsat})$$

(14)

where v_{sat} is given by (10). The current I_{dsat} can also be calculated either by exact current equation (12) or by its approximate value (without velocity saturation and $\alpha=1$). Both of these equations have been used in CAD models to calculate V_{dsat} from (14). Using current equation (12) for I_{dsat} in (14) and solving for V_{dsat} we have [4,5]

$$V_{dsat} = L_{eff}\,E_c\left[\sqrt{1 + \frac{2(V_{gs} - V_{th})}{\alpha\,L_{eff}\,E_c}} - 1\right] \quad (15)$$

Clearly V_{dsat} becomes function of channel length L_{eff} for short channel devices.

2.3.2 Saturation region current

For drain voltage V_{ds}, larger than V_{dsat}, the transistor is in saturation mode. When $V_{ds} > V_{dsat}$, the channel pinch-off point (or the velocity pinch-down point) P starts to move towards the source. This movement is referred to as channel length modulation. Point P still has a voltage V_{dsat} but the portion of the drain voltage beyond V_{dsat} is now dropped across a depletion region of length l_d from point P to the drain. Thus in saturation channel region splits into two different regions - one on the source side where the GCA is valid and second on the drain side where GCA is violated. For long channel devices, the second region is approximately identical to the pinch-off region and in such devices l_d can be neglected compared to L_{eff}. However for short channel devices, channel length modulation significantly affects the output conductance in saturation.

The approach normally used to model this effect is to replace the normal channel length with decreased channel length, L' ($= L_{eff} - l_d$), where GCA is valid so that

$$I_{ds} = I_{dsat}\frac{L_{eff}}{L_{eff} - l_d} \qquad (16)$$

To compute an exact value of l_d requires a two dimensional solution of Poisson's equation near the drain

$$\frac{\partial^2 V}{\partial x^2} + \frac{\partial^2 V}{\partial y^2} = -\frac{(qN_b + Q_c)}{\epsilon_{si}} \qquad (17)$$

Various models proposed and in use for l_d are based on approximations used to solve equation (17) and are well summarized by Engel et al [20].

Since the exact solution of Poisson's equation (17) is complex, various empirical relations that match the device behavior in saturation have also been proposed [2 and references therein].

Apart from channel length modulation, I_{ds} in saturation also increases due to the DIBL effect discussed earlier. For CAD models this increase is accounted for through a decrease in V_{th}, effectively increasing the gate driving voltage as discussed in section 2.1.3.

2.3.3 Subthreshold region current

Subthreshold or weak inversion conduction occurs when $V_{gs} < V_{th}$ or $\phi_f \leq \phi_s \leq 2\phi_f$. This region of operation is important because leakage current affects dynamic circuits and CMOS standby power. The potential ϕ_s or the band bending is nearly constant from source to drain because the inversion charge density Q_c is several order of magnitude smaller than the bulk charge density Q_b. Because ϕ_s is constant, carrier transport occurs primarily through diffusion. The source potential barrier and the drain depletion edge are regarded as minority carrier source and sink respectively, with minority carrier density being determined by the surface potential. The subthreshold diffusion current for large channel devices has been calculated on this basis and is given by [4-5,21 and references therein]

$$I_{ds} = \beta m n V_t^2 exp\left[(V_{gs} - V_{th})/nV_t\right] \quad (18)$$

where $V_t = kT/q$ is the termal voltage. The parameter n is a measure of the rate of exponential increase of I_{ds} with V_{gs} and is given by

$$n = 1 + \frac{C_d}{C_{ox}} + \frac{C_{ss}}{C_{ox}} = n_0 + \frac{C_d}{C_{ox}} \quad (19)$$

where the surface depletion layer capacitance C_d is the depletion capacitance. The surface state capacitance C_{ss} is normally regarded as an adjustable parameter through n_0 and is used to fit the value of n to measured characteristics. The fitting parameter m is inserted to correct the various approximations made in the derivation of equation (18). Equation (18) indicates that in weak inversion I_{ds} varies exponentially with V_{gs} and for $V_{ds} > 3V_t$, the current becomes independent of V_{ds}.

For short channel devices, and particularly at large V_{ds}, the surface potential ϕ_s is not constant along the length of the channel and is higher than for long channel devices. Although the current remains exponentially dependent on the gate voltage, various physical arguments used in the derivation of (18) no longer apply. Nevertheless this equation is often used for CAD models provided m and n are regarded as fitting parameters.

3 Capacitance characteristics

The capacitive characteristics of a MOSFET are the sum of two conceptually distinct terms: the intrinsic capacitance, which comes from the channel region of the device, and the extrinsic capacitance which includes the source and drain p-n junction capacitances plus the overlap capacitances of the gate over the junctions and the field. These extrinsic capacitances are often called parasitic capacitances.

The Meyer model [1] for the intrinsic capacitances of a MOSFET has been almost universally used in circuit simulators until very recently. In the Meyer model, the active gate-channel capacitance is considered to be voltage dependent, since it is a series combination of the geometrical oxide capacitance C_{ox} and the space charge capacitance C_{sc} for the channel. This distributed capacitance is split into three lumped capacitances, C_{gs}, C_{gd} and C_{gb} as shown in figure 2.

The Meyer model, though simple, leads to errors in some capacitances of short channel devices, and generally charge is not conserved. Thus the model becomes inadequate for simulating circuits like dynamic RAMs and switched capacitor circuits, which are sensitive to the capacitive components of the MOSFET currents. However, by using charges associated with the device terminals as the state variables, charge conservation is guaranted. This implies that we need to calculate the charges Q_G, Q_D, Q_S and Q_B as functions of the node voltages. The gate and the bulk charge can easily be obtained by integrating the corresponding charge densities over the area of the active gate region, thus

$$Q_G = \frac{W_{eff}^2}{I_{ds}} \int_0^{V_{ds}} \mu_{eff} Q_g(x) Q_c(x) \, dV \quad (20)$$

$$Q_B = \frac{W_{eff}^2}{I_{ds}} \int_0^{V_{ds}} \mu_{eff} Q_b(x) Q_c(x) \, dV \quad (21)$$

where I_{ds} is given by (12), Q_c is given by

$$Q_c(x) = C_{ox}(V_{gs} - V_{th} - \alpha V) \quad (22)$$

and Q_g is given by

$$Q_g(x) = C_{ox}(V_{gs} - V_{fb} - 2\phi_f - V) \quad (23)$$

so that $Q_b = -(Q_g + Q_c)$. However, the source and drain charges (Q_S and Q_D) can only be determined from the channel charge Q_G. It is thus necessary to partition the channel charge into charge associated with the drain terminal Q_D and charge associated with the source terminal Q_S such that $Q_G = Q_S + Q_D$.

Various approaches have been used in the literature to partition Q_G into Q_S and Q_D [5-7]. However, the one which can rigorously be shown to be correct and which has been demonstrated to agree with experiment is that proposed by Ward [6]. According to this approach

$$Q_S = -\frac{W_{eff}^2}{I_{ds}} \int_0^{V_{ds}} \mu_{eff}\left(1 - \frac{x}{L_{eff}}\right) Q_c^2 \, dV, \quad (24)$$

and

$$Q_D = -\frac{W_{eff}^2}{I_{ds}} \int_0^{V_{ds}} \mu_{eff} \frac{x}{L_{eff}} Q_c^2 \, dV. \quad (25)$$

To express x in the above equation in terms of V_{ds} we integrate (11) from $x = 0$ to an arbitrary point in the channel

$$x = -\frac{W_{eff}}{I_{ds}} \int_0^{V_{ds}} \mu_{eff} Q_c dV \qquad (26)$$

Figure 6 shows the total charges associated with the gate, bulk, source and drain of a MOSFET in all four regions. Note that at any V_{gs} total charge is zero. Once the charges are known, the corresponding capacitances can be obtained by taking partial derivatives of the charge equations with respect to different terminal voltages. Thus

$$C_{ij} = -\frac{\partial Q_i}{\partial V_j}, \quad i \neq j \;\; i,j = g,b,d,s \qquad (27)$$

and

$$C_{ij} = \frac{\partial Q_i}{\partial V_j}, \quad i = j \qquad (28)$$

Physically C_{ij} is the sensitivity of the stored charge associated with terminal i to the voltage applied to terminal j with all other voltages held constant. The above equations yield sixteen capacitances, four for each terminal of the device. One of these four capacitances, corresponding to each terminal of the device, is the self capacitance which is the sum of the remaining three capacitances. So in fact only twelve capacitances need to be evaluated. Only nine out of these twelve are independent. These twelve capacitances (also called transcapacitances) are the intrinsic capacitances of a MOSFET and in general are non-reciprocal.

4 Parasitic Elements

The role of parasitic resistance and capacitance in the determination of the transient response of integrated circuits becomes more dominant as the device size becomes smaller and the circuit density increases. For VLSI simulation, it is thus important to properly model the effect of these parasitics.

4.1 Parasitic Source-Drain Resistance

The parasitic resistance associated with the MOSFET is the series resistance of the source-drain (S-D) region. For short channel devices this plays an important role in limiting the transconductances of these devices [22]. It consists of S-D diffusion layer sheet resistance, contact resistance and spreading resistance. The latter occurs in the S-D region under the gate due to crowding of the current flow

lines in that region and constitutes major portion of the S-D resistance. Modeling this parasitic S-D resistance is important for proper fitting of the output conductance in the linear region [22].

For VLSI simulation, the current equations are often modified to include the series resistance R_t of the source and drain [23]. This results in θ being modified as $(\theta + \beta R_t)$ in (12).

4.2 Extrinsic Capacitances

Among the capacitive components, extrinsic to the active device, the capacitances of the source and drain diffusions are modeled as the sum of the sidewall (periphery), and bottom-wall (area) voltage dependent capacitances [24]

$$C_j = \frac{C_{joa} A_{sd}}{\left(1 + \frac{V_b}{\phi_A}\right)^{m_j}} + \frac{C_{jop} P_{sd}}{\left(1 + \frac{V_b}{\phi_P}\right)^{m_{jsw}}} \qquad (29)$$

where C_{joa} is the zero bias depletion region capacitance per unit area, C_{jop} is the zero bias depletion region capacitance per unit periphery, V_b is the reverse bias junction voltage, ϕ_A built-in potential for area capacitance, ϕ_P built-in potential for periphery capacitance, m_j area junction capacitance gradient factor, m_{jsw} periphery junction capacitance gradient factor, A_{sd} and P_{sd} are source/drain window area and side-wall perimeter length respectively.

Although in some structures, like the lightly doped drain the physical overlap of the gate and source-drain is essentially eliminated, in other devices these overlap capacitances can be significant fraction of the gate-oxide capacitance. For a device fabricated with a self-aligned process the value of these overlap capacitance C_{gso} and C_{gdo} differ significantly from the parallel plate capacitance value which to a first order approximation is given by

$$C_{gso} = C_{gdo} = C_{ox} W_{eff} d \qquad (30)$$

where $d(\sim 0.7 X_j)$ is the overlap. When d is small, the fringing effect can no longer be neglected and must be taken into account [25].

Since the channel length, width and gate electrode thickness are comparable, the fringing capacitance C_{fr} associated with the gate becomes a significant portion of the total gate capacitance. Even with reduction in the oxide thickness in concert with reduction in channel length, the need to maintain a thick electrode to reduce interconnect parasitics results in an increased coupling between the vertical edge of the gate electrode and the semiconductor surface [26]. Thus C_{fr} must be evaluated in order to obtain an accurate circuit simulation.

For CMOS technology, normally one of the transistors is inside the well. Therefore one must consider the capacitance due to the p-n junction between the well and the common substrate on which the well sits and is calculated from equation similar to (29).

5 Parameter Extraction

The accuracy of a model in predicting the device characteristics is fully dependent on the accuracy of the values extarcted for the model parameters. In general, model parameters are determined from technological process data, current-voltage (I-V) and capacitance-voltage (C-V) characteristics for certain device geometries. There are basically two ways to extract model parameter values from the I-V data. One way is to extract the parameter from data local to that region using some linear regression methods [2,5,27]. The second approach is to consider all of the parameters and all of the data simultaneously and to solve for the best fit of the model to the data using a nonlinear least-square optimization technique [28-29 and refrences therein]. Because there is sufficient interaction between the parameter values, several combinations of values will provide a working fit to the measured characteristics. Thus, it is not always clear, which are the correct values. However, this method produces a better fit to the data over the entire data space at the sacrifice of some physical insight.

It turns out that it is more practical to extract the parameters utilizing both approaches simultaneously. Thus several parameters are extracted from one set of local data using the optimization method in conjunction with the relevant model equations. Those parameters are then frozen while determining other parameters from a different local data set. In one example of this approach the MOSFET parameters are split into four groups. The first group of parameters (C_{ox} and X_j) is generally known from the technological process data. The second group of parameters (V_{fb}, N_b, γ_l etc) is determined from I-V characteristics in the strong inversion region at low V_{ds} (data set A), the third group (μ_0, θ, R_t etc) from the general $I_{ds}(V_{ds}, V_{gs}, V_{sb})$ characteristics in strong inversion (data set B), and the forth group (m, n and σ) from the subthreshold characteristics (data set C).

The extrapolated threshold voltage as a function of channel widths for two different substrate biases for n-channel devices is shown in figure 7a. The square are experimental points while the con-

tinuous and broken lines are those calculated based on equation (4). The value of the narrow width fitting parameter, γ_w, is 0.38 which shows that the increase in V_{th} with decreasing width is not significant down to 2 μm width. The variation of V_{th} with channel length is shown in figure 7b. Here again the short channel fitting parameter γ_l is 0.33 showing that the roll-off in V_{th} with channel length down to 0.5 μm is not significant.

Once parameters in group II are known, the current in the subthreshold region (data set C), is then used to obtain parameters m, n and σ (group IV parameters) using an optimizer and the equation (18). The remaining parameters (group III) are then obtained by using already determined parameters in the full model and fitting this by optimization to data obtained from the linear and saturation region of operation (data set B) using different channel length and width devices. For this we use equations (12), (15) and (16). The output characteristics of a n-channel CMOS device with channel length of 1 μm (L_{eff} = 0.76μm) and width of 3 μm (W_{eff} = 2.1μm) is shown in figure 8. Simulated data are plotted with solid lines while measured data are shown with squares. The average error is less than 3% for V_{bs}=0V in the linear and saturation region of operation, however, the error does increase as V_{bs} increses. Thus for V_{bs}=-2V, error is about 5%. This is mainly because of DIBL effect dependence on V_{bs} is not properly taken into account in the model. Since the model is not very accurate in the subthreshold region, therefore, the average error rises to 10-15 % for $I_{ds}(log)$ vs V_{gs} curves (Fig. 8b). Similar plots are obtained for p-channel CMOS devices of the same dimension. These various results demonstrate the accuracy and continuity which can be obtained with CAD models over a range of channel lengths in which significant short-channel effects are occurring.

It should be pointed out that gate oxide capacitance C_{ox} is not a fitting parameter. Its value is determined from direct measurement of a MOS capacitor or directly from a large transistor biased in accumulation mode. The overlap capacitance is measured from the bias condition in which either the source or the drain is pinched-off with respect to the gate. The source and drain junction capacitance per unit area is obtained from special test structure large area devices [24].

6 Summary

In this review of MOSFET modeling, the most commonly used models for circuit simulators are described. These models provide an accurate description of enhancement-mode MOSFET characteristics throughout the subthreshold, linear and saturation region of operation. Many other models are conceivable which are variations of the one reported here, but each model is a compromise between accuracy and simplicity. A short discussion of parameter extraction concludes this review.

7 References

1. J.E. Meyer, RCA Rev. 32,42(1971).

2. F.M. Klassen, in "Process and Device Modeling for Integrated Circuits",(Wiele, Engl and Jespers, Eds.) p. 541 Noordhoff Publishing, Leyden, 1977.

3. S. Liu and L.W. Nagel, IEEE J. Solid State Circuits, SC-17,983(1982).

4. G.T. Wright, IEE Proc. Pt. 1, 132,187(1985).

5. B.J. Sheu, D.L. Scharfetter, P.K. Ko and M.C. Jeng, IEEE J. Solid State Circuits, SC-22,558(1987).

6. D.E. Ward and R.W. Dutton, IEEE J. Solid-State Circuits, SC-13,703(1978).

7. P. Yang, B.D. Epler and P.K. Chatterjee, IEEE J. Solid State Circuits, SC-18,128(1983).

8. N.D. Arora, Solid State Electron., 30,559(1987).

9. T.N. Nguyen and J. D. Plummer, IEEE IEDM Tech. Dig. 596(1981).

10. C. R. Viswanathan, B.C. Burkey, G. Lubberts and T.J. Tredwell, IEEE Trans. Electron Devices, ED-32, 932(1985).

11. L.D. Yau, Solid State Electron., 17,1059(1974).

12. W. Fichtner, and H.W. Potzl, Int. J. Electron. 48,33(1979).

13. L.A. Akers and J.J. Sanchez, Solid State Electron., 25,621(1982).

14. S.C. Chamberlain and S Ramanan, IEEE Trans. Electron Devices, ED-33,1745,(1986).

15. S.C. Jain and P. Balk, Solid State Electron., 30,503(1987).

16. A.G. Sabnis, and J.T. Clemens, IEEE IEDM Tech. Dig. 18(1979).

17. S.C. Sun, and J.D. Plummer, IEEE Trans. Electron. Devices ED-27,1497(1980).

18. J.A. Cooper,Jr, D.F. Nelson, S.A. Schwarz, and K.K. Thornber,"VLSI Electronics: Microstructure Science," (N.G. Einspruch, and R.S. Bauer, eds.) Vol. 10, p.323. Academic Press, New York, 1985.

19. N.D. Arora, and G. Sh. Gildenblat, IEEE Trans. Electron. Devices ED-34,89(1987)

20. W.L. Engl, H.K. Dirks and B.Meinerzhagen, Proc. IEEE, 71,10(1983). 19,51(1976).

21. P A. Muls, G. J. Declerck and R. J. Van Overstraeten in "Advances in Electronics and Electron Physics" (ed.) Vol. 47, p.197. Academic Press, New York, 1981.

22. K.K. Ng and W.T. Lynch, IEEE Trans. Electron Devices, ED-34,503(1987).

23. P.I. Suciu and R.L. Johnston, IEEE Trans. Electron Devices, ED-27,1846(1980).

24. B.A. Freese, and G.L. Buller, IEEE Electron Device Lett. EDL-5,261(1984).

25. P. Vitanov, U. Schwabe and I. Eisele, IEEE Trans. Electron Devices, ED-31,96(1984).

26. E. W. Greeneich, IEEE Trans. Electron Devices, ED-30,1838(1983).

27. F. H. De La Moneda, H.N. Kotecha and M. Shatzkes, IEEE Electron Device Lett. EDL-5,491(1984).

28. D.E. Ward and K. Doganis, IEEE Trans. Computer-Aided Design, CAD-1,163(1982).

29. W. Maes, K.M.De Meyer and L.H. Dupas, IEEE Trans. Computer-Aided Design. CAD-1,320(1986).

134

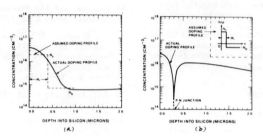

Fig. 1. Vertical doping profile of implant region under the gate for (a) n-channel and (b) p-channel devices. An idealized step profile of width W_s has the same active dose as the actual implanted profile.

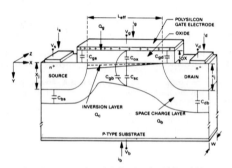

Fig. 2. Device crosssection showing device dimensions, voltage definition and device coordinates.

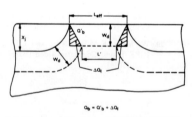

$$Q_b = Q'_b + \Delta Q_l$$

Fig. 3. Yau's charge sharing model for short channel MOSFET at threshold condition. The charge in the shaded area controlled by the source and drain.

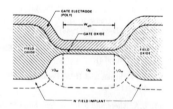

Fig. 4. Cross-section of a narrow-width device showing fringeing field component in the depletion charge (After Arora, Ref. 13)

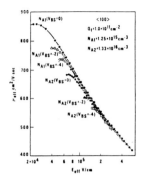

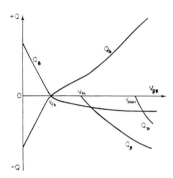

Fig. 5. Inversion layer electron mobility data of Sun and Plummer for Si at room temperature for four different substrate dopings. (After Ref. 17)

Fig. 6. Variation of total chrage associated with the gate (Q_G), source (Q_S), drain (Q_D) and bulk (Q_B) with V_g, in alll the region of operation of the device.

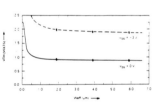

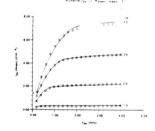

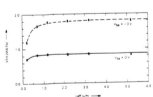

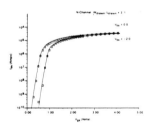

Fig. 7. Threshold voltage as a function of substrate bias for a n-channel CMOS for different (a) drawn channel width and (b) drawn channel length for $V_{ds}=0.1$ V. Dots are experimental points while the continuous line is calculated values

Fig. 8. Output characteristics of a W/L=3.1 n-channel CMOS (a) I_{ds} Vs V_{ds} at $V_{bs}=0$V (b) log(I_{ds}) vs V_{gs} at $V_{ds}=0.1$V and $V_{bs}=0,-2$V (L=, 150λ, λ=0.15 μm). Squares are experimental points while the continuous line is calculated values.

PULSE ANNEAL PROCESSING IN THREE-DIMENSIONAL
MICROELECTRONICS

Aleksandrov L.N.
Institute of Semiconductor Physics, Siberian Branch of
the USSR Academy of Sciences, 630090 Novosibirsk, USSR

New ideas on increasing the integrated circuits density
and shortening the bond lengths between large scale integration elements are connected with progress in three-dimensional
(3D) microelectronics /1, 2/. The 3D integrated circuits technology is based on the process of deposition of semiconductor
single crystalline layers on an amorphous dielectric substrate.
For its realization the recrystallization or melting-crystal-
lization of the amorphous layers in the annealing conditions
are required, so that do not affect the underlying earlier
formed floors of the integrated circuits. Therefore, the
usual furnace methods of heating as well as alloying by impurities diffusion are not applicable and pulse annealing becomes the base of the 3D microelectronics technology. The
pulse annealing in combination with ion implantation permits
crystallizing selectively and doping only the upper layer of
the integrated circuits.

The extended use of the pulse radiation effects in scientific research and engineering applications is connected with
necessity of conducting the computer simulation for a wide
interval of pulse duration ($10-10^{-15}$ s) and energy densities
($0.1-10$ J/cm^2) for ion, electron, light, laser beams. The
comparison of the relaxation time for the base interaction
processes of radiation pulses with semiconductors (electron

subsystem, excitation, electron-phonon interaction, electron-
-hole plasma recombination) gives a possibility to estimate
a relative contribution of the thermal and plasma processes
by the pulse action.

The experimental investigations and the theoretical analysis have showed that for the pulse duration more than 10 ns
and structural transformations on the basis of the thermal
model are described by the solution of the heat conductivity
equation with sources and sinks /3, 4/.

By this it is possible to make a complete thermodynamic
description of the proceeding processes and calculations of
the temperature field in dependence on energy and duration
of the pulses of different kinds that interact with a semi-

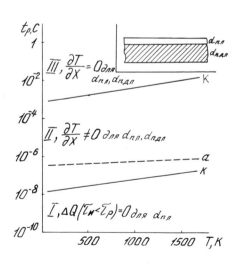

Fig. 1. Regions of the
thermal processes in silicon
(the film-substrate structure) vs the time of equal-
ization of temperature and
time of pulsed heating:
I - adiabatic process in
the amorphous (a) or crystalline (c) film; II - nonuniform heating of the film
and the substrate with
temperature gradients;
III - isothermic heating
of the film and the substrate.

138

conductor. The main attention was paid to the heating of
two-layer structure of the film-substrate type forming after
film deposition or after ion implantation of semiconductor,
with the thickness of the upper layer d being near 1 μm.

For d $>$ λ_T (thermal conductivity diffusion length) adiabatic heating is realized, for d $<$ λ_T isothermal one in
solid phase. In the adiabatic conditions where the pulses time
τ_p is shorter than 1 μs ($\lambda_T = \sqrt{\tau_p \cdot \alpha}$; $\alpha \approx$ 0.1 cm^{2}s^{-1}
for Si), layer melting does not affect the substrate. This
provides the possibility of the liquid-phase epitaxial crystallization of the melted layer on the single crystalline
substrate (Fig. 1).

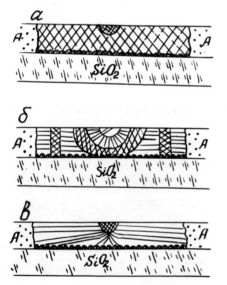

Fig. 2. Development of the
quasi-periodic polycrystalline (a), fine-grained
polycrystalline (b) and
coarse-grained quasi-mono-
crystalline (c) structures
during the self-sustaining
crystallization of an
amorphous film on SiO$_2$ from
the initial molten section.
In all cases a polycrystalline layer is formed near
the substrate.

In the case of silicon-on-insulator structure (SOI) we observe not epitaxy but change of the cooling regime gives a possibility to control grain sizes, low-angle interfaces in the layer after solidification and to obtain large-grained silicon regions. We have developed the methods of analytical solution of thermal problem, probability-statistical analysis of transformation, numerical solution by finite-difference form of equation /5/. This permits computer simulation studies of the process of amorphous silicon layer crystallization during pulse heating, elucidation of the conditions of the beaginning and end of self-sustained crystallization, the conditions of the order of $A \rightarrow 6$ or $A \rightarrow L \rightarrow C$ processes, rate of the crystal--melt interface motion, the impurity redistribution and capture, the large-grained and periodical structure formation conditions. By the method of computer simulation the accelerated influence of mechanical stresses on the film crystallization kinetics by pulse heating was shown /6/, the condition of preparation of epitaxial dislocation-free layers and superdoped semiconductor films and sharp p-n junctions with MBE and pulse annealing combination were determined. It is possible to take into account molecular beam ionization by film deposition, to determine effect of laser, light, electron, and ions beams on the films.

The method of computer simulation determined the control conditions for the molten zone movement by self-sustaining crystallization and gave possibility to receive the extended laterally growing regions of Si on SiO_2 over the whole substrate length (Fig. 2). The pulses are used of nanosecond duration for crystallization and of millisecond one for substrate heating. For the electron beam with 10 keV energy and pulse duration 500 ns the crystal growth rate is 3-6 m/s by SiO_2 warming up to 600-750 K /3/. The high crystallization rate guarantees complete capture of the implanted impurities.

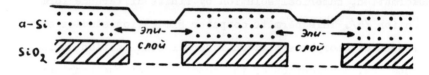

Fig. 3. Lateral oriented crystallization of silicon amorphous film on SiO_2 and single crystalline substrate by pulse (or scanning) heating beginning from the epitaxial film regions formed in the SiO_2 layer window at the initial stage of melted silicon layer crystallization.

The system of rectangular cuts in SiO_2 permits a regrowth of the cap a-Si layer by light pulse heating (duration time 30 ms, energy density 60-70 J/cm^2, and substrate temperature 1000 K). The single-crystalline regions obtained (size of 40-50 um) had electrophysical properties of bulk material, that permits to make silicon devices on the base of this SOI structure (Fig. 3) /7/.

The study of the formation processes of 3D integrated circuits by pulse heating confirms that by the pulse duration of more than 1 ns the computer simulation may be carried out by

thermal model. The Si layers obtained by pulse methods are
quite suitable for application in microelectronics.

References

1. Valiyev K.A. Microelectronics: Progress and Future.
   Moscow, Nauka, 1986, 141 p.
2. Aleksandrov L.N., Dagman E.E. Microelectronics as a
   Catalyst of the Progress in Engineering. - Economics and
   Industry Organization (USSR) 1984, No. 1, p. 39.
3. Balandin V.Yu., Dvurechensky A.V., Aleksandrov L.N.
   Computer Simulation of Structure Transformation in
   Amorphous Silicon Layers by Pulse Heating. - Surface:
   Physics, Chemistry, Mechanics (USSR) 1986, No. 1, p. 53;
   Phases of Self-Sustained Crystallization in Amorphous
   Layers. - Autometriya (USSR) 1987, No. 1, p. 64.
4. Aleksandrov L.N. Thermal and Plasma Models of Pulse
   Heating of Thin Films. Proc. of the 7th Czechoslovak
   Conf. on Electronics and Vacuum Physics, 1985, v. 2, p.415;
   Vacuum (GB) 1986, v. 36, p. 463.
5. Aleksandrov L.N. Kinetics of Crystallization and Regrowth
   of Semiconductor Films. Novosibirsk, Nauka, 1985, 224 p.
6. Aleksandrov L.N. The Accelerated Influence of Stresses on
   the Crystallization Kinetics of Ion-Implanted Silicon
   Layers by Pulse Heating. - J. of Techn.Phys.Letts. (USSR)
   1985, v. 11, No. 5, p. 286.
7. Kolyadenko S.N., Melnik I.G., Mokeyev M.V., Smirnov L.S.
   et al. Lateral Epitaxy in SOI Structure with Crystallization from Single-Crystalline Silicon by Pulse Heating.
   The Abstracts of the 7th All-Union Conf. on the Processes
   of Growth and Synthesis of Semiconductor Crystals and Films.
   Novosibirsk, USSR, 1986, v. 1, p. 159.

Measurement and Control of Mechanical Properties of Thin Films

H. Guckel, D. W. Burns

Wisconsin Center for Applied Microelectronics, Department of Electrical and
Computer Engineering, University of Wisconsin, Madison WI 53706, USA

Introduction

Semiconductor devices employ various deposited and chemically induced thin films. These films exhibit strain fields which are sensitive to processing conditions and can lead to internal stresses of several thousand atmospheres. Consequently, electronic characteristics which depend on lifetime or surface states or bulk and surface mobility change because all of these quantities are influenced by stress. If this problem is ignored during processing, larger undesireable tolerances result and, in extreme cases, which are typically in the VLSI device area, device failure is a consequence. If the problem is understood strain can be used to improve device performance. Thus, mobility increases due to tensile strain can and have been used to improve the performance of N-channel transistors.[1]

Micromechanics is a field which is maturing. The use of thin films to produce devices such as pressure transducers is attractive and has been shown to be feasible.[2] The design of micromechanical components which typically involve beam and plate deflections requires accurate measurements and process control of mechanical quantities such as the built-in strain field, Young's modulus, the Poisson ratio and the tensile strength of the films which are to be used. The measurement task becomes complicated because typical substrates are large in comparison to micromechanical devices. This implies that one is not only interested in the average value of the above mechanical properties but also in their deviation and their local value. This observation normally excludes techniques such as radius of curvature measurements. Furthermore, this type of measurement and many other techniques produce data which cannot be used easily to measure individual mechanical quantities. Thus, mechanical deflections depend typically on the built-in strain, Young's modulus and the Poisson ratio.

Experience with pressure transducers and fully supported plates as used in X-ray mask blanks, and the actual processing of various thin films result in the conclusion that the dominant mechanical quantity for thin films is the built-in strain field. The tensile strength is in second place and Young's modulus together with the Poisson ratio are needed if mechanical device performance is to be calculated, i.e. for design rather than feasibility studies.

The Measurement Technique

X-ray mask blanks require low atomic number materials in order for them to be reasonably transparent to x-rays produced by synchrotron radiation. This consideration and the fact that low pressure chemical vapor deposited films grow on all exposed surfaces and are not line of sight restricted, as for instance evaporated films, have resulted in a major effort to understand LPCVD films such as polysilicon and silicon nitride. Metal films such as sputtered aluminum alloys and electroplated gold, an x-ray absorber, and nickel have also received some attention. In all cases the built-in strain field, its magnitude and deviation, has been the dominant measured mechanical property.

The basic technique for measuring strain is readily understood by considering polysilicon on a for example basis. This material is produced by thermally decomposing silane, SiH_4, under vacuum conditions. Reactor temperatures below 580°C produce amorphous films, higher temperatures result in polycrystalline deposits. In either case as-deposited films on silicon dioxide covered wafers will exhibit built-in strain which is compressive and can be as large as 0.4%. The film is readily patterned by plasma etching. If the geometric features are those of a dumbell resistor with oversize pads, immersion into hydrofluoric acid will produce a free standing, doubly supported beam. This beam will tend to buckle if it is sufficiently long and if the strain field is large and compressive. Conversely, a short beam in a small compressive field will be straight. There are two important aspects of buckling: it depends only on geometry and does not depend on Young's modulus or the Poisson ratio, and it is very geometry dependent which implies a sharp threshold between not buckling and buckling. Since interference contrast microscopy is very sensitive to displacement, visual detection is simple and accurate.

The buckling issues are treated in detail in reference (3). It is possible to show that doubly supported beams with clamped supports buckle at a compressive strain level which is given by

(1) $\varepsilon = \pi^2 h^2 / 3 l_{cr}^2$

where the critical beam length is l_{cr} and the film thickness is h. By employing a mask which leads to beams of identical cross-section and varying length an accurate rapid determination of the strain field becomes possible. Furthermore, since strain fields as low as 0.05% cause buckling at 160 micron length and 2 micron film thickness the size of the needed micromechanical test pattern is quite small. Therefore, local strain level variation become measureable by scanning over the wafer surface.

A word of caution is in order. Equation (1) is based on the assumption that the built-in strain is uniformly distributed throughout the film thickness. In order to verify the validity of this assumption experimentally it is necessary to include another useful set of diagnostics: cantilever beams. If these structures bend up or down non-uniform strain or surface tension effects play a role and must be accounted for properly.

Silicon nitride is a material which can be produced from dichlorosilane and ammonia under LPCVD conditions at temperatures near 800°C. If this reaction is adjusted so that the film is essentially stochiometric, Si_3N_4 and a large built-in tensile field result. The consequence of this strain would be doubly supported beams which are straight and eventually rupture. Thus, a different type of test structure must be designed. Since buckling has the advantage of a direct strain measurement devices which buckle under tensile conditions are of interest. Lateral buckling as for instance in a doubly supported right angle beam or U-shaped beams is a possibility and has indeed been examined. It is difficult to use for quantitative data. A better structure involves a stiff ring which is rigidly clamped at both end points of a diameter. This type of structure contracts when it is etched free and the film is under tensile strain. The deformation maximizes for ring coordinates which are orthogonal to the support axis. If a weak beam with a small width is attached to the ring at the two maximum deflection points, buckling can be induced for specific ring diameters at a given strain level. The detailed mechanical analysis for this type of diagnostic structure is complicated, but has been completed. Experimental data from the structures, a set of ring and beams, allows quantitative evaluation of tensile fields.

Figure 1 summarizes the above ideas. It is the mask which is used to analyze strains which are typical for LPCVD films and acts as a single but very powerful diagnostic tool for thin film evaluation.

The ability to measure strain allows process optimization for repeatable, controllable and predetermined strain levels. In polysilicon this implies perfect grain size control. If this has not been achieved the measurement of Young's modulus and the Poisson ratio will not be very

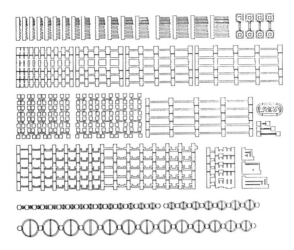

Figure 1. The Strain6 diagnostic mask contains doubly supported beams for measuring compressive strain fields, and ring and buckled beam structures for measuring tensile strain fields. Also included are structures that buckle laterally in either compression or tension.

successful. The argument for this is as follows. In single crystal silicon Young's modulus is given along the x' axis by

(2) $E' = 1/s_{11}'$

where the prime denotes an arbitrary orientation of the rotated coordinate system and

(3) $s_{11}' = s_{11} + (s_{11} - s_{12} - s_{44}/2)(l_1^4 + m_1^4 + n_1^4 - 1)$

where the s_{ij} are the compliance coefficients and l_1, m_1 and n_1 are appropriate direction cosines.

In polycrystalline silicon, Young's modulus becomes

(2a) $<E'> = <1/s_{11}'>$

The averaging occurs over all directions in the wafer plane. Structural information for the film can be included by using a texture function $G(\phi,\theta)$ which is of course an experimentally determined quantity. Then

(2b) $<E'> = (1/4\pi)\iint G(\phi,\theta)(1/s_{33}')\sin\theta d\theta d\phi$

It is obvious that a change in the texture function, a change in film morphology, leads to a change in Young's modulus. The same comment applies to the Poisson ratio which can be calculated by similar but more complex techniques. In either case data consistency can only be obtained with films for which the morphology is constant from process run to process run. A necessary condition for this is a process controlled strain field.

Stabilized films can be examined by a second set of micromechanical test structures. These are fundamentally doubly supported beams and plates. Both structures can be analyzed for clamped boundary conditions, an assumption which can be verified experimentally, and resonances. Experimental data from a series of structures lend themselves to an accurate measurement of strain, Young's modulus and the Poisson ratio. It is clear that this technique is much more complex than the simple strain measurement of the earlier discussion.

Material Results

The diagnostic structures have had dramatic influences in the understanding of polysilicon. This material is typically reacted at 640°C from pure silane. The film for this type of deposition is polycrystalline with large grain sizes. The grains typically change size with doping and subsequent anneal cycles. The application to controlled mechanical situations are therefore in doubt. This observation was used to re-examine the LPCVD process for silicon in some detail.

Reduction of the deposition temperature leads to reduced grain sizes. The transition from amorphous deposits to fine grained polysilicon occurs at 580°C. Films grown at this temperature on very clean oxidized silicon substrates exhibit compressive strains of up to 0.4%. This implies that the strain energy for 2 micron thick films on 3 inch wafers is approaching 0.1 joule. This energy is responsible for recrystallization and grain growth during subsequent anneal cycles. If the film and the substrate are defect free very rapid recrystallization with grain sizes as low as 100Å can be achieved. This grain size will not change significantly for doped or undoped samples even after heat treatment at 1150°C for several hours in nitrogen. The morphology of the film, its texture function, becomes essentially process independent. Typical results on 3 inch wafers are less than 50 defects at 0.5 micron aperture with a surface haze less than 200 Angstrom. Figure 2 contrasts standard polysilicon and fine grained polysilicon structures.

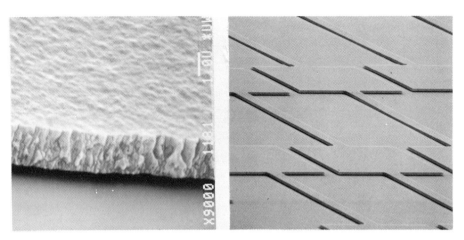

Figure 2. Micrographs of the sides of LPCVD polycrystalline silicon structures show the columnar grains in polysilicon deposited at 640°C (left), and virtually no texture visible in fine grained polysilicon grown at 580°C (right).

The strain field for fine grained polysilicon is of a very surprising nature. Until July 1987 all results indicated compressive behavior with starting strains of -0.4%. Annealing produced reductions to 0.001%. However, the films remained compresive. In July 1987 new annealing cycles were examined which convert the as-deposited compressive strain to tensile strains of up to 0.3%. Furthermore, process technologies were achieved which via the strain diagnostic resulted in the conclusion that any tensile strain level below 0.3% was indeed achieveable. Thus, free standing doubly supported beams which for compressive fields were limited to 160 micron in length at 2 micron film thickness because of buckling, were produced with over 1000 micron lengths. Furthermore, very large area fully supported diaphragms of 2 micron thickness with 1cm x 1cm lateral dimensions were produced and tested in a 3 x 3 array without failure and pinholes. The implication of this progress on x-ray lithography and micromechanics is very significant.

A second material which has received attention is silicon nitride. It is produced from dichlorosilane and ammonia. Two observations are useful: stochiometric silicon nitride normally has a built-in strain which is tensile; polysilicon which is obtained here by setting the ammonia flow to zero is an as-deposited compressive film. This implies that the ammonia to dichlorosilane ratio can be used to control the strain level in the film. In particular, zero strain silicon rich "silicon nitride" films are achievable and produce pin-hole free situations in electronic device processing.

Figure 3. A 3 x 3 array of 1 cm^2 square by 2 micron thick plates has been processed using a tensile film of polycrystalline silicon. The plates were fabricated by depositing and heat treating the film, then using an anisotropic etch to remove silicon in the substrate by back-etching. The etch was stopped on a thin layer of silicon nitride between the poly and the substrate, which was eventually removed.

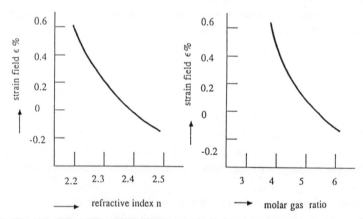

Figure 4. The refractive index and intrinsic stress varies as the deposition conditions for LPCVD silicon nitride are changed. Increased dichlorosilane to ammonia ratios result in lower tensile strain films, and will even produce compressive nitride films.

Attempts to understand the interplay between strain and grain size and electroplating conditions are currently in progress. This work is particularly important because gold is readily plated into vertical flank photoresist structures with submicron dimensions. Since gold, a high atomic number material, is a good x-ray absorber the application area is x-ray masks.

Applications

All of the mechanical results have had a significant benefit in electronic device applications. Thus, by carefully considering stress concentrations and of course very clean processing leakage currents in silicon devices can be reduced to less than 10 electrons per second per square micron at 15V reverse bias and 25°C. Perhaps just is important is the fact that the leakage current distribution becomes a very tight Gaussian with very small deviation. The understanding of the influence of polysilicon strain on the channel mobility is just beginning. However, very significant increases have already been measured for silicon on insulator structures.

Mask construction for x-ray lithography is a difficult task which requires mechanical considerations. The strain compensated silicon nitride work has led to mask blanks with 3 cm x 3 cm areas at film thicknesses of only 1 micron. Polysilicon blanks, which potentially are better thermal conductors, have been produced only because the available strain diagnostics allowed a quick evaluation of experimental data to guide process development towards tensile configuration of this important film. This in turn removed all size restrictions which certainly exist in a compressive situation.

In the micromechanics area the progress has been significant enough to conclude that feasibility has been demonstrated and product design can start. This conclusion is particularly true for the miniature pressure transducer area. The device, a sealed pill box, is constructed by using fine grained polysilicon over oxide posts to define the pillbox. Lateral etching of the oxide leads to an open box. Sealing via a new concept, reactive sealing, is a batch process which produces vacuum conditions in the pillbox by etch channel closure and oxygen gettering for oxide seals. Dielectric isolation for polysilicon sensing resistors which are fabricated from a second, doped polysilicon deposition is achieved by positioning these devices on a silicon nitride film which covers the entire cavity and also acts as a diffusion barrier. The end result is an absolute pressure transducer which for a one atmosphere design involves a pill box of 120 micron x 120 micron with a 7500Å overpressure gap. This implies that a 3 inch wafer contains more than 12,000 devices of this type. The implications towards sensor arrays for tactile sensing, redundancy, extended range and statistical averaging is significant.

150

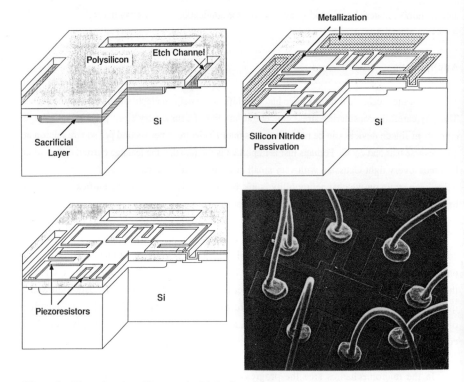

Figure 5. These drawings illustrate the fabrication of the planar processed pressure transducer. This transducer design features piezoresistive sensing, an effective over-pressure stop, a built-in vacuum reference, and utilizes a low-strain fine grained polysilicon plate material.

References

(1) J. C. C. Fan, B.-Y. Tsaur and M. W. Geis, "Graphite-Strip-Heater Zone-Melting Recrystallization of Si Films," Journal of Crystal Growth 63, 1983, pp 453-483.

(2) H. Guckel, D. W. Burns, C. R. Rutigliano, D. K. Showers and J. Uglow, "Fine Grained Polysilicon and its Application to Planar Pressure Transducers," Transducers '87, The 4th International Conference on Solid State Sensors and Actuators, Tokyo, Japan, June 1987.

(3) H. Guckel, T. Randazzo and D. W. Burns, "A Simple Technique for the Determination of Mechanical Strain in Thin Films with Applications to Polysilicon," Journal of Applied Physics 57 (5), 1 March 1985, pp 1671-1675.

FLICKER NOISE IN SEMICONDUCTOR DEVICES

R. Sharan

DEPARTMENT OF ELECTRICAL ENGINEERING
INDIAN INSTITUTE OF TECHNOLOGY,KANPUR-208016

Abstract:

A stochastic process with a power spectral density given by $Kf^{-\alpha}$ is called flicker noise. Many engineering groups have observed a relationship between large amplitude of flicker noise and high failure rates of devices and this effect has been used for screening out failure-prone devices. A simple phenomenological model to explain this widely observed relationship is presented and the limitations of this approach are examined.

1. Introduction :

A stochastic process with a power spectral density given by $Kf^{-\alpha}$ is called flicker noise and has been studied for about six decades. During this period this research has branched in several directions. The noteworthy directions are the following:

(a) Flicker noise has been observed in several [1-4] electronic systems (resisters, diodes, bipolar junction transistor, MOSFETS, amplifiers,oscillators etc.) and also in some [1] non-electronic systems (biological systems, economics data, hydrological data etc.). There has been a search for a unified theory which can explain all the effects but this, so far, has not yielded result. For example even the controversy whether the stochastic process which models flicker noise should be stationary or non stationary does not appear to be settled. It is felt by some that the process should be stationary because power spectral density is rather reproducibly measured by different groups at different times. The proponents of nonstationary nature of flicker noise, on the other hand, point out to the theoretical singularity at the origin and to the experimental observation that no bending of power spectral density has been observed up to very very low frequencies of observation. Some clarification of this controversy has been provided by Keshner [1] who considers this stochastic process to be evolutionary(i.e quasi-stationary).

(b) Although a unified theory of flicker noise has not been developed, there are several theories that have been proposed, each one of which explains the flicker effect in some devices quite well but not in other devices. For example, McWhorter's theory [5] explains surface effects, Hooge's theory and its extensions [6] explain bulk effect, and there are theories [7,8] to explain flicker noise of time-keeping oscillators. The result of these theoretical models seem to agree with experimental results on specially designed, carefully fabricated devices.

(c) The deleterious effect of flicker noise on amplification at low frequencies has been appreciated in the field of instrumentation. The problem is overcome either by using heterodyning [9] (i.e by frequency shifting, say, by lock-in-amplifier) or by designing specially tailored filters [10]. This area has gained considerably by borrowing ideas from control systems, communication systems and from system-theoretic studies. It is worth noting that there has been very little interaction between the physical electronics models mentioned in (b) and system theoretic models.

(d) For about two decades it has been observed in several reliability studies of semiconductor devices [11,12] that there is a direct relation between high amplitude of flicker noise and early failure of devices. This relationship has been used to screen out failure-prone devices and is of engineering importance. The published literature on this topic is sparse and there appear to be several gaps in the understanding of this relationship. Some aspects of this have been studied in the present work.Before we conclude this section, it is important to emphasize that our concern here is not with specially designed,carefully fabricated devices to test the theories of flicker noise, rather with the commercially produced devices which are used to fabricate instruments and systems.

2. Flicker noise and reliability.

Two well-known boat-shaped curves for semiconductor devices are of interest. One is a typical plot of power spectral density (p.s.d.) versus frequency [2] and the other is a typical reliability plot of the failure rate versus time [11]. Let us first consider the initial portions of these curves which are known as the flicker noise region of p.s.d. curve and the infant mortality region of the reliability curve,respectively. One finds that some devices from an ensemble of similarly processed devices show high amplitude of flicker noise and fail early contributing to infant mortality[11,12]. Care must however be exercised in drawing inferences from this observation because of the reasons elaborated below.

(i) The theoretical definition and experimental methods of measurement of p.s.d. are rather unambiguous, but there is scope of ambiguity in the case of reliability curve. This ambiguity does not arise from the definition of the failure rate (because once the probability density function is known the failure rate can be uniquely obtained [13]) but resides in the choice of the variable and the criteria which determine when a failure has taken place. These (variable and criteria) depend on engineering judgment and, if possible, should be standardized. In the absence of such standardization one has to make an empirical choice, some guidelines for which are given below.

(ii) Failure can be catastrophic, gradual or intermediate between the two. At the early stage of development of a device catastrophic failures take place and the avoidable causes of this are eliminated, hence our interest is not in these failures. An example of intermediate case is the degradation of beta of a transistor each time the base-emitter junction is reverse-biased, leading to failure of circuits like astable multivibrators [11]. In this case it is definitely and reproducibily established [11] that the degradation of beta is accompanied with a substantial increase of flicker noise. This intermediate effect is well-studied. Our interest, here, is in the gradual failure where the choice of criteria is that when the value of a variable has exceeded a certain limit, failure would take place. The chance of this happening is larger,larger the mean square fluctuation of the variable. Hence one should find a relation between the amplitude of flicker noise and the mean-square value of a chosen variable. Let us call this variable $x(t)$ and define the time of first failure $\mathcal{T}$ in an ensemble as where the value of $x(t+\mathcal{T})$ of one device has drifted and exceeded the threshold value so that failure has taken place. Our interest is in finding a relation between the mean-square value of the fluctuation $[x(t+\mathcal{T}) - x(t)]$ and its power spectral density.

3. Stationary or non-stationary process ?

The question of stationarity or non-stationarity of flicker noise has received most unambiguous treatment in the study of time-keeping oscillators [7,8]. There it has been noted that whereas the variable itself is nonstationary, the finite differences of the variable are statistically well behaved quantities. In other words, the flicker process is nonstationary, so that the autocorrelation $R(t,\mathcal{T})$ is both a function of the time (t) elapsed from the origin (i.e., aging or memory) and the time-difference $(\mathcal{T})$ between samples. Because of the t-dependence and ambiguity about the origin (t=0) the process is ill-behaved. It can be shown, however, that $R(t,0)-$

$R(t, \tau)$ is independent of t [14] hence is well behaved. To illustrate this point, let us consider a system having an impulse response $h(t) = 1/t^{\frac{1}{2}}$ and drive it with white noise. Then the autocorrelation function of the output is given by [14]:

$$R(t,\tau) = \int_{\delta}^{t-\tau} h(u)\, h(u+\tau)\, du = \int_{\delta}^{t-\tau} u^{-\frac{1}{2}} (u+\tau)^{-\frac{1}{2}}\, du$$

$$\approx \ln 4 + \ln (t/\delta) - \ln (|\tau|/\delta) \qquad (1)$$

Eqn (1) is valid for $\delta \ll \tau$. The lower limit δ is due to a high-frequency cut-off that has been introduced by a "smoothed" impulse response $h(t,) = (\frac{1}{\delta}) \int_{t}^{t+\delta} h(u)\, du$. It is obvious from Eqn (1) that $R(t,\delta) - R(t,\tau)$ is independent of t and only dependent on τ. Also:

$$E[\Delta x^2] = E\{[x(t+\tau) - x(t)]^2\} \qquad = 2[R(t,0) - R(t,\tau)] \qquad (2)$$

Hence now the mean square fluctuation can be considered to be stationary.

4. Mean-square value and flicker noise

In the previous section, it has been shown that for a filter with impulse response $h(u) = u^{-\frac{1}{2}}$ and driven by white noise, the autocorrelation $R(t,0) - R(t,\tau)$ can be considered stationary. We conjecture that this remains true for the physical stochastic signals of devices corresponding to flicker noise. Hence we can use Weiner-Khinchine theorem to obtain $R(t,0) - R(t,\tau)$ for a process with the power spectral density $S_x(\omega) = K\omega^{-\alpha}$. This gives [15]:

$$E[\Delta x^2] = (4K/\pi) \int_{0}^{\infty} \omega \sin^2(\frac{\omega \tau}{2})\, d\omega \qquad (3)$$

Using a change of variable $z = \omega\tau/2$ and integrating Eqn (3), one obtains [15] :

$$E[\Delta x^2] \approx (2k/\pi)\, \tau^{\alpha-1} \left\{ \Gamma(1-\alpha) \cos\left[\frac{(1-\alpha)\pi}{2}\right] \right\} \qquad (4)$$

This equation is valid for $1 < \alpha \leq 3$ but is not valid for $\alpha = 1$. For $\alpha = 1$, one gets [15]:

$$E[\Delta x^2] \approx (2K/\pi) [.57 + \ln(\omega_h \tau) - ci(\omega_h \tau)] \qquad (5)$$

where the last term is negligible for large $\omega_h \tau$. Here ω_h is an upper cut-off frequency.

A comparison of Eqn (4) and (5) is interesting. The amplitude of $E[\Delta x^2]$ is dependent on K in both the cases but when $\alpha > 1$, the amplitude increases in proportion to $\gamma^{\alpha-1}$. Since γ is large, the amplitude of $E[\Delta x^2]$, hence the drift of $x(t)$ and ensuing failure, is sensitively dependent on the value of α. It is customary in failure analysis to measure the power spectral density at spot frequencies, thereby only concentrating on the amplitude of the flicker noise. Our consideration here suggests that an accurate determination of α is also of vital importance in predicting the failure of devices. It is important to emphasize here that in the past an accurate determination of α would have been very tedious and time consuming but with the use of personal computer controlled measurement, data acquisition and analysis, now the value of α can be determined with comparative ease.

5. Conclusion

It has been conjectured that the difference $(R(t,0)-R(t,\gamma))$ in autocorrelation of physical signal of devices corresponding to flicker noise is independent of t, hence is stationary. In this case, the mean square fluctuation is dependent on the amplitude and the slope of the power spectral density of flicker noise. Hence the drift and failure due to mean square fluctuation can be predicted by measuring the flicker noise. This will be of help for those devices and systems where burn-in tests are not feasible due to either the cost or the complexity of the system.

References

1. M.S. Keshner, "1/f noise", Proc IEEE, 70, 212, 1982(Mar).

2. A. Ambrozy, "Electronic noise", p.122, McGraw Hill, New York, 1982.

3. D. Wolf, "1/f noise" in Noise in Physical Systems, p122, Ed. D. Wolf, Springer Verlag, Heidelberg, 1979.

4. A. Vanderzeil, "Flicker Noise in Electronic Devices", Advances in Electronics and Electron Physics, vol.49, Academic Press, New York, 1979.

5. S. Christenssan, et.al., "Low Frequency Noise in MOS Transistors - I and II, Solid State Elecr.,11,797 and 813, 1968.

6. M. Nelkin, "1/f resistance fluctuation" in Chaos and Statistical Methods, Ed. Y. Kuramato, Springer Verlag, Heidelberg,1984.

7.D.W. Allan,"Statistics for atomic frequency standards", Proc IEEE 54, 221,1966.

8.J.A. Barnes, "Atomic timekeeping and the statistics of precision signal generators" Proc IEEE 54, 207, 1966.

9.P.H.Sydenham, Handbook of Mesurement Science, vol.I, p.431, John Wiley, New York, 1982.

10.J. Llacer, "Optimum filtering in the presence of 1/f noise" NUclear Instr. Meth. 130, 565, 1975.

11.C.D.Matchenbacher, F.C. Fitchen, "Low-noise Electronic Design" p.87, John Wiley, New York, 1973.

12.L.K.J. Vandamme et.al., "1/f noise as a reliability estimation of solar cells" Solid St.Electr. 26, 671, 1983,(July).

13.S.M. Ross, "Introduction to Probability Models", Academic Press New York, 1972.

14.V. Radeka, "1/f Noise in Physical Measurements" IEEE Trans. Nucl. Sci., vol NS-16, p.17,1969 (Oct).

15.J.A. Ringo, P.O. Lauritzen, "Voltage drift related to low-frequency noise in reference diodes", Proc IEEE,60,236, 1972 (Feb).

JUNCTION TERMINATION - A TECHNIQUE FOR NEAR IDEAL BREAKDOWN IN PLANAR JUNCTIONS

S. Ahmad
Semiconductor Devices Area
Central Electronics Engineering Research Institute
Pilani (Rajasthan) 333031

Abstract

A brief review of various techniques available for junction termination in planar junction is undertaken here to find out an optimum method from breakdown point of view. Advantages and drawbacks associated with each method are examined for practical realisation of the associated structures. It is shown here that a near ideal breakdown can be accomplished in a planar junction if another junction of appropriate parameters is used for termination.

1. Introduction

Earlier development /1/ of high voltage power devices employed deeper junction and mesa type structures for achieving higher breakdown voltages. Techniques /1/ used for modifying junction termination in these were useful only in deeper junction devices as a fine control on physical removal of semiconductor material was generally required. However, in design of modern power devices, breakdown voltage alone is not the factor to be considered. Forward voltage drop associated with the involved p-n junction is another important /1/ parameter. Though higher breakdown voltage can be obtained by lowering doping in the low doped substrate, but this affects the forward drop adversely /1/. Therefore, a trade off has to be found between reasonable breakdown voltage and tolerable forward voltage drop. This indicates towards preferred use of structures involving relatively shallower junctions and restricted extension of low doped substrate i.e. punch through type devices. In this connection, planar junctions appear to meet these requirements with added features of high degree of control on fabrication processes and reproducibility. However, there is a serious drawback that planar junctions have very poor breakdown voltage due to junction curvature effect and surface charges present in passivation /1/. In order to utilize attractive features of the planar process, breakdown voltage of planar junction must be improved. This is the motivation behind search for high breakdown voltage planar junctions using different forms of junction termination techniques. A considerable amount of investigation has been carried out in this direction. Several methods

/1/ have been suggested on the basis of results obtained from two dimensional numerical simulations of the field distributions. Some recent studies /2, 3/ indicate that a near ideal breakdown i.e. breakdown corresponding to parallel plane region, free of curvature effect, is possible to achieve from a planar junction with appropriate junction termination.

In the present paper, various methods of junction termination suggested for planar devices are examined in brief. Advantages and drawbacks associated with each technique are looked into from the point of view of their practical realisations. The basic aim of this comparative study is to find out a simple and reliable method for achieving near ideal breakdown in a planar junction.

2. Planar Junction and Terminations

In a planar junction, as shown schematically in Fig.1, p^+-diffusion is done through an opening in the oxide layer. Junction formed by low doped substrate i.e. n^- region, has a parallel plane portion surrounded by edge region formed due to lateral diffusion under the diffusion mask. Junction contour in edge region is formed by geometry of the opening in oxide layer. For a rectangular opening, junction is spherical at the corners and cylindrical along rest of the edges. In case of circular diffusion window junction contour is cylindrical throughout.

The term "near ideal breakdown characteristics" /2/ used in this paper refers to the breakdown associated with the parallel plane region. This depends upon doping profile and related dimensions of high and low doped regions of the junction.

In a planar junction edge region causes maximum reduction in breakdown voltage. Influence of junction curvature, caused by side diffusion for a known geometry of diffusion opening, has been studied in detail and computation results /4/ are already available. Further simulations in two dimension for such devices reveal that the overall effect of junction curvature is not a result of junction contour alone but a combination of junction contour and doping profile in high doping side of the junction. This basic fact led different workers to propose various kinds of junction terminations to reduce electric field strength in edge region.

Termination methods considered in this paper are divided into four groups. Group A consists of techniques like field plate /1/ and floating field rings /1/. In group B, junction termintion extension (JTE) /5/ and variable lateral doping (VLD) /6/ methods are kept because a typical ion implantation with tight control on dosage is common to both. Technique like reduced surface field structure /7/ comes under group C. Lastly, methods / 8,9,10,11 / based on termination employing another planar junction with parameters different from main junction are considered in group D.

Group A

A. Field Plate

Field plate /1/ has been used to control electric field across the junction at surface as shown schematically in Fig. 2. Field plate may be either connected to a separate supply to control bias on it or it may be left floating. In another configuration, it may be connected to metallisation of the main junction. Field plate has also been used in conjunction with other techniques. But improvement in breakdown voltage due to field plate is relatively less, hence not much effective alone. Another disadvantage here is the requirement of thick oxide layer /6/ to support bias, if used in high voltage devices. This may put severe restriction on its practical realisability if thickness requirement runs in few micrometers.

B. Floating Field Ring

Location of a floating field ring /1/ surrounding a planar junction is schematically shown in Fig. 3. Presence of a floating field ring in vicinity of the edge region of a planar junction reduces field crowding and thus improves breakdown voltage. Instead of one, a number of such rings /1/ can be employed for this purpose but at the cost of increased wafer area usage. However, location and dimensions of field rings should be chosen such that proper sharing of applied bias is affected. The aim here is to create field distributions in edge region of the main junction and around the field ring, such that breakdown occurs simultaneously. For this, a careful design has to be made using numerical simulations. However, it has been noted that numerical solution of Poisson's equation in presence of field rings is rather time consuming. Thus it adds another drawback. Otherwise fabrication of field rings is quite simple. In case field ring junction is required to be same as the main, some additional openings in the same mask used to define diffusion window are sufficient to realise them in one step. This advantage is found quite useful in cases /1/ where marginal increase in wafer area usage can be tolerated against the substantial improvement in the breakdown. This is more valid in those cases where reduction in doping level in substrate for increasing breakdown voltage may offset forward drop.

GROUP B

Two methods have been proposed using the technique of ion implantation. One is known as junction termination extension (JTE) /5/ and other is called variation of lateral doping (VLD) /6/. In either of these, some percentage of the critical charge, defined as the product of peak electric field at breakdown in plane region and dielectric constant of silicon, is implanted in the edge region. The details follow.

A. JTE Method

In junction termination extension /5/, a controlled implantation of the same dopant as used in main junction, is done in the region shown in Fig. 4. An improvement of the order of 95% of the parallel plane breakdown voltage can be obtained if charge in implanted region is about 60-80% of critical charge defined earlier. Instead of one, two implantation approach was also suggested as alternative to control fields in surface and bulk regions. The main disadvantage of this technique is the observation of increased leakage current due to ion implantation at or near surface in experimental devices. Further, effectiveness of this technique is rather sensitive to quality of the passivation. Thus, these two factors contribute towards lesser applicability of this technique in fabricating devices with practical reproducibility.

B. VLD Method

This method /6/ was employed by Stengl et al to design and fabricate high breakdown devices using multiple implants in the region shown schematically in Fig. 5. Here, the usual junction curvature due to planar junction formation is increased by making the junction to spread over laterally. This added extension using low doping along the edges reduces electric field. Once again, to reduce surface fields, integrated ion implantation dose was related to the critical charge as in the case of JTE method. From fabrication point of view, this method seems to be attractive. Openings in diffusion mask for both main and terminating junction are needed to fabricate this structure. Locations and areas of a number of smaller openings required for VLD can be chosen in such a manner that after drive-in step, the overall profile produces desired gradual change. Considerable improvement occurs when compared to JTE technique. This structure is relatively less prone to positive oxide charges when compared to floating field ring structure. Further, the design is much simpler than those of the floating field ring.

GROUP C

RESURF Structures

Concept of reduced surface field (RESURF) /7/ was used by Appels et al to realise a high voltage junction for various applications. Here, starting material is a thin n^- epitaxial layer on p^--type substrate. A deep p^+ diffusion followed by a relatively shallower n^+ diffusion, for making contacts to p^--substrate and n^--epi-layer respectively are used to fabricate RESURF device as shown schematically in Fig. 6. Though different from conventional planar junction considered so far, this device uses two junctions to support applied reverse bias. One is along the surface involving n^--p^+ which is critical from breakdown considerations. Other is formed between n^- and p^--substrate. By controlling doping and thickness of epitaxial layer, horizontal junction at surface can be influenced by vertical junction in the plane region. When epitaxial layer thickness d_{epi} is such that $d_{epi} \cdot N_{epi} = 10^{12}/cm^2$, where N_{epi} is doping in epitaxial layer, the peak electric field at surface is lower than that across the vertical junction. Similarly, distance between the n^+ and p^+ contacts i.e. width of the RESURF layer, is also critical because if it is smaller, uneven sharing of voltage may occur between reverse biased n^+ - n^- and n^- - p^+ junctions. This gives rise to corner breakdown at much lower bias due to curvature associated with shallow n^+ diffusion.

RESURF structure has been used in making high voltage bipolar, junction field effect and MOS transistors /7/. Because of its configuration more suited to conventional IC structures, maximum use has been made there for integraton of above devices.

Design is rather simpler here than those involved with field ring. It has been studied by Walker et al /3/ in detail where the influence of surface charges upto $5 \times 10^{11}/cm^2$ has been considered. RESURF is found less sensitive to surface charges as compared to field ring.

Fabrication involves all conventional techniques hence easier but it does need two diffusions besides special type of n^- on p^--epitaxial wafers.

GROUP D

Methods of termination employing another planar junction to modify field distribution in the edge can be divided into two further subgroups. In one, a p^+- π - n^--type of structure is employed. In the second, the terminating junction is chosen to be deeper and has relatively low doping. The details follow.

A. $\underline{P^+\text{-}\pi\text{-}N^-\text{-Structures}}$

Here /8/, as shown in Fig. 7 schematically, the junction consists of two regions. As the doping in π region of π-n^-- junction is relatively lower, severity of the junction curvature is reduced. Further, π-n^- junction is deeper than p^+-π junction, therefore associated larger radius of curvature does not reduce the breakdown as would have been the case if heavily doped p-region was to form the same junction. The doping in the π-region and its extension beyond the edge of the main junction were taken as parameter for optimisation purposes. The proposed structure removed the peak field from the surface to the edge region giving substantial improvement in the breakdown voltage. Further, effect of fixed oxide surface charges Q_{ss} was also considered in this work. However, because the terminating junction forms practically a part of the main junction, the adverse effect of Q_{ss} could not be counter balanced completely.

Practical realisation of p^+-π-n^--type of structure is simple and therefore the proposal is attractive. However, since breakdown appears to be confined to the edge region below surface the structure is reported to be influenced by oxide charges if the value is higher than $10^{11}/cm^2$.

B. Pocket Termination

In another proposal /2,9/ a low doped junction was employed to terminate the main junction as shown schematically in Fig. 8. Even with rectangular junction and constant doping assumed in earlier simulation /2/ some very useful results were noted. When the terminating junction was taken deeper than the main junction and sufficiently wider in the lateral direction, a large improvement in breakdown voltage was observed. Even the breakdown shifted from edge regions of either of the junctions to the plane region below the main junction. Though such results were obtained in case of rather extremely simple structures which are difficult to fabricate but on the other hand represented the worst case type from breakdown point of view. Rectangular junction contours are the wrost to reduce breakdown voltage. Thus if breakdown shifts to near ideal situation for such structures, the behaviour in case of normal rounded corner type diffused junctions would be still more easy to observe. This was confirmed in a later study /9/ employing diffused junctions with finite overlap as shown in Fig. 9. Additionally, the largest avalanche multiplication path (LAMP) was computed for each combination of terminating junction parameters. Shift in the position of LAMP affected by changes in the parameters was used to look for the optimum structure. Exercise of optimisation was carried out for cases with and without the presence of fixed oxide charges Q_{ss}. It was further seen that upto fairly large magnitude of Q_{ss} (i.e. $10^{12}/cm^2$), breakdown could be shifted to the plane region by considering combined influence of terminating junction parameters along with oxide charges.

This method is very simple from fabrication point of view. Conventional methods of predeposition and drive-in can be used to realise pocket junction followed by formation of p^+-n junction allowing for the required overlap. Ion implanted predeposition would be attractive for controlling the precise doping profile in the pocket junction.

Two other groups /10,11/ also analysed termination similar to the one described above. When compared, it is found that in these two cases breakdown shifted from surface to edge regions due to relatively shallower junctions used. However, in one case /11/, the indication towards possible optimisation was given while discussing the effects of oxide charges Q_{ss}. Incidentally, in these analyses /10,11/ mostly high breakdown devices were considered where large dimensions are involved. Whereas in the case of Ahmad et al /2,9/ a 100V device was analysed which involved structural dimensions in the range of few micrometer. Because of this a fairly large range of parameter variations could be tried out and additional results could be seen.

3. Discussions

For accomplishing the ideal breakdown condition in a planar junction, the breakdown should take place in the parallel plane region. During junction termination, it can be noted that improvement in breakdown might occur due to shift of peak electric field from surface to edge region in the bulk. But this is not yet the ultimate goal to look for. When, largest avalanche multiplication path along field line is confined to the edge region, the structure

164

is sensitive to the presence of positive oxide charges Q_{ss}. This would cause problems of stability and reproducibility in the performance. From oxide charges affect as well it is preferable to have breakdown in parallel plane region. Besides these effects, influence of the geometry of diffusion window is also taken care of if the parallel plane region is containing the LAMP.

Keeping in view the goal of achieving ideal breakdown, the methods of termination described earlier are examined below.

Methods in group A are such that there is hardly any possibility of complete shift of breakdown. Similarly the techniques of group B employ ion implantations confined to the regions close to surface. Therefore the possibility of getting ideal breakdown appears to be remote here as well. RESURF devices of group C do indicate towards this possibility. But only disadvantage with this structure is an unwanted influence of high field present at the n^+-n-junction. This ultimately limits the utility of an otherwise attractive method. In group D, the method based on p^+-π-n^--type structure looks quite promising. But because of the presence of low doped π-region in series with p^+-region along the parallel plane part of the main junction, its utility becomes limited due to an expected larger forward drop. The methods given in last subgroup of group D appear to be optimum. With right kind of junction parameters used for termination, a near ideal breakdown can be accomplished without putting any further restriction on forward drop. Taking relatively higher doping in p^+-region, an ideal breakdown can be realised using conventional techniques of planar processing.

4. Conclusion

A careful comparison of the methods available for improving the breakdown voltage of a planar junction enabled to find out an optimum technique as described above. The method, thus arrived at, is free from adverse affect of junction curvature due to side diffusion and surface charges present in the passivating layer.

Acknowledgements

Author is thankful to his colleagues Mr. J. Akhtar and Mr. ASV Sarma for their kind help. Constant encouragement from Dr. G.N. Acharya, Director, is gratefully acknowledged.

References

/1/ B.J. Baliga, IEE Proc., 129, Pt I, No. 5, p 173,82.
/2/ S.Ahmad and J. Akhtar, IEEE Elec Dev Lett, EDL-6(9), p 465, 85.
/3/ P. Walker, J.T. Davies and K.I. Nuttal, IEE Proc. 132, Pt I, No. 6,85 and K. Board and M. Darwish, ibid, No. 4, p 177, 85.
/4/ J. Akhtar and S. Ahmad, IEEE Trans Elec Dev, ED-31, No. 12, p 1781, 84.
/5/ V.A.K. Temple, IEEE Trans Elec Dev, ED-30, No. 8, p 954, 83.
/6/ R. Stengl, U. Gosele, C. Fellinger, M. Beyer and S. Walesch, IEEE Trans Elec Dev, ED-33, No. 3, p 426, 86.
/7/ J.A. Appels, M.G. Collet, P.A.H. Hart, H.M.J. Vaes and J.F.C.M. Verhoeven, Philips J. of Res., 35, No. 1, p 1, 80.
/8/ K. Hwang and D.H. Navon, IEEE Trans Elec Dev, ED-31, No. 9, p 1126, 84.
/9/ J. Akhtar and S. Ahmad, to be published.
/10/ V. Boisson, M. Le Helley and J-P Chante, IEEE Trans Elec Dev, ED-33, No.1, p 80, 86.
/11/ S. Georgescu, T. Dunca, D. Sdrulla and I. Ghita, Solid State Electronics, 26, No. 10, p 1035, 86.

Figure Captions

Fig 1. Schematics of various regions in a planar junction used in the text.
Fig 2. Field Plate Termination of a planar junction.
Fig 3. Cross Section of Floating Field Ring Configuration.
Fig 4. Junction Termination Extension in a planar junction.
Fig 5. Details of VLD Structure for termination.
Fig 6. High Voltage RESURF Device on an Epitaxial Wafer.
Fig 7. P^+-π-N^- Diode Termination.
Fig 8. Pocket Termination Using Rectangular Junction.
Fig 9. Diffused Junctions Pocket Termination.

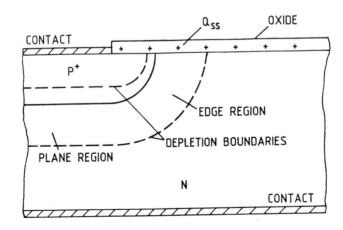

FIG. 1.

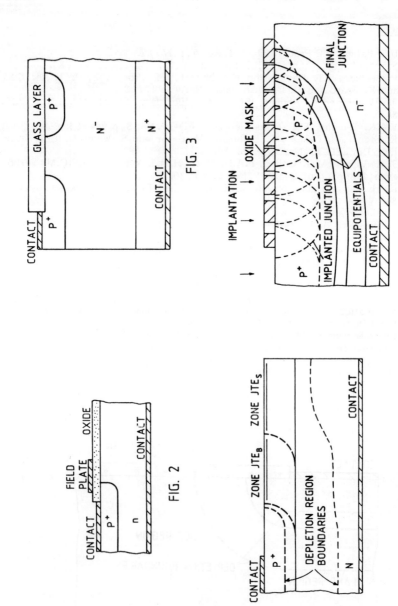

CONTACT P⁺ GLASS LAYER P⁺

N⁻

N⁺

CONTACT

FIG. 3

IMPLANTATION OXIDE MASK

FINAL JUNCTION

P⁺ IMPLANTED JUNCTION

P⁻

n⁻

EQUIPOTENTIALS

CONTACT

FIG. 5

FIELD PLATE OXIDE

CONTACT P⁺

n

CONTACT

FIG. 2

ZONE JTE_B ZONE JTE_S

CONTACT P⁺

DEPLETION REGION BOUNDARIES

N

CONTACT

FIG. 4

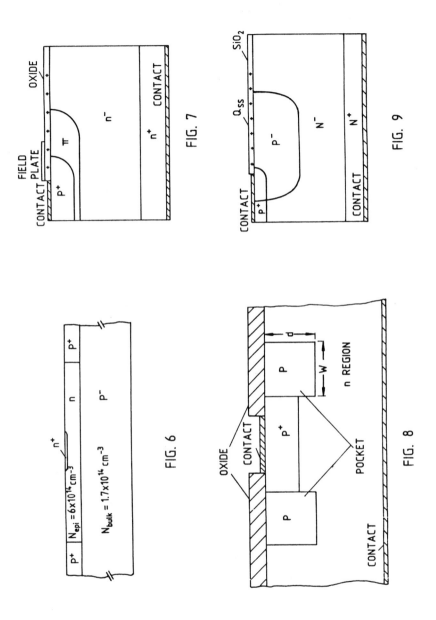

FIG. 7

FIG. 9

FIG. 6

FIG. 8

SOME IMPLICATIONS OF MINIATURISATION OF SEMICONDUCTOR DEVICES

M. Satyam, K. Ramkumar, K.S. Gurumurthy
Department of Electrical Communication Engineering
Indian Institute of Science
Bangalore-560 012, India

Abstract

The tendancy to design smaller and smaller devices for very large scale IC's through "Scaling" does not seem to yield the expected device performance when one goes to very small dimensions. One of the problems is the fringe effects which become more and more prominent at these dimensions. These effects are not taken into account, generally, in the scaling. Another aspect of the problem is the behaviour of the various layers of the semiconductor device when they become very thin. In particular, when the gate oxide of an MOS device becomes thin (thickness of the order of a few hundred angstroms), various phenomena connected with the transportation of carriers in and across the oxide come into picture. These phenomena alter the behaviour of the device and many times the device becomes non functional.

In this paper, some aspects of scaling of MOS devices are dealt with first and their implications on small geometry devices are brought out. One of the problem areas, viz., the electrical behaviour of thin oxides is later discussed in detail. There seem to be alternate approach to miniaturisation of electronic circuits and systems through the reduction in the number of components unlike in the case of MOS integrated circuits where each device is miniaturised. This approach is briefly discussed in the end.

Section I

Introduction to the physics of small geometry MOS devices

(i) Scaling: In the attempt to reduce the dimensions of MOS devices, a scaling process is adopted which aims at reducing the dimensions without disturbing the electric field pattern in the device. This process involves changing of conductivities and operating voltages as well. The scaled and original parameters of a typical device are given below.

Parameter		Scaled value
Length of channel	L	L/K
Channel width	W	W/K
Oxide thickness	t	t/K
Resistivity of semiconductor	ρ	ρ/K

Characteristics of Original device		Characteristics of Scaled device
Drain voltage	V_D	V_D/K
Threshold voltage	V_T	V_T/K
Transconductance	g_m	g_m
Current	I_D	I_D/K
Capacitance	C	C/K
Breakdown voltage	BV	$BV/\sqrt{K}$
Power disipation	P	P/K^2

K is the scaling factor.

From the table, it may be noted that

(1)   reduction of the width W by a factor of K does
       not have anything to do with keeping the electric
       field constant.

(2)   the field in the oxide of the scaled device will be
       same as that in the original device, only under
       accumalation condition.

(3)   The reduction in W essentially reduces the current
       and transconductance of the device.

(4)   The breakdown voltage of the scaled device is
       lowered.

Therefore, scaling all the parameters by a constant factor gives rise
to undesirable effects as far as breakdown voltage, maximum current
and transconductance are concerned. Thus, it is necessary to think
of operating the devices at low currents, low voltages which is not
always possible.

The need to operate devices at low currents calls for the
use of large value resistors which may be difficult to realise through
the conventional MOSFETs of comparable lengths. This problem
of getting high value resistors can perhaps be tackled, to some extent,
by using poly silicon resistors. Poly silicon resistor does not occupy
additional space since it can be deposited on the passivating oxide
on another MOSFET itself. However, these resistors exhibit
negative resistance characteristics that may lead to instabilities.

Further, reproducibility of these resistors is poor. In view of these,

poly silicon has been studied by the devices group at I.I.Sc and the

results of these will be discussed in a later section.

Another problem encountered in small geometry devices is

that of predicting the values of threshold voltage of the devices.

Normally, one dimensional analysis is used. However, because of the

fringe effects, prediction of the threshold voltage for small devices

needs a two dimensional analysis of the electric field.

(ii) **Electrical conduction in thin oxides.**

When the gate oxide of an MOS devices becomes very thin

(thickness $\sim$ 100 A° to 800 A°), it is no more an insulator layer

capable of blocking flow of current across it. At the operating voltages,

the electric field across this oxide is sufficiently large to support (i)

tunneling of carriers and (ii) avalanche multiplication of carriers in the

insulator. Further, these processes get complicated because of the

carrier, trapping inside the oxide layer. The time constants associated

with these processes are sometimes large and they give rise to time

dependent behaviour of the oxide layer. It is very much essential

to understand the electrical behaviour of these thin oxides so that

they can be used for devices. The various investigations that have

been carried out on thin oxide MOS structures are discussed in

Section II.

(iii) **Functional Approach to Miniaturisation - An alternative?**

The problems to be overcome to get devices with reasonable characteristics are; many fold - They range from designing devices based on two dimensional analysis, getting at better fabrication tools, getting at better materials and new processes etc. On the other hand, one can think of a new approach in which the same goal of miniaturisation can perhaps be achieved by developing new device and circuit configurations which will involve less number of components. This approach is generally referred to as "Functional Approach". A blend of functional approach with the conventional approach to miniaturisation through reduction in sizes of individual devices might provide an ultimate solution to the problem of miniaturisation without undue effort. The devices group at I.I.Sc has been putting considerable effort in this area and some aspects of the work in this direction will be discussed in Section III.

Section II

(i) **Electrical conduction in thin silicon oxide.**

Generally, the electrical behaviour of thin oxides is studied by making MOS structures as shown in fig. 1. When the thickness of the oxide is in the range of a few hundred angstroms, it is observed that the oxide conducts and the conduction depends on the time for which the electrical stress (voltage or current) is applied. A typical V-I characteristic as measured by Holland et al is given in fig 2. From this figure it may be seen that there is a shift of the characteristic after it is stressed. This has been attributed to the charging of the insulator due to carrier trapping (1). As the voltage across the MOS structure is increased, the current at some stage starts increasing steeply. At this point various additional processes enter into the process of conduction and these are responsible for the increase in current. They are - avalanche multiplication of carriers and the trapping of multiplied carriers in the insulator. The following is a description of the effect of multiplication and trapping of carriers which finally leads to the time dependent breakdown.

Fig. 3a indicates the electric field distribution in the oxide initially, When the multiplication and trapping are negligible. The few carriers that tunnel into the insulator get multiplied in the region beyond x_o. The multiplied carriers in turn get accelerated towards the respective electrodes and in the process create additional carriers by further multiplication. On account of this, the electron concentration goes on increasing from the metal side to the semiconductor side while the hole density increases from the semiconductor side to the metal. This gives rise to increased electron trapping in the oxide on the metal side and increased hole trapping on the semiconductor side. This trapped charge on both sides of the oxide distorts the electric field in this layer in such a way that there is considerable increase in the field near the metal as well as near the semiconductor. A typical variation of the electric field after the avalanche multi- plication–trapping processes is shown in fig 3b. The increase in electric field changes the tunnel current; invariably the tunnel current gets reduced. However, the multiplication in the high field regions increases and this increase makes up some of the decrease in the tunnel current. Ultimately, a stage is reached when the current flow can be maintained without the presence of any tunnel current, but fully supported by the multiplication process. At this stage, the device becomes self sustained in maintaining current flow and it may be considered to be under breakdown. This breakdown gives rise to different characteristics depending on the condition under which it is produced, viz., constant voltage or constant current. The process of build up of the electric field near the electrodes involves some time and it depends upon the initial voltage or initial current, the multiplication coefficient of carriers and the carrier trapping coefficient. This process leads to a delay in the breakdown of the oxide from the instant at which the voltage or current is applied. This is known as the "time dependent breakdown". This general view of the breakdown has been put forward by various investigators. However there are differences in the manner in which the problem is attacked.

Harari, (2) observed that the characteristics of the time dependent breakdown is an index of the quality of the oxide. Extensive measurements have been carried out on oxides in the thickness range $30-300\overset{\circ}{A}$, in the temperature range $77^{\circ}K$ to $300^{\circ}K$. He reports that the time to breakdown decreases with increase in temperature. He proposes that the mechanism of breakdown is intimately related to electron trapping and electron trap generation and their dependence on temperature. He argues that impact ionisation may not be responsible for the breakdown and it may be caused by the continuous build up of internal electric fields because of the generation of extra electron traps. According to him, the time to breakdown is lower at higher temperature because the traps that may be generated at higher temperatures for a given electrical stress is higher and this gives rise to increased electron trapping which in turn gives rise to higher fields near the electrodes.

Chenming Hu et al (3) have looked at the problem from a different angle. They have computed the total stored charge in the oxide by integrating the current with respect to time. They describe the oxide in terms of "weak areas" and "robust areas". According to them, the time for "Breakdown" is arbitrarily defined as the time needed for the current to increase by ten fold. Their observations are confined to constant voltage operation. In their model they have considered hole trapping and they argue that hole trapping decides the breakdown in the silicon oxide films.

Yamabe K and Taniguchi K (4) studied the time dependent dielectric breakdown in thin silicon oxides of thickness around 200 A$^{\circ}$ on a very large number of devices. Their analysis is one of statistical nature and they have grouped their samples into three categories - one failing at very low fields, another failing at fields of about 5 MV/cm and the third group withstanding upto 10 MV/cm. The failure of the first group is apparently caused by a defective oxide structure on account of defects like pin holes. The failure of the second group which has a reasonable breakdown strength is attributed to defects in oxides like microprecipitates, metallic contamination etc. The last group perhaps has defect free oxide. It is also reported that the time to breakdown at a particular stress decreases with increase in temperature.

Chen C.F. and Wu. C.Y. (5) have carefully studied the various phenomena that take place at the Si-Sio$_2$ interface and bulk Si0$_2$ and propose a theoretical model which takes into account various factors like trapping of electrons, positive charge generation by impact ionisation, dangling bonds, etc. They too describes the oxide in terms of "weak spots" and "robust areas". Their contention is that the positive charge aggregation in the weak spots produces a distortion in the electric field and this increases the impact ionisation which finally leads to electrical breakdown. They have gone through the detailed mechanisms of trapping and avalanche multiplication of carriers and have developed a computer program which may be useful for computer Aided Design.

Chen D.N. and Cheng. Y.C. (6) have pointed out that no single mechanism of breakdown can explain all the observed phenomena like presence of molten metal, molten silicon and molten oxide in the vicinity of breakdown region. They propose two main mechanisms, viz., Avalanche multiplication through small regions in materials with large band gap and hot filament transportation. In films with small band gap, pure avalanche multiplication can take place and non destructive breakdown can occur. In materials with large band gap, initially, weak avalanche multiplication occurs and then, because of the high fields, current transportation through filaments which are at high temperature occur. This type of breakdown is a destructive one. The defect density influences the onset of filaments and hence the destructive breakdown.

Weinberg. Z.A. and Nguyen T.N.(7) have examined the relation between the positive charge accumalation at the $Si-SiO_2$ interface and the occurence of breakdown. They prepared oxides having different hole-trapping properties through a short rapid thermal annealing in oxygen. Measurements on these oxides involving testing of hole trapping, high field stressing, the initial current transients at constant, gate voltage, and breakdown statistics. Their conclusion is that the simple model of positive charge feedback mechanism is not the main cause of breakdown, but is one of the processes occuring during high field stressing. The positive charge can cause large current increase but the current quickly decays because of its neutralisation. The exact details of the processes of positive charge generation and its neutralisation is not clear.

The work carried out in this area at I.I.Sc. is indicated briefly in Section III.

Section III

A. **Electrical conduction in Polycrystalline Silicon-Poly silicon resistors**

The electrical behaviour of a polycrystalline material is greatly influenced by the grain boundaries present in it. It is well known that the crystal structure breaksdown at these grain boundaries and this has several implications on the properties of the polycrystalline material. At the grain boundary a potential barrier gets built up due to the breakdown of the crystal structure and the trapping of carriers by the dangling bonds. The dependence of this barrier potential on the dynamics of trapping and release of carriers from the grain boundary has been studied (8)   The barriers at the grain boundaries impede the motion of carriers and thereby affect the carrier mobility. The variation of mobility with parameters like grain size and the characteristics of the grain boundary have been theoretically computed (8). Typical variations are shown in figure 4.   . Because of the presence of the potential barrier at the grain boundaries, one expects the density of carriers to vary from the centre of the grain to the grain boundary. The way in which this concentration varies has been studied and its implications on the variation of conductivity in a polysilicon wafer has also been investigated, by looking at the diffusion and trapping of carriers at the grain boundaries. A typical variation of conductivity along a grain is shown in fig 5. . The overall effect of the potential barriers in a polycrystalline semiconductors from the point of resistance has been analysed. Polysilicon resistors exhibit a negative resistance in the voltage-current characteristic and this has been explained based on the self heating of the resistor and its effect on the grain boundary barrier (9). The voltage-current characteristics of a typical poly silicon resistor is shown in fig. 6.

B. **Electrical conduction in thin oxides**

One of the important phenomenon associated with the electrical conduction in thin silicon oxides, namely, "time dependent breakdown" has been investigated. As indicated earlier, various investigators have expressed the general view that charge carrier trapping, and trap generation are perhaps responsible for the large time involved in the breakdown. However, no attempt seems to have been made to really analyse the multiplication of carriers along the insulating oxide and the consequent trapping which inturn influences the multiplication. This process has been studied through the following equations governing the motion, multiplication and trapping of carriers.

$$\frac{\partial n(x,t)}{\partial t} = (\alpha_n - \eta_n) \cdot v_n \, n(x,t) + (\alpha_p - \eta_p) \cdot v_p \, p(x,t) - \frac{\partial (n \cdot v_n)}{\partial x}$$

$$\frac{\partial p(x,t)}{\partial t} = (\alpha_n - \eta_n) v_n \, n(x,t) + (\alpha_p - \eta_p) v_p \, p(x,t) - \frac{\partial (p \cdot v_p)}{\partial x}$$

$$\frac{\partial D}{\partial x} = q \left(\eta_p p - \eta_n n \right)$$

α_n, α_p — ionisation coefficients,

η_n, η_p — carrier trapping coefficients.

v_n, v_p — drift velocities

Solution of these equations with appropriate initial conditions
yields the variation of voltage or current in the device, with
time. This analysis appears to picturise clearly the various
processes that take place in a thin oxide film and lead to
breakdown. This model can perhaps help in prolonging the
time to breakdown by appropriate control of trap densities at
different points in the oxide.

C.   Functional approach to microelectronics.

As explained earlier, functional approach attempts to
achieve miniaturisation through reduction in the number of
components per function and hence the required chip area.
To illustre the strength of this approach, new functional blocks
synthesised to perform various functions are shown in figs. 7-8.

References

1. S. Holland, et al, IEEE Electron. Dev. Lelt, Vol.5, P.302 (1984)
2. E. Harari, Journ. Appl.Phys., Vol.49, P.2478 (1978)
3. Chenming Hu et al, IEEE Trans. Electron Dev., Vol.32,P.413 (1985)
4. K.Yamabe and K.Taniguchi, IEEE. Trans.Electron Dev,Vol.32,423 (1985)
5. C.F. Chen and C.Y. Wu, Journ. Appl.Phys, Vol.60, P.3926 (1986)
6. D.N.Chen and Y.C.Cheng, Journ.Appl.Phys, Vol.61, P.1592 (1987)
7. Z.A.Weinberg and T.N.Nguyen, Journ.Appl.Phy,Vol.61, P. 947 (1987)
8. K. Ramkumar and M.Satyam, Appl.Phys.lett, Vol.39, P.898 (1981)
9. K. Ramkumar and M.Satyam, Journ.Appl.Phys, Vol.62, P.174 (1987)

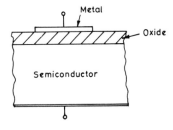

Fig. 1

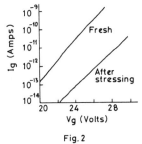

Fig. 2

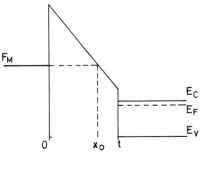

Fig. 3 (a)

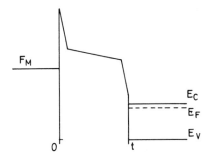

Fig. 3 (b)

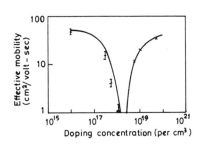

Fig. 4

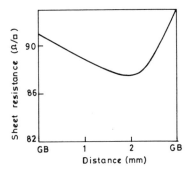

Fig. 5

182

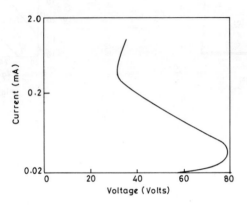

Fig. 6

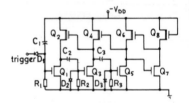

Fig.7 Three stage monostable
multivibrator.

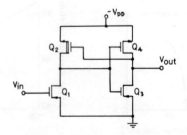

Fig. 8 CMOS Schmitt trigger.

SUBMICRON CMOS TECHNOLOGY

Dinesh K. Sharma
S.S.E. Group, Tata Institute of Fundamental Research
BOMBAY : 400 005 (INDIA)

1. INTRODUCTION

Over the history of Integrated Circuits, there has been an effort to reduce the device size in order to pack more functions per package, to reduce costs and to enhance performance. As a result, device sizes have shown a steady downward trend over the years. In the case of MOS devices, the scaling theory has been extremely useful in optimizing the design of small geometry transistors [1]. For successful device scaling, device design has to be accompanied by advances in the capabilities of processing equipment and procedures. In particular, finer lithography and etching techniques have to be developed for each round of device shrinking. Through a series of device scaling steps, channel lengths of about one micrometer are now becoming quite common[2-5]. At this level of circuit density, CMOS has been the technology of choice because of its low power consumption, rail to rail logic swing and high noise margins.

The next logical step in device miniaturisation would be to scale down the device dimensions even further and develop submicrometer CMOS technologies. The term submicrometer covers a wide range of processes. Processes with channel lengths in the vicinity of 0.9 micrometers, 0.5 micrometers and even 0.25 micrometers or less are in development simultaneously and all of these qualify, of course, to be called submicrometer processes. The problems, optimization techniques and design compromises made at these different levels of scaling are necessarily quite different. In this paper, unless specified otherwise, we shall refer to processes with 0.5 to 0.9 micrometer channel length as submicrometer processes.

2. THE CHALLANGE OF SUBMICROMETER CMOS

As we approach submicrometer geometries, the task of continued device shrinking presents a formidable challange. This is because several problems need to be overcome before further scaling. These problems can be roughly classified as process related, device related or system related. In this paper I shall briefly describe these problems and the solutions which are being adopted to counter these.

2.1 Process Related Problems

The main process related problems include (in addition to the obvious need for fine line lithography and etching), the difficulty of creating extremely shallow junctions required for these technologies, the need to make good ohmic contacts to these shallow junctions without causing shorts, the requirement of growing extremely thin but pinhole free gate oxides and the need to provide low resistance and electro-migration free interconnects.

2.1.1 Fine line lithography

This is of course the basic requirement for any submicrometer process. Several techniques such as electrn beam direct write on wafer, deep UV projection, X-ray lithography and ion beam lithography are capable of providing the kind of resolution required by submicrometer processes. Of these, electron beam and deep UV techniques have emerged as the most practical for immediate use. Work continues, however, on X-ray lithography and ion-beam lithography for use in future processes.

Optical techniques have the advantage of providing high throughput while keeping the cost and complexity of equipment at moderate levels. Therefore, optical techniques are often preferred over others as long as these can provide the required resolution. It is believed that deep UV projection with direct step on wafer will be adequate for feature sizes upto 0.6 micrometer or so. For feature sizes smaller than this, electron beam direct write on wafer is expected to be the technique of choice.

Electron beam lithography is capable of producing lines and spaces down to about 0.1 micrometer. The main drawbacks of this technique are its low throughput and the cost and complexity of the equipment required. One advantage of this technique is that it does not require any masks and can therefore be quite cost effective for those designs where the pattern to be exposed needs to be changed frequently (for prototyping runs and semicustom circuits, for example).

In addition to the technique for exposing the resist, the type of resist used is quite important by itself. In order to provide consistent results over all kinds of topography and reflectivities, tri-layer resists are being used extensively.

In these systems, the base layer serves to provide a planarizing layer and may include absorbing dyes to overcome internal reflections and interference effects in the resist. The intermediate layer provides a hard mask to avoid resist erosion during subsequent etching steps. The top layer is the photo-sensitive or imaging layer. It can be quite thin now to provide the best resolution possible, because planarization and resistance to etchants are provided by other layers. The imaging layer is exposed and developed as usual. The intermediate and the planarization layers are then etched away using reactive ion etching (R.I.E.).

2.1.2 Dry Etching Techniques

Only dry etching techniques can provide the kind of resolution required by submicrometer processes. Very careful control of etching slopes is required because too steep a slope may sometimes result in problems with step coverage for subsequent layers and too gradual a step would blow out the minimum line and spacing kind of design rules. As we approach submicrometer geometries, the post-etch cleaning of polymer deposits from the plasma process becomes quite tough. The R.F. power level has to be optimized, so that there is no polymer deposit in the etched holes. It is possible to clean some of the deposit by oxidizing it away in an Oxygen plasma. This, however, oxidizes the underlying silicon as well and if the junctions are very shallow, one may not be able to afford the loss of material from the contact areas.

2.1.3 Creation of Shallow Junctions

In order to keep short channel effects under control, the source-drain junctions need to be quite shallow. A rough rule of thumb is that junction depths need to be about a quarter to a fifth of the effective channel length. This means we need junctions which are no deeper than about 0.2 micrometer. For such short channel devices, the parasitic elements account for a major portion of the circuit delay. To keep the parasitic delays down, the contact resistance has to be minimized by making the surface concentration as high as possible. The twin requirements of high surface concentration and small junction depth make it mandatory to use the least amount of thermal budget after the source/drain junctions have been implanted. This can be achieved by the use of Rapid Thermal Annealing techniques. The source/drain implants have to be made at low energies (less than 10 KeV for boron) to keep the junctions this shallow. This presents considerable difficulty because the ion beam tends to de-focus at such low energies, the sputtering rate is rather high and channeling probability is also quite high. The problem is particularly difficult for p+n junctions since boron is a small atom and exhibits a large implant range at a given energy and its diffusion constant is also relatively high. The problem of channeling can be controlled to some extent by pre-amorphizing silicon before impurity implants. This is accomplished by implanting silicon or germanium to produce sufficient damage to cause amorphization [6]. The same effect can be achieved by the use of Boron Fluoride as the implanted specis in place of elemental boron. The crystal is grown back after impurity implantation by solid phase epitaxy through a low temperature anneal. The impurity can then be activated through Rapid Thermal Annealing (typically about 10 seconds at about 1000 C).

An alternative to this recipe is to give up implanted junctions and to go back to solid source diffusions. This alternative will become all the more attractive for channel lengths in the 0.25 micrometer range. An interesting variation of this approach is to use the contacting metal (say a silicide) itself as the source of impurity and to carry out contact alloying and source/drain diffusion in one Rapid Thermal Annealing step [7]. The diffusion and seggregation behaviour of impurities at silicon-silicide interfaces has to be better characterized before this approach becomes practical [8].

2.1.4 Interconnects

In the ideal scaling theory, voltages, currents and device dimensions are scaled down by the same factor (say K). This has the net effect of increasing the current density by a factor K. This presents a serious problem because electromigration would limit the amount of scaling possible. This difficulty is partially solved by using special alloys such as aluminium-silicon-copper to reduce electromigration. There is a price to pay for this choice, however, since this increases the resistivity of the interconnect metal. The desired properties of low resistivity, high resistance to electromigration, good step coverage and fine geometry are unfortunately not compatible with each other and compromises have to be made in choosing the composition and the thickness of the interconnect metal to be used.

Since the interconnect delay does not scale, while the device delay is reduced by scaling, the total circuit delay starts being dominated by interconnect delays. If this happens, further device scaling would bring diminishing returns. Therefore, the conductivity of the gate material and interconnect metal must be improved. In case of gate material, this can be done by paralleling the polysilicon with silicides. For gate lengths well below 0.5 micrometers, this may be inadequate and refractory metal clad polysilicon as well as direct use of refractory metals as gate material may be

required. The recent discovery of high temperature
superconductors holds great promise for reducing interconnect
delays for systems which can be cooled to liquid Nitrogen
temperatures. As we shall see later, cooling to liquid Nitrogen
temperature has many other advantages as well [9].

2.2 Device Design Problems

The main device properties of concern are the
threshold voltages of n and p channel devices, subthreshold
slopes of these devices, field turn on voltages in the n and p
regions and latchup characteristics of the parasitic pnpn
device.

2.2.1 Threshold Voltages

The ideal scaling theory demands that supply voltages
as well as threshold voltages be scaled down by the same
factor. Unfortunately system considerations make it very
difficult to follow this rule. Power supply and threshold
voltages are generally not changed continuously with scaling,
but are fixed at some standard values (for example, power supply
voltage may be kept fixed at 5.0V) and revised only once in a
while. (The new proposed supply voltage is 3.3 V). As a result,
the threshold voltage is generally higher in magnitude than the
optimum dictated by scaling theory. This means that the
threshold correction implant, and therefore the impurity doping
in the channel, is much higher than the optimum. This leads to
lower carrier mobilities due to impurity scattering. This
problem is all the more acute for submicrometer CMOS, since the
oxide thickness is low, which means gate oxide capacitance is
large and therefore much larger correction doses need to be
implanted to shift the threshold voltage by a given amount. It
appears necessary that one may have to change over to the new
supply voltages of 3.3 V and correspondingly scaled down
threshold voltages of about + and - 0.5 V for submicrometer CMOS
designs.

2.2.2 Subthreshold slopes

The subthreshold slope is the rate of current
reduction for a device as the gate voltage is taken below the
threshold voltage. Generally, the inverse of this slope is
specified for a device as the number of millivolts (below
threshold voltage) per decade of reduction in drain current.
Thus a lower value (in milliVolts/decade) signifies a steeper
turn off characteristic. This parameter is of some concern
during device scaling because its value remains unaffected by
scaling laws. If the threshold voltage is scaled down, it may

not be possible to turn a transistor off very effectively.
Values of about 80 to 100 mV/decade are typical for this
parameter. (In fact for submicrometer MOS design, one of the
difficult design goals is to keep the subthreshold slope below
100 mV/decade for all operating conditions.) Therefore, if the
threshold voltage is 0.5 V, only about 5 decades of current
reduction may be possible from the on to off state.
(Contemporary designs demand a six decade ratio between on and
off resistances.) It is clear, therefore, that as we go towards
smaller channel lengths, the circuit designers will have to keep
in mind that the 'off' devices actually draw much more current
than they used to at longer channel lengths.

One way for improving the subthreshold slope is to cool
down the transistor (say to liquid Nitrogen temperatures) since
the slope scales directly with absolute temperature [9]. This
approach is very attractive because it also improves the channel
mobility, latchup characteristics and electomigration properties
of the interconnect metal. However, the decision to incorporate
liquid Nitrogen cooling in any system is not trivially taken and
therefore, this approach has not been very popular inspite of
its considerable advantages.

2.2.3 Field Turn-on Voltages

In a properly designed technology, the parasitic
devices need as much attention as the active devices. The
isolation between active devices requires a careful optimisation
of lateral and vertical doping profiles. Trade offs between
junction capacitances, body effect on threshold voltages (both
requiring low substrate doping) and field turn on voltages
(requiring high substrate doping) are necessary. The short
channel effects on field devices are important because these
determine how close various active devices may be placed to each
other.

2.2.4 Latchup

Latchup has been a source of concern for CMOS
processes all along. It becomes a serious problem as we go to
submicrometer CMOS since we would like to place the n and p
channel devices as close as poissible and this reduces the base
widths of lateral parasitic bipolar transistors, increasing the
likelyhood of latchup to occur. It is possible to design a
structure in which the holding voltage of the parasitic pnpn
structure is actually higher than the supply voltage. If this
is done, then the circuit becomes latchup proof since even if a
transient causes the pnpn structure to turn on, there is not
enough voltage to hold it in the on state and so the structure
unlatches as soon as the transient is gone. High holding

voltages can be achieved by the use of epitaxial substrates as the starting material and retrograde doping in the n-well. This task is made somewhat easier if the power supply is scaled down since the holding voltage has to exceed just 3.3V then rather than the full 5.0V. A rule of thumb is that the minimum n+ to p+ spacing is of the order of the epi thickness. For a submicrometer CMOS process, this would require p on p+ substrates with an epi thickness of about 2 to 3 micrometers.

Cooling the circuit to liquid Nitrogen temperatures is beneficial for latch up too, since low temperature operation reduces the gain of parasitic bipolar devices.

2.3 System Related Problems

Certain solutions proposed in the last sections require decisions at a system level. For example, the decision to choose a supply voltage of 3.3V may have serious implications of compatibility for the systems. Similarly, the decision to employ liquid Nitrogen cooling has to be taken depending on cost-benefit trade offs. Most of these considerations are beyond the scope of this paper and we shall not discuss these at any great length. The only comment we make at this stage is that the experimental results reported in the following section assume a power supply voltage of 3.3V and no liquid Nitrogen cooling.

3. A Submicron CMOS Technology

The technology presented in this section barely qualifies as a submicron CMOS technology. It was originally designed as an agressive 1 micron CMOS process, having been optimised for a channel length of 0.9 +or- 0.25 micrometrs. This means that the transistor with the worst case channel length of 0.65 micrometers would still be within specifications. Clearly further scaling is necessary to get a half micrometer like CMOS process but these results are being included to give a flavour of the design decisions taken.

The process uses a 10 ohm-cm p on p+ epitaxial starting substrate and retrogradely doped n wells formed with 700 KeV implants. The p and n type channel stoppers are self-aligned to active area edges as well as well boundaries, allowing close spacing of n and p channel transistors. The retrograde well design in combination with thin epitaxial starting substrate assured a holding voltage higher than 3.3V for an n+ to p+ spacing of 4.0 micrometers (fig. 1). The source drain junctions were formed with and without germanium preamorphisation for comparison and the p+n junction was 0.2 micrometers deep with preamorphisation (fig. 2). The shallower preamorphized junction gives improved short channel characteristics for the p channel

190

transistor (fig.3). The process provides unloaded gate delays of about 220 pico seconds.

4. ACKNOWLEDGEMENTS

The technology described above was developed at MCNC, North Carolina, U.S.A. in collaboration with S. Goodwin-Johansson, D. Wen and C.M. Osburn of M.C.N.C. and C.K. Kim of KAIST, seoul, S. Korea.

REFERENCES

[1]   R.H. Dennard, F.H. Gaensslen, H.N. Yu, V.L. Rideout, E.L. Bassous, and A.R. LeBlanc; "Design of ion-implanted MOSFET's with very small physical dimensions"; IEEE J. Solid State Circuits, Vol. SC-9, p.256, (1974)

[2]   Y. Taur, W. Chang and R. Dennard; "Characterisation and modelling of a Latchup free 1 μm CMOS Technology"; IEDM Technical Digest, pp.398-401, (1984)

[3]   R. Martin, A. Lewis, T. Huang and J. Chen; "A new process for one micron and finer CMOS", IEDM Technical Digest, pp. 403-406, (1985)

[4]   K. Lee, B. Jones, C. Burke, L. Tran, J. Shimer and M. Chen; "Lightly doped drain structure for advanced CMOS (Twin Tub IV)"; IEDM Technical Digest, pp 242-245, (1985)

[5]   D. Sharma, S. Goodwin-Johansson, D-S. Wen, C.K. Kim and C.M. Osburn; "A 1μm CMOS technology with low temperature processing"; 1st International. conf on ULSI, Electrochemical Soc. , 1987

[6]   J. Liu and J. Wortman; "Optimization of Germanium preamorphization implant shallow junctions"; Extended abstracts, Vol 86-1, Electrochemical society meeting, Boston, pp 347-348, May 1986.

[7]   F. Shone, K. Saraswat and J. Plummer; "Formation of 0.1μm N+/P and P+/N junctions by doped silicide technology"; IEDM Technical Digest, pp 407-410, (1985)

[8]   C.M. Osburn, T. Brat, D. Sharma, N. Parikh, W.K. Chu,
        D. Griffis and S. Lin;
        "The   effects   of Titanium Silicide formation on dopant
        redistribution";
        1st   International.   conf on ULSI, Electrochemical Soc.
        , 1987

[9]   J.Y.-C. Sun, Y. Taur, R.H. Dennard ans S.P. Klepner;
        "Sub - micrometer   channel   CMOS   for   low   temperature
        operation ";
        IEEE trans. ED - 34, pp. 19-27 , (1987)

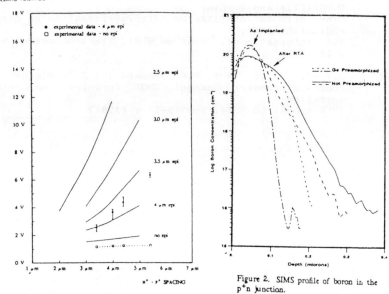

Figure 2. SIMS profile of boron in the $p^{+}n$ junction.

Figure 1 . Measured and simulated holding voltage as a function of epi thickness and n^{+} to p^{+} spacing.

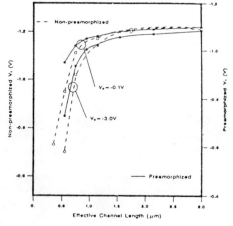

Figure 3. Measured short channel effect for p channel transistors for germanium preamorphized junctions and non-preamorphized junctions.

COMPUTER INTEGRATED MANUFACTURING SCIENCE AND TECHNOLOGY FOR VLSI

Prof. Krishna C. Saraswat

The Center for Integrated Systems
Stanford University
Stanford, CA 94305 USA

To make a state-of-the-art integrated circuit process more manufacturable, we must be able to understand both the numerous individual process technologies used to fabricate the complete device as well as the important device, circuit and system limitations in sufficient detail to monitor and control the overall fabrication sequence. Specifically, we must understand the sensitivity of device, circuit and system performance to each important step in the fabrication sequence. Moreover, we should be able to predict the manufacturability of an integrated circuit before we actually manufacture it.

The salient objective of this program is to enable accurate simulation and control of computer-integrated manufacturing of very large scale integrated (VLSI) systems, through creative application of computer software and hardware engineering.

The Stanford research program in semiconductor manufacturing science is working to improve the nature of semiconductor manufacturing. The breadth of modern semiconductor products depends on mass production of DRAMs, to debug and refine new processes before they are utilized for lower volume but higher profit products. This research program is developing methodologies to permit the rapid development and engineering of new processes so that highly innovative products depending on new processes can be fielded more rapidly.

A key goal of this research is the *programmable factory*, an integrated system of manufacturing equipment, sensors, and computer hardware and software. Just as a programmer can rapidly modify and debug a complex computer program, so a process engineer should be able to modify and debug the complex processes which controls the manufacturing of semiconductors. To make the factory easier to program, we are developing a powerful process CAD system.

A second key goal of our research is the *virtual factory*, a factory that can be run in simulation. Just as VLSI circuits are simulated before they are cast in silicon, so VLSI manufacturing processes should be simulated before they are run in the factory. We are developing a powerful computer-based simulation system that will permit us to simulate semiconductor manufacturing processes in their entirety, using knowledge about equipment, processes, materials, devices, and circuits to predict critical measurements of manufacturing performance such as yield, electrical performance, throughput, and equipment utilization.

The programmable factory and the virtual factory are highly related and interdependent. Both the programmable and the virtual factories must be based on a large common knowledge base that captures knowledge about equipment, processes, materials, schedules and other aspects of semiconductor manufacturing. The different software systems needed to support process development tasks -- e.g., design, debugging, execution, data acquisition/interpretation, control, updating -- need to have access to similar knowledge. For efficiency of development and ease of maintenance, it is crucial that information about a piece of equipment, for example, not be encoded one way for process design and another way for process debugging. Much of the knowledge and much of the software that supports one will also support the other. For example, both the programmable factory and the virtual factory need graphical interfaces giving the user access to equipment, communications, operations data, and schedules. The interfaces ought to be as similar as possible, the differences between the programmable and the virtual factory appear at a low level; most of the software will run the same whether the factory is being run "for real" or in simulation.

Accurate simulation and control of a complete manufacturing line for VLSI systems demands a tightly integrated set of research tasks. Development of programmable and virtual factories require on one hand development of knowledge based computer software and hardware and on the other hand development of techniques to acquire the knowledge about manufacturing technology. This knowledge should be fundamental and not empirical in order to ensure portability. Computer control of a programmable laboratory that fabricates working prototypes of VLSI systems will verify the simulation programs and convincingly demonstrate the essential unity of all research tasks.

The focus of the **FABLE: Manufacturing Automation** project is a growing body of knowledge about semiconductor manufacturing together with a set of techniques and tools for encoding and using that knowledge to support the automated factory. The development of the knowledge base and the tools are driven in part by questions such as the following:

What does the programmable factory need to know?

How can such knowledge be encoded for use by the automated factory?

How can the knowledge be efficiently acquired and updated?

These questions are stimulated by FABLE-related projects in factory simulation, equipment modeling, intelligent processing equipment automation, automated test structure interpretation, and graphical equipment interfaces. The research on *knowledge tools* is developing a common knowledge base capturing knowledge about equipment, processes, materials, schedules and other aspects of semiconductor manufacturing, using techniques from artificial intelligence. This knowledge base will be used to facilitate the development of expert systems to help design, debug, execute, monitor, control, and modify processes. The research on *equipment automation* will improve the reproducibility of semiconductor manufacturing by developing programmable processing equipment. The research on *equipment communications* will improve the flow of information between host computers and manufacturing equipment using advanced versions of the SECS protocols and local area networks. The *user interface* research will improve user access to the programmable factory by supplying powerful, graphics-based user interfaces to processing equipment. Together, these four projects will provide a consistent set of methodologies to support an integrated factory automation system. With the FABLE system we will be able to describe fabrication processes so that they are understandable, designable, debuggable, portable -- and runnable, by a suitable computer-based automation system.

The **Factory Modeling and Management** project is focussing on factory level productivity relationships, as opposed to the wafer level concerns of physical scientists. In this project we are fundamentally concerned with the relationships that connect throughput, manufacturing cycle time, and various aspects of managerial policy. The ultimate goal of this project is to show how manufacturing resources can be used more efficiently in the semiconductor industry. We hope to provide insights and methods of anlaysis that can guide managerial decision making with regard to both system design (capital allocation) and operating policies.

The **Simulation** project is developing a sophisticated VLSI manufacturing line simulator which will be able to predict device, circuit, and yield behavior describing key parametric distributions of a complex VLSI chip based on a specification of the process, the characteristics of the fabrication equipment and laboratory environment, and a description of the geometric pattern information. This simulator is incorporating equipment, process, device and circuit models in a manufacturing environment. An example of this work is SHIPS, short for Stanford High-Level Incremental Process Simulation. The most pervasive and recognizable form of centralized knowledge in a wafer fabrication line is the process runsheet. While this form of process specification embodies only a part of the complete body of knowledge required for wafer processing it nevertheless remains largely isolated from and underutilized by other tools such as simulation programs.

To address the traditional simulation area we have developed SHIPS -- a flexible Unix-based system that aids process specification and simulation at the level of the SUPREM family of process simulation tools. The purpose of this tool is to allow a process engineer to specify and simulate a VLSI process using physical process models so as to facilitate statistically significant simulation results. This requires computationally intensive process simulation and the statistical analysis of results. We refer to the methodology embodied in this tool as *manufacturing-level process specification* and *simulation*. Similar efforts are in progress for device and circuit simulators.

In the **Technology** project we are developing a new modeling discipline-semiconductor manufacturing equipment modeling. Perhaps the greatest obstacle to simulating a semiconductor manufacturing line is an almost total lack of physical models to describe process-parameter variations resulting from characteristics (e.g. the geometry) of particular machine designs. Existing models used in process simulators such as SUPREM are highly generic and assume that a specific local environment is replicated across an entire wafer, from wafer-to-wafer in a batch and from batch-to-batch. This assumption is an invaluable simplification in developing process models per se but neglects *the* fundamental manufacturing problem of parameter distributions and their causes. Equipment modeling will address this problem. The Etching project is focussing on modeling plasma, reactive ion and other etching equipment. The Rapid Thermal Processing project is concerned with modeling of the thermal environment of a lamp heated system for accurate measurement and control of the wafer temperature for applications, such as, annealing, oxidation, nitridation and CVD. Ion Implantation project is dealing with modeling of an ion implanter keeping in mind the large number of internal and external variables. Strong interaction with the FABLE group is taking place to use knowledge based techniques. Sputtering equipment is focussing on developing an equipment model with emphasis on step coverage. At the same time we are developing an expert sputtering system by working with the FABLE group. The Lithography project is concerned with modeling optical, electron beam and other lithography equipment. The general approach to each project begins with the development of new test structures and measurement tools to improve understanding of manufacturing parameter distributions. This experimental data is then used as a guide to the formulation of predictive physical models for various machines. Agreement between measured and theoretical distributions will verify the models and serve as a basis for defining in concert with manufacturers new machine concepts for future generations of equipment.

The **Testing and Diagnostics** project is developing new tools for in-situ, in-process and end-of-process monitoring and evaluation of IC processes. These activities include the development of new approaches to test structures, measurement techniques, data reduction and display, and data interpretation techniques. The goals of these activities are the integration of the above techniques into "expert" systems for providing accurate assessment of the integrity and capabilities of individual IC processing steps, as well as entire IC processes.

Acknowledgements

This work is being supported by SRC contract 84-01-046 and DARPA contract MDA 903-80-C-0432.

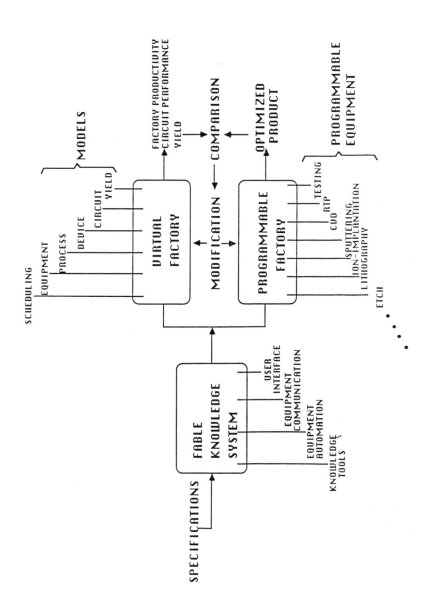

Non-equilibrium Carrier Transport in Hot Electron
and Heterojunction Bipolar Transistors

John R. Hayes[*]

Bell Communications Research Inc.,
Red Bank, NJ 07701-7020, USA.

ABSTRACT

In this article I will review our understanding of non-equilibrium carrier dynamics in both hot electron and heterojunction bipolar transistors. The results, obtained using "Hot Electron Spectroscopy", enable us to measure the injected electron momentum distribution. In the case of hot electron transistors we can observe both ballistic electron transport and electron activation from the Fermi sea in the hot electron spectra. These results can be explained by considering a coupled electron/phonon system. In the case of heterojunction bipolar transistors (HBT's) the spectra is closer to equilibrium and the carrier distribution can be characterized by an effective electron temperature far in excess of equilibrium.

TEXT

The advent of thin film epitaxial semiconductor crystal growth techniques have enabled researchers to use variable energy band gap regions to conceive of and hopefully improve electronic and optoelectronic devices. Of particular interest to people concerned with hot electron effects are the hot electron and heterojunction bipolar transistor. Such devices enable one to investigate the interactions of hot electrons with a cool electron gas (hot electron transistors) and with a cool hole gas (n-p-n heterojunction bipolar transistor). The experimental technique used for this study is referred to as hot electron spectroscopy and the measured quantity is the hot electron spectra. Although this technique will be described here in terms of a hot electron transistor it can be easily understood and equally well applied to heterojunction bipolar transistors as will be discussed.

The results described in this paper were obtained by means of a planar GaAs structure, a schematic diagram of which is shown in Fig. 1. The structure was grown by Molecular Beam Epitaxy, at 650 C, on <100> oriented semi-insulating GaAs substrates. After the growth of a thick (1um) buffer layer, which served as the collector contact layer two bulk triangular potential barriers were formed either by sheet charge doping[1] (planar doped barriers) or compositionally grading with AlGaAs[2]. They were bounded on each side by n^+ layers. The barriers were designed to inject below the threshold for intervalley scattering so that electron transport only occurs in the direct T minimum. The wafers were fabricated into two-level mesa structures in order that the three n^+ regions could be contacted individually. The first etch, which reveals the transit region is critically depth dependent, and was made by the successive growth and removal of an anodic oxide grown in a solution of $H_2O:H_3PO_4$ having a Ph of 2.5. The third n^+ region was revealed using a standard chemical etch with $100H_2O:10H_3PO_4:2H_2O_2$, which has an etch rate of about

800Å/min at room temperature. Ohmic contacts were formed on the three n^+ regions by rapidly thermal annealing an evaporated Au-Sn alloy.

A schematic diagram of the energy band structure used for hot electron spectroscopy studies is shown in Fig. 2. Since the structure resembles a unipolar transistor, standard transistor notation is used to describe the currents and voltages involved. In order to perform hot electron spectroscopy[3-4] the transit region (base) was grounded and a negative bias ($-V_{be}$) was applied to the emitter enabling a near mono-energetic beam of electrons to be injected into the transit region. With the collector potential (ϕ_{bc}) greater than the injected electron energy no electrons will be collected; none have sufficient energy to surmount the barrier. With increasing base/collector bias (V_{bc}), the barrier ϕ_{bc} is lowered and electrons satisfying the necessary criteria to surmount the barrier will be collected. Hence by continuously varying the base/collector voltage we establish a means of spectroscopically resolving the resulting injected electron distribution after interaction in the transit region.

Those electrons which are injected into the base will have a component of momentum parallel to the emitter (P_p), which is conserved, and a component of momentum normal to the emitter plane (P_n) which is dramatically increased. With the typical injection energies used here ($=0.25eV$) the electrons become strongly collimated and a maximum angle exists for electron injection (Θ_{max}) that is given by

$$\Theta_{max}=\tan(P_n/P_p) \qquad\qquad 1$$

For GaAs at low temperatures typical of those used in these experiments a Θ_{max} of less than 10^0 would exist.

Electrons injected into the base suffer both elastic and inelastic collisions whilst interacting with the electron/phonon and ionized impurity system. The resulting distribution, that is analyzed by the hot electron analyzer may be described in terms of the distribution of possible momentum values $n(P_p)$ of electrons having particular values of momentum normal to the analyzer barrier. The collector current, I_c, for a particular barrier energy O_{bc} is given by

$$I_c=-(e/m_e^*)\int_{P^0}^{\infty} P_n n(P_n)dP_n \qquad\qquad 2$$

were m_e^* is the effective mass, e the electron charge and $P_n^0=(2m_e^*\phi_{bc})^{1/2}$. Taking the derivative of I_c with respect to V_{bc} one can show that

$$dI_c/dV_{bc}=n(P_n) \qquad\qquad 3$$

Hence by differentiating I_c with respect to V_{bc} we may obtain the electron momentum distribution at the hot electron analyzer.

The results of four spectra[5] measured at 4.2K are shown in Fig. 3. Each sample had a hot electron injector injecting electrons into the transit region at an excess energy above the conduction band edge of 0.25eV. The doping in the transit region was similar for all samples ($n=1\times10^{18}cm^{-3}$). However they differed in their transit region widths; sample (a) 650Å, (b) 850Å, (c) 1200Å and (d) 1700Å. As can be seen in Fig 3. there is a significant change from spectra to spectra over small changes in the transit region width indicating that the injected electrons are being strongly scattered. The high energy peak (low voltage bias) results from unscattered electrons traversing the transit region. In thin samples the peak at low energies (high voltage bias) results predominantly from electron activation from the Fermi sea. As the transit region width increases there is an increased contribution from scattered electrons.

To understand the electron scattering mechanisms giving rise to the hot electron spectra shown in Fig. 3 we have developed a theory of non-equilibrium transport that takes account of scattering from the whole electron/phonon system. In GaAs at the carrier concentrations used in these experiments ($n=1\times10^{18}cm^{-3}$) the long wavelength collective oscillatory mode (plasmon) of the electron gas couples strongly with the LO phonons. These two oscillations do not exist independent of each other but couple together strongly to form a coupled electron/phonon system. In addition to scattering by long wavelength coupled plasmon/phonon modes, they also scatter by the electron/electron interaction. With decreasing wavelength the collective modes are damped and only optic phonon scattering remains. Details of calculating the scattering strengths for these mechanisms have been presented elsewhere[6] and in addition to obtaining the dispersion relationship one can also obtain the scattering rates. The calculated inelastic scattering rate from the electron/phonon system ($1/\tau_{in}$) together with the elastic scattering rate for elastic scattering from ionized impurities ($1/\tau_{el}$) is shown in Fig. 4. By considering scattering from both elastic and inelastic mechanisms we obtain a mean free path for hot electrons in n-type GaAs, doped $1\times10^{18}cm^{-3}$ and injected at 0.25eV, of 350Å. This is in excellent agreement with the value experimentally obtained by studying the magnetic field dependence of the hot electron spectra[7].

We believe we now have a complete understanding of the non-equilibrium electron dynamics in n-type material at low injection energies [$E_i < E_\Gamma L$]. To extend our knowledge of hot electron effects we have also investigated the interaction of hot electrons with a degenerate hole gas in heterojunction bipolar transistor structures having both uniform[8] and compositionally graded base regions. The first experiments to be described are transistors having emitters that inject hot electrons over a heterojunction step into a uniform band gap base. The second experiment describe the heating of thermally injected electrons into as region that has a uniform quasi-electric field (an electric field that only acts on the injected electrons) of 15kV/cm.

To study hot electron cooling in n-type material we used transistors that consisted of an n-type ($2\times10^{17}cm^{-3}$) AlGaAs emitter from which electrons could be injected into a Be-doped p-type ($2\times10^{18}cm^{-3}$) base and collected with an n-type ($2\times10^{17}cm^{-3}$) AlGaAs collector. Some compositionally grading adjacent to the base/collector heterojunction was incorporated in the structure to reduce quantum mechanical reflections at the base/collector interface. Transistors were fabricated using the same techniques as those previously described for hot electron transistors. Ohmic contacts were made to the transistor by rapidly annealing an evaporated Au-Sn alloy to the emitter and collector regions (n-type) and a Au-Be alloy to the base region (p-type). Results of measurements on two HBT's having uniform base regions are reported here both of which have the same base doping (p=2x10^{18}cm^{-3}) but different base widths: W_b= 900Å and 450Å. When the emitter is forwarded biased electrons are injected into the base with an excess energy above the conduction band edge (ϕ_{bc}=0.2eV) (sample with base width 900Å) and ϕ_{bc} (sample with base width 450Å). As in the case of the hot electron transistor the injected electrons scatter in the base and impinge on the base/collector junction where there momentum distribution is characterized. Unlike the case of the hot electron transistor the distribution is analyzed by forward biasing the collector.

As the base/collector junction is forwarded biased the heterojunction barrier raises and when a bias of 0.4V is applied one just begins to analyze the injected distribution. With increasing forward bias, the collector current decreases until finally at V_{bc}=-1.6V ϕ_{bc}>ϕ_{be} and no electrons are collected. Calculations show that the variation of ϕ_{bc} with V_{bc} is linear. The variation in collector band edge manifests itself in both the common emitter and common base characteristics. Typically transistors with 900Å (450Å) base widths had common emitter current gains of over 50 (100) at room temperature increasing to 500 (10,000) at 100K. The common base characteristics of the transistor having the 900Å base width is shown in Fig. 5 at two temperatures 300K and 77.8K. Current saturation is nonideal occurring between V_{bc} =-0.4 and -1.6V.

Thus measuring the hot electron spectrum, dI_c/dV_{bc} as a function of V_{bc} in the common base configuration gives direct information on the hot electron momentum distribution. Fig. 5 shows the hot electron spectra for two samples, at a temperature of 4.2K, having base widths of (a) 450Å and (b) 900Å. The injection energy ϕ_{be} and the conduction-band minimum in the base E_c are indicated in the figure. There is no evidence in either spectra of the initial injected distribution with most of the spectral weight occurring close to E_c (the conduction band edge). This indicates that the initial distribution has been strongly scattered to such an extent that one may infer an effective electron temperature T_e to describe the electron distribution at the collector. For samples having a 900Å base a T_e of 150K was inferred whereas the sample having a 450Å transit region a T_e of 500K was inferred. It should be noted that the conduction band edge (E_c) is not at the same position on each spectrum due to the different Al concentrations used in the collector of each sample.

Finally in this review I will describe the effect of acceleration of electrons in an electric field in the base of a graded gap bipolar transistor. The transistor has a base width of 900Å and differs from the previous structure by having the electrons thermally injected into the base region and then heated in the electric field. The sample was grown on <100> oriented GaAs substrates and consisted of a 5000Å buffer layer doped to 1x10^{18}cm^{-3} followed by an AlGaAs (Al=0.35) collector doped to 2x10^{17}cm^{-3} and terminated over 50Å at a compositionally graded base. The base was graded over 900Å from GaAs to AlGaAs (Al=0.2) over the p-type base (p=2x10^{18}cm^{-3}) giving a quasi-electric field in the base of 15kV/cm. The electrons were injected into the base from an AlGaAs

202

(Al=0.20) emitter that was compositionally graded over 150Å so that an equilibrium distribution could be injected into the base. Hence if a non-equilibrium distribution is measured it is entirely due to carrier heating in the base electric field.

The measurements, made at 4.2K, in order that thermal smearing effect be eliminated. At 4.2K the transistor had a current gain of 1600 an increase of a factor 5 over its room temperature value. A hot electron temperature of 650K was obtained indicating significant heating of the distribution occurs.

In conclusion we have measured the hot electron distribution function of electrons injected into both n and p-type GaAs. Two prominent features are observed in the distribution function for narrow n-type samples; a high energy peak attributed to ballistic electrons and a low energy peak attributed to electrons activated from the Fermi sea. In p-type samples no high energy peak is observed but one can describe the distribution in terms of an effective electron temperature far in excess of equilibrium. Recent studies of compositionally graded bases indicate that hot electron spectroscopy can be used to investigate transport at high electric fields.

* Work done in collaboration with A. F. J. Levi, P. M. Platzmann, A. C. Gossard, R. Bhat, J. P. Harbison, W. Wiegmann and J. H. English

REFERENCES

1. R. J. Malik, T. R. AuCoin, R. L. Ross, K. Board, C. E. C. Wood and L. F. Eastman. "Planar doped barriers in GaAs by molecular beam epitaxy," Elec. Lett., vol. 16, pp.836-838, 1980.

2. A. C. Gossard, C. L. Allyn, A. C. Gossard and W. Wiegmann, "New rectifing semiconductor structure by molecular beam epitaxy", Appl. Phys. Lett., vol 36, pp.36-38, 1980.

3. J. R. Hayes, A. F. J. Levi and W. Wiegmann, "Hot electron spectroscopy ," Electron. Lett., vol. 20, pp.851-852, 1984

4. --, "hot electron spectroscopy of GaAs," Phys. Rev. Lett., vol 54, pp. 1570-1572, 1985.

5. J. R. Hayes and A. F. J. Levi, "Dynamics of extreme nonequilibrium electron transport in GaAs," IEEE Jour. Quan. Elec., QE-22, pp.1744-1752.

6. A. F. J. Levi, J. R. Hayes, P. M. Platzman and W. Wiegmann, "Injected hot electron transport in GaAs", Phys. Rev. Lett., vol 55, pp. 2071-2073, 1985

7. J. R. Hayes, A. F. J. Levi and W. Wiegmann, "Magnetic field dependence of hot-electron transport in GaAs", Appl. Phys. Lett., vol 47, pp.964-966, 1985

8. J. R. Hayes, A. F. J. Levi, A. C. Gossard and J. H. English, "Base transport dynamics in a heterojunction bipolar transistor", Appl. Phys. Lett. 49, pp.1481-1483, 1986.

FIGURES

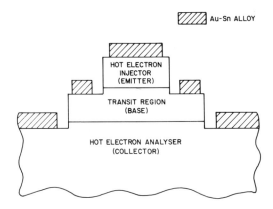

Figure 1 Schematic diagram of the fabricated hot electron transistor comprising the three active transistor areas (hot electron injector, transit region and hot electron analyzer). The ohmic contact regions are also shown.

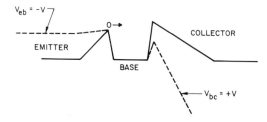

Figure 2 Schematic diagram of the energy band structure of a typical hot electron transistor. Typically the short arm of the potential barriers were 150Å and the long arms were 1500Å.

204

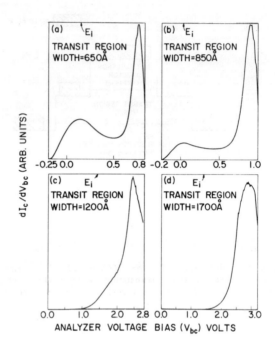

Figure 3 Measured hot electron spectra of four samples having the same transit region doping of $1 \times 10^{18} cm^{-3}$ but different indicated transit region widths. The injection energy for each sample ($E_i = 0.25eV$) is indicated on the upper axis.

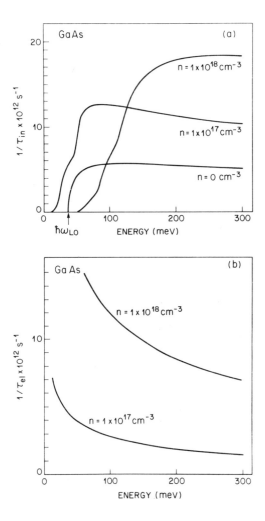

Figure 4 Total inelastic scattering rate calculated for GaAs (a) using the coupled electron/phonon model described in the text and elastic scattering rate from ionized impurities (b) as a function of injection energy for indicated carrier concentrations.

206

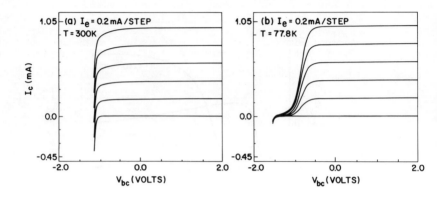

Figure 5 Common base characteristics of the heterojunction bipolar transistor having a 900Å base region.

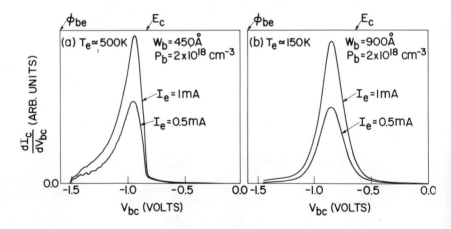

Figure 6 Hot electron spectra, measured at 4.2K, of two heterojunction bipolar transistors having base widths (W_b) of (a) 450Å and (b) 900Å. The inferred effective electron temperature (T_e) is indicated in the figures.

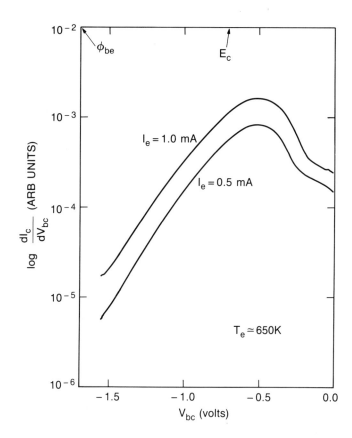

Figure 7 Measured hot electron spectrum for a heterojunction bipolar transistor having a 900Å compositionally graded base (the quasi electric field in the base was 15kV/cm). An effective electron temperature of 650K has been obtained.

SOME ASPECTS OF LIGHTLY DOPED DRAIN TRANSISTOR

W.S. Khokle and P.N. Andhare, R.K. Nahar, N.M. Devashrayee and Suresh Chandra

Central Electronics Engineering Research Institute, Pilani - 333031 (Rajasthan) INDIA

Introduction

As the minimum feature size is shrunk below 3 microns, following phenomena adversely affect the device performance. These phenomena are : reduction in enhancement threshold voltage with decreasing gate length, reduction in source-drain breakdown voltage with decreasing gate length, increase in hot electron induced instabilities in threshold voltage, transconductance etc. Conventional approach does not allow reduction of all the three degradations simultaneously, and optimisation tried within the conventional approach means optimisation at the cost of one of there degradations. By introducing a lightly doped n^- region between the highly doped n^+ region and the p substrate, it becomes possible to optimise all the three degradations simultaneously. This is the strength and crux of the lightly doped drain [LDD] transistor structure. Doping of n^- region and its length determine the tradeoff between source-drain breakdown [S/D Bd], hot electron degradation [HED] on one hand and drive capabilities of LDD transistor on the other. Details of this optimisation are given in R-1,2. An important technique developed for better and independent control of n^- region is known as oxide spacer technology, [R-3]. This paper, a part from showing other advantages of LDD, also shows the speed improvement obtained in an actual circuit. It should be noted that improvement in speed of LDD structures comes basically from reduction in overlap capacitance, leading to small RC constant. However, a tradeoff is involved. This improvement is possible only upto such n^- concentrations, that do not result in increase of R of the RC constant. Another property displayed by LDD structure is the asymmetry in Id-Vd characteristics, when source and drain are interchanged [R-4]. The reduction in Id is as much as 40% and is of serious nature. It was pointed out that asymmetry was the result of the inclination of n^- implant and would disappear if the implant angle is changed from the customery 7° tilt to 0° tilt. However, the implications of such an approach on channeling in implantation is not worked out. Asymmetrical Id, Vd characteristics remain an undesirable feature of LDD structure to this date. Measurements on LDD represent tough task. Optimum lengths of n^- region are smaller than 0.1 micron. R-5 gives the currently accepted method of measurement of n^- length but it is complicated and requires more than 3 transistors of different lengths & same width.

With this perspctive, following general trends could be noted.

For conventional LDD, substrate current peak goes down as concentration on n^- region is decreased from its n^+ value i.e. 10^{20} to 10^{17} atoms/cm^3. However this decrease in substrate current does not mean that the hot electron resistance of LDD improves with the continuous decrease in the substrate current. R - 6 & 7 show that LDD is inferior to conventional transistor by factor of 10^4 [$10^6/10^2$ for conventional/LDD], when HED is measured in linear mode. On the other hand, R-8 shows that when HED is measured in saturation mode LDD proves superior by a factor of 10 [$10^5/10^6$ conventional/LDD, degradation time]. This clearly brings about the complex nature of the problem.

It is shown in R-9 that for concentration in the range of $1-2 \times 10^{18}$ a/cm^3, LDD shows degradation time higher by a factor of 10^6 [$10^6/10^0$ LDD/conventional]. In this case, both Vth measurement and stressing were done in saturation region. It is thus seen that LDD shows much severe HED, when measured in linear mode and relatively less HED, when measured in saturation mode. However, instead of looking at HED of LDD being inferior [HED measured in linear mode] or superior, [HED measured in saturation mode] compared to conventional transistor, a better view would be to take a composite view of HED of LDD. This is necessary as an actual transistor working as enhancement driver has to go through both saturation and linear mode, when transition from logic 1 to 0 occurs. Thus an appropriate figure of merit should consist of sum of HED in linear and HED in saturation mode. This would be more realistic.

Novel LDD structures

Many novel device structures have been proposed to reduce the HED of conventional LDD. In buried LDD and graded buried LDD, the basic idea is to produce a n^-/n^+ region under the conventional n^- region, [R-10]. This creates a field which reduces hot electron injection in the oxide and leads to superior HED compared to conventional LDD. It has been shown that only graded buried layer LDD is superior to conventional transistor in terms of HED as measured by Δ Id/Id %. The buried layer LDD shows a degradation time of 1000 Sec. i.e. same as for conventional and normal LDD. But buried graded LDD shows under similar stressing conditions a degradation time of 10^4 Sec. i.e. 10 times better.

Another structure suggested for improved HED is inverse T LDD structure. Its basic features are an inverted [R-11] T shaped poly silicon gate. n^+ and n^- implants are off-set by the outer and inner walls of polly. Lower n^- doses are used due to the inherent reduction in spacer induced degradation.

Compared to HED of conventional LDD, it shows an improvement by a factor 10^4, (ratio of degradation time measured in seconds with % change in linear design current). Thus IT LDD has two features with regard to its HED. It is superior to conventional LDD by a factor 10^4 as against a factor of 10 for B.G. LDD in terms of degradation time. And at the same time, it is superior to conventional transistor by a factor of 10^3.

It is also superior in terms of asymmetrical $I_{sub}-V_g$ charactericties of LDD, [R-11].

Optimisation of HED, restricts n^- concentration of LDD between 1 to 2.5×10^{18} a/cm^3. Improvement in S/d breakdown has to be realised by some other technique. For example a p^+ region is put under n^- region of LDD. This p^+ region restricts the spread of depletion width near the source and drain and thereby improves both the Vth roll off with gate length and S/d breakdown. It is shown that transistor Vds can be maintained to 5 volts for .25 and .4 micron gate length devices by using a p^+ concentration of 6×10^{17} and 3×10^{17} 0/cm^3. However, the effect of this p^+ concentration on HED has not been studied, [R-12].

Apart from hot electron degradation V_{TH} roll off and S/d punch through effect, another consideration of importance is the speed or delay associated with each of the aforesaid LDD structures. [R-9] compares the ring oscillator delay of conventional transistors, LDD transistors and B.LDD transistors, as a function of drain to gate overlap capacitance Cdg and linear transconductance [Gm] for a gate length of .8 micron. A conventional transistor gives a delay of about 187.5 PS, compared to 200 PS for LDD [$n - 10^{13}$] and 180 PS for LDD of $n - 4x10^{13}$. Lowest delay of about 170 PS is obtained for buried graded LDD. While this reference does not give delay associated with IT LDD, one can make a rough estimate of its delay on the basis of its Cdg & Gm values. Cdg is expected to be higher by 20%, Gm by 50%. On these considerations, estimated delay of ITLDD comes out to be 150 P.S. i.e. lowest in the family of LDD structures.

IT LDD is, thus, seems superior in many respects but being the latest in the evolving family of LDD based structures, it is not as well understood as others. In a study of gate off-set with respect to drain [R-13], it is shown that overlap of source/drain with gate plays a crucial role on its HED related phenomena. It is shown that minimum overlap leads to maximum HED. The reason for this could be the reduction in peak field obtainable with different overlaps, as shown in [R-14].

The superiority of IT LDD can be explained from another point of view viz the reduced contribution of parasitic resistance. As reported in R-15,16 the parasitic resistance can be seen to be composed of four components viz. contact resistance of S/d contact, sheet resistance of n^{+} and n^{-} regions, the spreading resistance and the accumulation layer resistance. The last two components are gate voltage dependent. They also depend critically on the steepness of doping profile and use of Arsenic gives a steeper profile and less contribution from the last two components. In a conventional LDD, part of the n^- region is under CVD oxide which is much thicker than gate oxide, whereas in the ITLDD, the whole of n^- region is under the control of gate element with oxide thickness equal to gate oxide. As the concentration of carrier in accumulation layer is proportional to oxide capacitance and oxide capacitance for normal LDD being 10 times smaller than for ITLDD, the resistance contribution of ITLDD in accumulation layer will be 10 times less and hence the parasatic resistance would be much less for IT LDD compared to LDD . This is the factor responsible for the transconductance of ITLDD being not only larger than conventional LDD but also being greater than Buried graded LDD and conventional transistor.

LDD Development at CEERI

We have developed a simple technique to distinguish a conventional transistor from a LDD transistor without depending upon any of the functional difference between the two types. Our technique is based on capacitance voltage characteristics of $n^+ n^- p$ junction.

The source/drain junction of a LDD transistor with respect to substrate is a graded junction, which should give a straight line, when C^{-3} is plotted against reverse source voltage. However, as the graded junction of LDD on n side consists of two regions n^+ & n^-, C^{-3} versus reverse voltage should show two distinct slopes corresponding to n^+ and n^- regions. Fig. 1 shows

actual C-V variation of two batches of LDD transistor with
different n^- region length. The kink in the C-V curve clearly
shows the existence of n^+ n^- p junction. It should also be noted
that kink occurs at higher reverse voltage for a device with
larges n^- drives in time - implying a larger length of n^-
region. The kink in the C^{-3} - V curve can give a simple proof of
the existence of n^- region of the LDD transistor, which is
independent of any of its functional property. Implantation
energy of n^- seems to have a large effect on the length of n^-
region. This contention is based on SUPREM II simulation. For
the same drive in time of n^- impurity, smaller the energy of
implant, larger would be the n^- length. Fig. 2(a) shows this
effect in a quantitative way. In this figure, impurity profile
of n^+ n^- p junction obtained from SUPREM simulation is shown for
a typical 60 min drive in of n^- impurities, with energy of
implant being 120 KV and dose, 10^{13} a/cm^2. Profiles are
perpendicular to gate length i.e. in the substrate. From this
one has to estimate the lateral profile. Above profile brings
out one more fact distinctly i.e. length of n^- region is least at
the surface and increases as one goes deeper in substrate. This
is not, strictly speaking, due to two dimensional diffusion. It
is primarily due to higher background concentration at the
surface as a concequence of enhancement implant. Hence
describing LDD transistor by one length of n^- region is
unrealistic as is being done so far. This is important as length
of n^- region has its effect on on resistance, hot electron
injection and breakdown improvement. Fig. 2(b) shows the

215

schematic of LDD transistor traditionally depicted and as modified by the consideration of varying n^- length. Length of n^- region for the above profile would be 0.8 times the difference of vertical junction depths of n^- and n^+ regions. However, it should be noted that lateral diffusion under above conditions cannot be equal to 0.8 times vertical Xj of n^- impurity because vertical Xj consists of two parts. Xj implant + Xj diffusion, the former being a function of energy of implant i.e. its Rp. For lateral diffusion, however Xj implant term would be zero. Therefore, Lateral Xj of n^- impurity = 0.8 [Vertical Xj - Xj implant].

Table 1 summarises the length of n^- obtained with and without considering the effect of energy of implant.

Table - 1

Implant considera- tion	Energy of Implant	Rp	Vertical Xj	Length of n^- region
1. Without Rp effect	120 KV	0.15 micron	0.5 micron	0.12 micron
2. With Rp effect	120 KV	0.15 micron	0.5 micron	0.00 micron
3. With Rp effect	40 KV	0.04 micron	0.5 micron	0.08 micron

Oxide window shape seems to have significant effect on the length of n^- region. Fig. 2(c) shows state of a planar slice after source-drain window etching by wet etching and befor n^- implant. Considering the effect of oxide slope in Fig. 2(c), n^+ diffusion would take place through a window of length X, but the implantation would take place through a window of length

[X + Rp]. This would increase the length of lateral Xj by an amount equal to Rp for an oxide of slope of 45°. In practice, this would vary and this could change contribution from window slope. For angles > 45, additional to lateral Xj would be smaller than Rp and for angles <45, it would be larger than Rp. The former is closer to dry etching. It can thus be seen that the length of n̄ region is sensitive to many process parameters. Energy of implantation of n̄ impurity and slope of oxide window can change its magnitude significantly.

Conclusion

Factors influencing the performance of various LDD structures have been reviewed. A simple technique to distinguish LDD and conventional transistor has been developed. Energy of n̄ implant and slope of oxide window have been shown to affect the length of n̄ region significantly. A rough quantitative estimate of these two effects have been made.

References

1. "Design and Characterisation of LDD IGFET"
    IEEE Trans. on ED, Vol. 27, August 1980 page 1359.

2. "Optimization of Lightly doped drain MOSFETS using a new quasi ballastic simulation tool" - IEDM, 1984, page 770.

3. "Fabrication of LDD with oxide spacer technology"
    IEEE Trans. on ED. Vol. 29, April, 1982, Page 590.

4. "Asymmetrical characteristics in LDD and minimum overlap MOSFETS" Electron Devices Letters, EDL - 7, Jan., 1986, Page 16.

5. "Source and drain series resistance of LDD MOSFETS Electron Devices Letters, EDL - 9, 1984, page 365.

6. "Structure enhanced MOSFET degradation due to hot electron injection" Electron Devices Letters, EDL - 5, March 1984, page 71.

7. Comments on above R-5, Electron Devices Letters, EDL - 5, July 1984, page - 256 and reply by authors of R-5 on page - 258.

8. "Design and characterisation of LDD with self aligned Titanium Disilicide" IEEE Trans. on ED., Vol. 33, March 1986, page 345.

9. "Optimised and Reliable LDD structure for 1 micron NMOSFETS based on substrate current analysis", Internation Electron Devices Meeting, IEDM, 1983, page 392.

10. "Reliability and Performance of submicron LDD MOSFETS with buried As n Impurity profiles", IEDM, 1985, page 246.

11. "A novel submicron LDD Transistor with inverse-T Gate structure", IEDM, 1986, page 742 .

12. "Halo doping Effects in submicron DI-LDD Devices Design" IEDM, 1985, page 230 .

13. "Impact of source-drain overlap with gate on hot electron degradation rate", Electron Devices Letters, EDL-5 page 71, 1984.

14. "A theoretical study of Gate/Drain offset in LDD MOSFET", Electron Devices Letters, EDL - 7, March 1986, page 152.

15. "Parasatic Resistance Characterisation for Optimum design of half micron MOSFETS" IEDM, 1986, page 732.

16. "Analysis of the Gate voltage dependent series resistance of MOSFET's, IEEE Trans on ED, Vol. 33, July 1986, page 965.

218

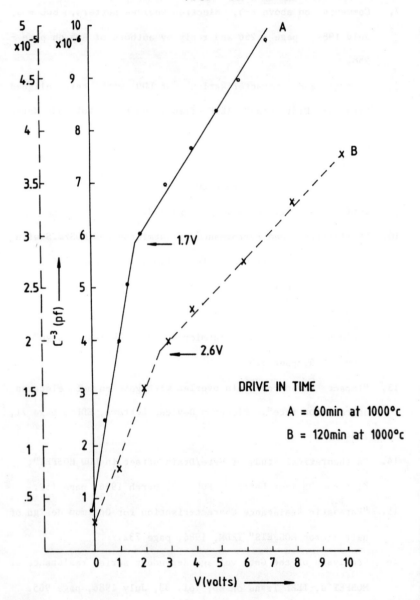

C^{-3} VERSUS V FOR LDD STRUCTURE

P IMPLANT DOSE = 5 x 10^{13} atoms cm^{-2} AT 120KV

Area = 5.7 x $10^{-4} cm^2$

A

B

1.7V

2.6V

C^{-3} (pf)

DRIVE IN TIME

A = 60min at 1000°c

B = 120min at 1000°c

V(volts)

FIG·1

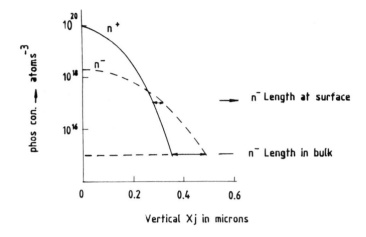

Fig.2(a) SIMULATED PROFILE OF $n^+ n^- p$ JUNCTION

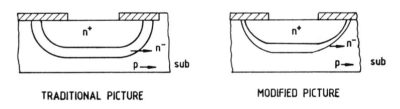

TRADITIONAL PICTURE   MODIFIED PICTURE

Fig.2(b) $n^+ n p$ JUNCTION IN BULK & AT SURFACE OF THE SLICE

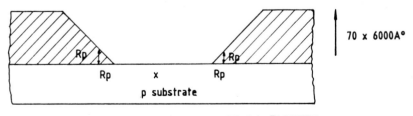

Fig. 2(c) EFFECT OF OXIDE SLOPE ON n^- LENGTH

Fig. Partial SURFACE LAYER OF STC... SLICE

TRADITIONAL PICTURE   MODIFIED PICTURE

Fig 2(ii) ...n-n JUNCTION IN BULK S AT SURFACE OF THE SLICE

Fig 2(i) - EFFECT OF OXIDE SLOPE ON n LENGTH

MATERIALS

RECENT ADVANCES IN SILICON DEVICES

S.C. Jain[*] and K.H. Winters

T.P. Division, B.424.4, Harwell Laboratory,
Oxon OX11 ORA, England.

[*](Permanent address: 39 New Campus
I.I.T Hauz Khas New Delhi 110016)

Abstract

The last ten years have seen great advances in the physics and technology of silicon devices. Dimensions have shrunk to values which we thought would be unattainable even by 1990. Supply voltages for MOSFET circuits have been reduced. Heavy doping effects are now better understood. Polysilicon contact technology has matured and the fabrication of advanced self-aligned bipolar transistors has become possible. Significant advances in analytical modelling of short channel MOSFETs have been made. Transport of carriers in the inversion layer at low as well as room temperature is now better understood. Some of these advances are discussed in this paper.

1.   INTRODUCTION

During the 1960's, the performance of bipolar transistors improved considerably but in the following decade emphasis shifted towards achieving a higher level of integration. This trend is continuing still today for both bipolar and MOSFET circuits. Many scientists believed, and some still believe that MOSFETs will displace bipolar transistors from their high speed role. However recent advances in bipolar technology and present trends do not seem to support this belief [1,2].

Many new devices have appeared during the last few years. Some of these devices are bipolar-like e.g. heterojunction emitter bipolar transistors and hot electron transistors [1]. Others are FET like devices, e.g. heterojunction FET(HEMT) and permeable base transistors. Considerable progress has been made in the understanding and technology of the static induction transistor.

In this paper we review some of the recent advances that have been made in silicon devices. The structure of the paper is as follows: we discuss scaling considerations in section 2, advances in bipolar devices in section 3 and advances in MOSFETs, in section 4.

2. SCALING AND COMPARISON OF BIPOLAR AND MOS DEVICES

The reduction of feature size and operating voltage since 1970 is shown in Fig. 1. It is clear that the rate of shrinking of dimensions is decreasing. Of course, the shrinking of feature size cannot continue indefinitely and limits must ultimately be encountered; these limits will be associated with tolerances of fine dimensions, electric breakdown and punch-through, parasitic capacitances and resistances, and breakdown of charge transport models. Work in the last 3 or 4 years has shown that earlier limits were too pessimistic. Scaling laws (Fig. 2) for MOSFETs require a reduction of operating voltage, and lower voltages reduce: (1) the break-down and punch-through effects, (2) power dissipation and (3) heat generation. They permit higher level of integration and this results in a reduction in weight, improvement in reliability and reduction in cost. These advantages of reduced operating voltage have been known for a long time but circuits have operated at 5 volts for years (see Fig. 1). Finally IBM in the USA introduced a 4-Mbit RAM and NTT Corp in Japan, a 16-Mbit RAM with 0.8 μm and 0.7 μm feature size operating at 3.3V, in February this year. Complex logic chips will benefit even more by the drop in supply voltage.

However all TTL (transistor-transistor logic) parts are designed for 5V interface standards and a reduction in voltage diminishes compatibility with the TTL components. These difficulties combined with a "Fear of Novelty" have delayed the progress in exploiting reduced operating voltages.

Bipolar scaling is more complex and different from MOSFET scaling. In bipolar devices, voltages are already low and cannot be scaled further. The delay is given by

$$t(delay) = \frac{CV}{I} \qquad (1)$$

The depletion layer widths W decrease on scaling down, the capacitance C increases and hence the current I must increase to keep t low. Doping levels are increased to reduce W and to handle larger currents. The increased base doping is also necessary to avoid punch-through and to reduce base resistance. The extrinsic base resistance becomes small and overall performance improves considerably by the self-alignment technique developed by Ning et al. [3]. The time constant $R_b C_c$ is proportional to the square of the emitter stripe width. Thus scaling dimensions by k results in $t(delay) \propto k$, $J \propto k^{-2}$, $P = $ constant and $Pt \propto k$, where J is the terminal current and P the power. Differences in the results of scaling between MOSFET and bipolar devices arise because V is scaled down in the MOSFET but not in the bipolar case and $Pt \propto V^2$.

The limits in bipolar scaling arise on several accounts. Increased base doping reduces emitter injection efficiency. Reduction in emitter junction depth also reduces current gain. In the limit of very small base thicknesses, diffusion equations may not hold because electrons may pass through the base without collisions. Monte Carlo studies of this problem have been made [1] which show that diffusion equations should hold in silicon down to 200 nm. As the current density increases, voltage drop across the contact can become serious. Electromigration also becomes a problem at high current densities.

At very small insulator thickness or base width, quantum mechanical tunnelling interferes with the performance of both bipolar and MOS devices. The quantum mechanical effects reduce the transconductance of the inversion layer. If the device size is very small, statistical fluctuations in the number of dopant atoms become significant and similar problems can arise with respect to the number of electrons.

Limits of bipolar performance are reached at 2.5 times the feature size as compared to that of a MOSFET. Bipolar devices continue to be superior in terms of transconductance per unit area. A MOSFET device has other advantages, however: it can be used in a dense and highly functional logic because it has an insulating gate and because it is a unipolar device; it does not suffer from the problem of saturation; unlike a bipolar device its input impedence is not low at large forward bias.

3. SOME RECENT WORK ON BIPOLAR DEVICES

We will discuss three areas in this field, (a) Heavy Doping Effects (b) Polysilicon Emitter Contact and (c) Extrinsic Base Contact in small and fast bipolar transistors.

(a) <u>Heavy Doping Effects</u>

When the dopant concentration increases to more than $10^{17} cm^{-3}$, the band gap of silicon, life time of minority carriers and carrier mobility become very sensitive to its value. It is only in the last 2 or 3 years

that reliable data for these quantities have become available [4-7].
Recent values for n type silicon are shown in Ref. 4 to 6; the studies on p
silicon are not as extensive [6].

In reference [7] a new method has been used to determine the band gap
shrinkage ΔEg and mobility in p type silicon. Jain and Murlidharan [4,8]
had shown that the earlier part of OCVD (open circuit voltage decay) is
very sensitive to the emitter dark saturation current and therefore to the
band gap shrinkage. The maximum time for which this emitter current effect
is large is about $t \lesssim 3 \tau_B$ where τ_B is the base lifetime. Jain and
Murlidharan could not use this method to determine ΔEg because τ_B was of
the order of a microsecond and a very fast oscilloscope was required to
measure the voltage decay for times smaller than one microsecond. Another
difficulty which arises at these small values of time is that a large part
of the voltage decay is due to resistive drop and to additional emitter
effects which are difficult to separate out. Totterdell et al [7] removed
this difficulty by using p n diodes with very low doping in the base so
that the life time τ_B is several milliseconds. The emitter doping is
$\sim 10^{19} \text{cm}^{-3}$ at the surface. The emitter is transparent and the surface
recombination velocity is infinite. The initial voltage decay for several
milliseconds gives the value of emitter dark current which can be used to
determine $e^{\Delta Eg/kT}/D$ as a function of impurity concentration. At the time
of writing, only preliminary results are available but these appear to be
in agreement with the results of Ref. 6.

(b) Polysilicon Emitter Contact
We mentioned in the previous section that the emitter efficiency
decreases because of heavy-doping effects and because of the scaling
requirement that junction depth should be small. This difficulty is
largely overcome in the modern polysilicon-contacted transistor.

In 1975 Graul et al. [9] showed that by introducing a thin doped
polysilicon layer between the metal emitter contact and single crystal
emitter, the current gain of a bipolar transistor can be increased by a
factor up to 10. The cut-off frequency also improves significantly. This
gave rise to intense activity in the field and a large number of papers
have appeared from several laboratories in many countries; at the time of
writing, the latest paper is by Ashburn et al. in June of 1987 [10].
Very high gains and a cut-off frequency f_T up to 10 GHz have been obtained

using polysilicon contacted bipolar transistors, which are now part of the production technology of most semiconductor industries in the world.

In their original paper, Graul et al. [9] suggested that the improvement is caused by the smaller band gap of polysilicon. De Graaf and de Groot [18] explained the current gain in terms of tunnelling currents through the interfacial (between poly and single crystal silicon) oxide layer. Ning and Isaac [see ref. 10] attributed the phenomenon to the reduced mobility of carriers in polysilicon. It is now known that all the above mentioned factors influence the performance of the polysilicon contact emitter, their relative importance depending upon the process and technology used in fabricating the transistor. Most references can be found in the paper by Ashburn et al. [10].

(c) Extrinsic Base Contact in Ultrasmall Bipolar Transistors

As the emitter stripe becomes narrower, the contribution of the extrinsic base (Fig. 3) to the total base resistance becomes larger. If the extrinsic base dopant diffuses into the intrinsic base, current gain and cut off frequency of the transistor are degraded. Chuang et al. [see Ref. 2 for references to earlier work] overcame this difficulty by a process that forms the intrinsic and the extrinsic base separately. Recent two dimensional simulations by the IBM group [2] have shown that the switch-on and punch-through characteristics of advanced self-aligned bipolar transistors also depend on the side wall thickness (Fig. 3) which determines the structure of the contact region between the extrinsic and intrinsic bases. This subject is being pursued with intense activity at the present time.

4. SOME RECENT WORK ON MOS DEVICES

Jain and Balk [11] have developed an analytical model to calculate the behaviour of short channel MOSFETs. It is a unified quasi two-dimensional model which calculates DIBL (drain-induced barrier lowering) as well as high electric field near the drain. Jain and Balk have shown that there is an inherent contradiction between the design requirements for optimizing short channel effects (i.e. reducing DIBL) and minimizing the high field effects. The model is in excellent agreement with a large body of experimental data as well as with numerical simulations.

Winters et al. [12] have examined the approximations used by Jain and Balk by exact numerical simulations. The important approximations on which the Jain and Balk model is based are

(i) The lower boundary of the depletion layer under the gate can be represented by a straight line (Fig. 4).

(ii) At the interface x = 0,

$$\frac{dE(x,y)}{dx} = \text{const.} \ \frac{V_G - \phi_s(y)}{W(y)}$$

where $W(y)$ represents the lower boundary of the depletion edge as discussed in (i) and ϕ_s is the interface surface potential.

(iii) The interface surface potential $\phi_L(y)$ for a short channel of length L can be represented in terms of $\phi_\alpha(y)$ applicable to a long channel MOSFET as follows

$$\phi_L(y) = \phi_\alpha(y) + \phi_\alpha(L-y)$$

(iv) In the short channel the current is assumed to be controlled by diffusion as well as by thermionic emission.

Some of these approximations have been used by other authors and very recently by Mayaram et al. [13].

The work of Winters et al. shows that (iii) is valid to a very good approximation. The assumptions (i) and (iv) are also acceptable. The most serious difficulty arises in assumption (ii). Numerical simulations show that near the drain junction, $\frac{dE(y)}{dx}$ is not constant. In fact E(x,y) starts with a very low value at x=0, increases to a maximum and then decreases approximately linearly. This behaviour is not consistent with assumption (ii). Winters et al. suggested that since the current does not flow at the interface but somewhat below it, where E(x,y) reaches a maximum, assumption (ii) becomes valid along the line of maximum current flow.

Another area where significant advances have been made concerns the transport of carriers in the inversion layers at low (up to 2°K) as well as high temperatures. The low temperature results are interesting both from

practical as well as academic points of view and interest in low
temperature circuits is growing. Since the inversion layer forms a quantum
mechanical two dimensional gas at low temperatures, its investigation is of
fundamental importance [14,15,16].

The measured values of conductivity as a function of 1/T are shown in
Fig. 5 for different values of inversion layer carrier density n [14].
Mott was the first to provide the interpretation of these observations in
terms of a mobility edge model (Fig. 6) [see Mott's review in Ref. 15]. At
low values of n, all carriers are in localised states. This localisation
occurs because of the inhomogeneity in interface charge distribution and
the carriers have to be thermally activated to energies more than E_{cm} to be
able to contribute to conduction. The energy of activation W is the
difference $E_{cm}-E_F$. Since E_F increases as n increases, W decreases for
larger values of n. When n becomes sufficiently large, $E_F > E_{cm}$ and
conduction becomes metallic. All plots extrapolated to 1/T = 0 meet at one
point as predicted by Mott's mobility edge model.

Subsequent work showed, however, that many experimental results do not
agree with this model. In particular, the plots did not meet at one point
at 1/T = 0. The increased conductivity at very low temperatures could be
explained in terms of Mott's variable range hopping, but no explanation
could be found for the increase and deviation from straight lines above
5°K. Lastly Hall carrier concentration was found to be independent of
temperature, contradicting the prediction of the mobility edge model.
Arnold tried unsuccessfully to explain these results on the basis of a
macroscopic inhomogeneity and percolation model [16].

Several Soviet authors [16] and Adkins [14] have suggested that the
correlation effects in the electron gas are so large that Wigner
condensation must take place. At T=0, the electrons 'freeze' into a glass,
while at finite temperatures, the electrons move collectively like a
viscous fluid. Since all electrons move in this model, the carrier
concentration is constant and mobility now becomes thermally activated.

At room and higher temperatures the behaviour is similar but
activation energies are now of the order of 100 meV as compared to ~ 5 meV
at low temperatures. Rawlings, Jain and Leake [17] have analyzed a large
body of experimental data and have developed an improved theory based on
the macroscopic inhomogeneity model; unlike the earlier theories, Rawlings

model is applicable to large inhomogeneity. Rawlings et al. have also taken into account the interactions between neighbouring inhomogeneities. Their preliminary results show that in many cases, fixed oxide charges are not Gaussian distributed and their effects on I-V curves may be similar to that of interface states. These results are important in interpreting experiments on irradiated MOSFETs.

FIGURE CAPTIONS

Fig. 1 Decrease of feature size and operating voltage over the past 15 years [S. Chou, Spectrum April 1987].

Fig. 2 Scaling Laws for MOSFETs and their impact on gate delay and power dissipation [S. Chou, Spectrum, April 1987].

Fig. 3 Structure of an ultra small and ultra fast bipolar transistor. The contact region between extrinsic p^+ base and intrinsic p base is critical in determining the performance of the device [C.T. Chuang et al, Ref. 2].

Fig. 4 Depletion edge in a short channel MOSFET. In the analytical model of Ref. 11, vertical field is assumed to vary linearly and become zero along the line FH.

Fig. 5 The plots of observed conductivity vs 1/T [C.J. Adkins ref. 14]

Fig. 6 Mott's mobility edge model.

REFERENCES

1. P.M. Solomon, Proc. IEEE 70, 489 (1982).

2. C.T. Chuang, D.D. Tang, G.P. Li and E. Hackbarth, IEEE Trans.Elect. Dev. ED34, 1519 (1987).

3. T.H. Ning et al. IEEE Trans.Elect.Dev. ED28, 1010 (1981).

4. S.C. Jain, E.L. Heasell and D.J. Roulston. Progress in Quantum Electronics, 11 pp.105-204 (1987).

5. J.A. Del Alamo and R.M. Swanson, IEEE Trans. Elect.Dev. ED34, 1580 (1987).

6. S.E. Swirhun, Y.H. Kwark and R.M. Swanson IEDM 1986 p.24 (1986).

7. D. Totterdell, J.W. Leake and S.C. Jain to be published.

8. See the review V.K. Tewary and S.C. Jain, Advances in Electronics and Electron Physics 67, 329 (1986).

9. J. Graul et al. IEDM 1975 p.450.

10. P. Ashburn, D.J. Roulston and C.R. Selvakumar IEEE
     Trans.Elect.Devices, ED34, 1346 (1987).

11. S.C. Jain and P. Balk, Solid State Electronics, 30, 503 (1987).

12. K.H. Winters, S.C. Jain, A.S. Valerie and J. Hohl, unpublished work.

13. K. Mayaram, J.C. Lee and Cheming Hu, IEEE Trans.Elect.Dev. ED34, 1509
     (1987).

14. C.J. Adkin, J.Phys.C: Solid State Phys., 11 851 (1978).

15. N.F. Mott, "Conduction in Non-Crystalline Materials", Clarendon Press,
     Oxford (1987).

16. S.C. Jain "A review of Carrier Transport in Silicon Inversion Layers"
     to be published.

17. K.J. Rawlings, S.C. Jain and J.W. Leake, to be published.

18. H.C. de Graaf and J.G. de Groot, IEDM 1978 p.333.

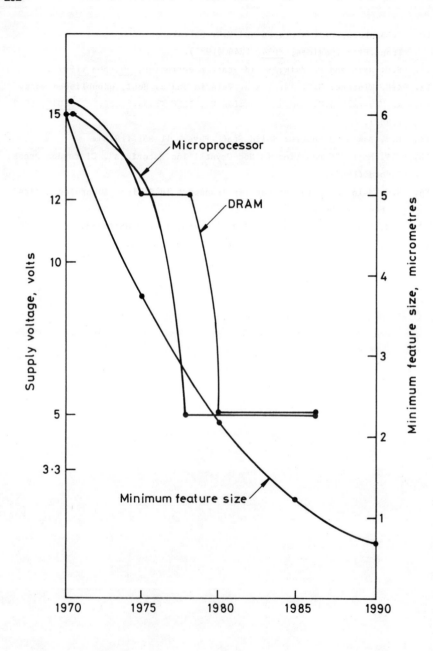

Fig. 1.

A

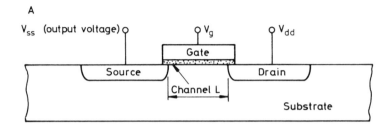

V_{ss} (output voltage) V_g V_{dd}

Gate

Source   Drain

Channel L

Substrate

B

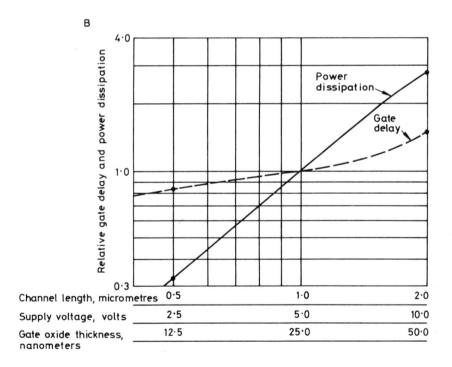

Relative gate delay and power dissipation

Power dissipation

Gate delay

Channel length, micrometres	0·5	1·0	2·0
Supply voltage, volts	2·5	5·0	10·0
Gate oxide thickness, nanometers	12·5	25·0	50·0

Fig. 2.

234

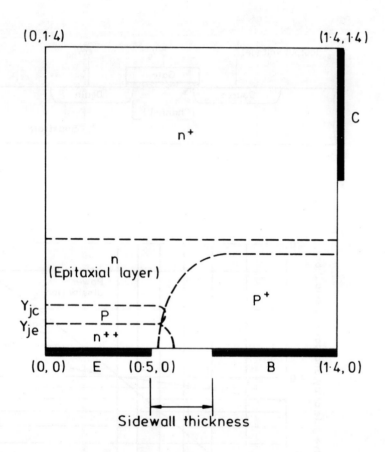

Fig. 3.

235

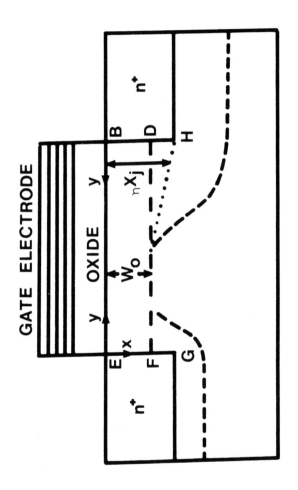

Fig. 4.

236

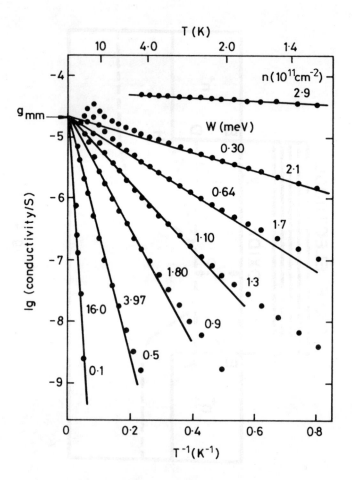

Fig. 5.

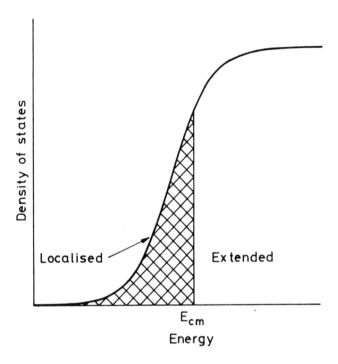

Fig. 6.

MODELING SEMICONDUCTOR DEVICES WITH
POSITION-DEPENDENT MATERIAL PARAMETERS

Alan H. Marshak

Department of Electrical and Computer Engineering
Louisiana State University
Baton Rouge, LA 70803 USA

Abstract

An overview of the transport model describing electron and hole
motion and density in solids with nonuniform band structure is presented.
This includes materials with nonuniform composition, like heterostructures, and devices with highly doped regions, like the emitter region of
modern bipolar transistors and solar cells. Effects due to carrier degeneracy, changes in the energy band edges, and changes in the density
of states produce terms in the carrier- and current-density equations in
addition to those found in the conventional Shockley model. These new
terms are discussed.

The general energy-band diagram relating the electrostatic potential, electron affinity and bandgap is given. The current densities are
written in terms of gradients of quasi-Fermi level and the carrier densities in terms of normalization integrals. The concepts of generalized
drift and diffusion are discussed. The transport equations applicable
to parabolic bands, nondegenerate material and the rigid-band model are
presented. The special case of minority-carrier flow in quasi-neutral
material is given. Physical concepts are emphasized.

1. INTRODUCTION

The band structure of a solid becomes nonuniform when the crystal potential is not strictly periodic, but which varies for different unit cells. This occurs in materials with nonuniform composition, or with nonuniform temperature or strain. The emerging technologies based on metallo-organic chemical vapor deposition (MOCVD) and molecular beam epitaxy (MBE) promise routine manufacture of advanced compound semiconductors. Material parameters like m_e, ε, χ and E_g can become position dependent. Impurity doping also affects the crystal potential, thereby causing a change in the band structure. In highly doped material, in excess of 10^{18} cm^{-3} for Si, the situation is complex. Changes in bandgap, electron affinity and density of states occur. These changes are due to several mechanisms: impurity band widening, band tailing, and many-body effects. These mechanisms produce unequal shifts from intrinsic band edges because they affect electrons and holes differently. A good review of these effects is given by Mertens et al. [1].

The purpose of this paper is to review the general transport equations describing electron and hole motion and density valid for solids with nonuniform band structure. Analogous transport effects occur in devices with nonuniform composition, like heterojunctions, or with highly doped regions, like the emitter region of modern bipolar transistors and solar cells. The concepts of generalized drift and diffusion are established and various special cases are presented. Size effects, such as those in short-channel MOS transistors, are not considered.

2. THE TRANSPORT EQUATIONS

Device analysis utilizes equations that express the flow and conservation of electrons and holes in regions of the material. Proceeding from the semiclassical $\underset{\sim}{k}$-space Boltzmann transport equation, Marshak and Van Vliet derived current-density expressions for degenerate materials with nonuniform band structure [2]. For isotropic material with constant temperature the electron current density is given by

$$\underset{\sim}{J}_n = n\mu_n \underset{\sim}{\nabla} E_{fn} \ . \tag{1}$$

Here E_{fn} represents the electrochemical potential or quasi-Fermi level for electrons in the conduction band. The hole current density is

$$\underset{\sim}{J}_p = p\mu_p \underset{\sim}{\nabla} E_{fp} \tag{2}$$

where E_{fp} is the quasi-Fermi level for holes. Recent work by Van Vliet et al. indicates that (1) and (2) remain valid for non-Bloch states [3]. Thus these equations can be used to describe electrical current through localized states, such as in a magnetic field or in amorphous material or in highly doped materials below the mobility edge. The form of (1) and (2) is consistent with that of irreversible thermodynamics.

The continuity equations are usually written as

$$\frac{\partial n}{\partial t} = G_n - R_n + \frac{1}{e} \nabla \cdot \underset{\sim}{J}_n \tag{3}$$

$$\frac{\partial p}{\partial t} = G_p - R_p - \frac{1}{e} \nabla \cdot \underset{\sim}{J}_p . \tag{4}$$

These equations are independent of the material and thus are quite general. Here G is the generation rate caused by external excitation. If $G = 0$, then $R_n = R_p$.

The carrier densities n and p can be derived by integrating in energy space the product of the number of allowed states per unit volume and the probability that a state is occupied. Carrier density in a band, as well as the mobility μ, involves a predetermined group of states characterized by a common quasi-Fermi level. The electron density in the conduction band can be expressed by

$$n = \int_{E_c}^{\infty} f \, g_c(W, \underset{\sim}{r}) \, dE \tag{5}$$

where $W = E - E_c$ is the electron kinetic energy. Here the bottom of the conduction band E_c is interpreted as an effective band edge (or mobility edge) separating localized from extended states, and the density of allowed states g_c is assumed to be an arbitrary function of energy and position. The probability f that a state of energy E is occupied by an electron is given by the Fermi function

$$f = \left[1 + \exp\left(\frac{E - E_{fn}}{kT} \right) \right]^{-1} . \tag{6}$$

Likewise, the hole density is given by

$$p = \int_{-\infty}^{E_v} f_h g_v(W_h, \underset{\sim}{r}) \, dE . \tag{7}$$

Here $f_h = 1 - f$, with $E_{fn} \to E_{fp}$, $W_h = E_v - E \geq 0$ is the hole kinetic energy, and E_v is the top of the valence band.

For materials with position-dependent band structure, the relationship between the band edges and the electrostatic potential ϕ needs
clarification. Fig. 1 shows the general energy-band diagram of a nonuniform semiconductor for the 1-D case. The local vacuum level E_ℓ represents the energy of an electron at a given point if it were at rest and
free from the influence of the crystal potential. Hence E_ℓ indicates
only additional potential energy. For convenience, let $E_\ell = E_o$ when
$\phi = 0$, where E_o denotes a reference level. Thus $E_o - E_\ell = e\phi$ and it follows that the gradient of E_ℓ is proportional to the electric field. The
electron affinity χ is defined as that energy which is required to excite
an electron from E_c to E_ℓ. This definition for χ includes the effect of
the image force seen by an electron just outside the surface of the material. Here $\chi = E_\ell - E_c$ and $E_g = E_c - E_v$ are treated as basic material
parameters. From Fig. 1, which was constructed using the above definitions, we observe that

$$E_c = E_o - \chi - e\phi \tag{8}$$

$$E_v = E_o - \chi - E_g - e\phi \ . \tag{9}$$

Note that the total potential energy E_c of the conduction-band electrons
consists of two parts: energy $E_o - \chi$ due to the crystal potential and
energy $- e\phi$ due to the electrostatic field. For the valence band, energy
$E_o - \chi - E_g$ is due to the crystal potential. Since ϕ is a continuous
function (excluding zero-width dipole layers), any discontinuity in χ or
E_g, such as that occuring in abrupt heterojunctions, must be reflected as
a change in the band edges E_c and E_v. We further note from (8) and (9)
that the often used definition [4] of ϕ in terms of a mid-gap energy
value E_i is inapplicable for materials with nonuniform band structure
since $\underset{\sim}{\nabla}\phi$ is not proportional to $\underset{\sim}{\nabla}E_c$ or to $\underset{\sim}{\nabla}E_v$.

Finally, ϕ is related to the carrier densities by Poisson's equation

$$\epsilon\nabla^2\phi + \underset{\sim}{\nabla}\epsilon \cdot \underset{\sim}{\nabla}\phi = -\rho = -e(p - n + N) \tag{10}$$

where $N = N_D - N_A$ is the net ionized donor impurity density. Equations
(3), (4), (5), (7), and (10) define a transport model with variables n,
p, E_{fn}, E_{fp}, and ϕ for degenerate material with nonuniform band structure.

3. GENERALIZED DRIFT AND DIFFUSION

Equations (1) and (2) are the basic forms for the current density. Since E_{fn} and E_{fp} are not directly measurable, the current equations are usually rewritten in terms of the variables n, p, and ϕ. Generalized drift and diffusion can be derived in a number of ways. Here I consider n in energy space and solve for $\underset{\sim}{\nabla}E_{fn}$, as was done in [5]. Using (6) in (5) yields

$$n = \int_0^{\infty} \frac{g_c(W, \underset{\sim}{r}) dW}{1 + \exp\left(\frac{W}{kT}\right) \exp\left(\frac{E_c - E_{fn}}{kT}\right)} \quad . \tag{11}$$

From (11) we observe than $n = n(E_{fn} - E_c, T, \underset{\sim}{r})$. Solving $\underset{\sim}{\nabla}n$ for $\underset{\sim}{\nabla}E_{fn}$ and substituting into (1) yields the electron current density

$$\underset{\sim}{J}_n = eD_n\underset{\sim}{\nabla}n + n\mu_n\underset{\sim}{\nabla}E_c - n\mu_n\underset{\sim}{\nabla}\Gamma_n \tag{12}$$

where D_n is related to μ_n by the generalized Einstein relation

$$eD_n \frac{\partial n}{\partial(E_{fn} - E_c)} \equiv n\mu_n \tag{13}$$

and

$$n\mu_n \underset{\sim}{\nabla}\Gamma_n \equiv eD_n \int_0^{\infty} f \underset{\sim}{\nabla}^W g_c(W, \underset{\sim}{r}) dW \quad . \tag{14}$$

The symbol $\underset{\sim}{\nabla}^W$ denotes the gradient operator for variable W held constant. The second term in (12) denotes generalized drift, the other two terms diffusion. From (8), we have

$$\underset{\sim}{\nabla}E_c = - \underset{\sim}{\nabla}\chi - e\underset{\sim}{\nabla}\phi \quad . \tag{15}$$

The development for the hole current density follows along similar lines. From $p = p(E_v - E_{fp}, T, \underset{\sim}{r})$ it follows that

$$\underset{\sim}{J}_p = - eD_p\underset{\sim}{\nabla}p + p\mu_p\underset{\sim}{\nabla}E_v + p\mu_p\underset{\sim}{\nabla}\Gamma_p \tag{16}$$

where

$$eD_p \frac{\partial p}{\partial(E_v - E_{fp})} \equiv p\mu_p \tag{17}$$

$$p\mu_p\underset{\sim}{\nabla}\Gamma_p \equiv eD_p \int_0^{\infty} f_h \underset{\sim}{\nabla}^{W_h} g_v(W_h, \underset{\sim}{r}) dW_h \tag{18}$$

$$\underset{\sim}{\nabla}E_v = - \underset{\sim}{\nabla}(\chi + E_g) - e\underset{\sim}{\nabla}\phi \quad . \tag{19}$$

The gradients ∇E_c and ∇E_v contain two effects of different physical origin, namely the variations in the band caused by the electrostatic field and by the changes in material parameters which are inherent in the nonuniform structure. Thus the drift force $F_n = -\nabla E_c$ experienced by an electron in the conduction band and the drift force $F_p = \nabla E_v$ experienced by a hole can be quite different. Only the electric field $E = -\nabla\phi$ is the same for both the conduction and the valence band. Besides the usual diffusion term, there is an extra term in (12) and (16) caused by the position-dependent density of states.

Substituting the current equations (12) and (16) into the continuity equations (3) and (4) yield general transport equations. They along with Poisson's equation (10) represent a set of three partial differential equations in three unknowns, namely n, p, and ϕ. In principle, this set of equations can be solved given appropriate boundary conditions and specific models for the material parameters ε, χ, g_c, g_v, and E_g. The following special cases of general interest elucidate these results.

Parabolic Bands

When the $E - k$ relation is quadratic, as in low-doped Ge and Si, the density of states is given by

$$g_c = \frac{\sqrt{2}\, m_e^{3/2}}{\pi^2 \hbar^3}\, W^{1/2} \tag{20}$$

where m_e is the electron density-of-states effective mass. Using (20) in (11) we obtain

$$n = N_c\, F_{1/2}\!\left(\frac{E_{fn} - E_c}{kT}\right) \tag{21}$$

with

$$N_c = 2\left(\frac{m_e kT}{2\pi\hbar^2}\right)^{3/2}. \tag{22}$$

Note that N_c is a function of position if m_e depends on position. Evaluating (13) and (14) yields

$$\frac{D_n}{\mu_n} = \frac{kT}{e}\, \frac{F_{1/2}(\eta)}{F_{-1/2}(\eta)} \tag{23}$$

$$\nabla\Gamma_n = \frac{3e}{2}\, \frac{D_n}{\mu_n}\, \nabla\ln m_e. \tag{24}$$

Here $F_{1/2}(\eta)$ is the Fermi integral of order 1/2. Equation (24) indicates that electrons will tend to move in the direction of increasing effective mass. Similar results follow for holes. The analysis can be extended to the nonparabolic case (e.g. InSb and GaAs) when the density of states can be expressed as a series $g = \Sigma \, a_\nu(\underset{\sim}{r})W^\nu$ [6].

Nondegenerate Material

In the nondegenerate limit a number of simplifications are possible. The current densities are given by (12) and (16); however the carrier densities (5) and (7) reduce to

$$n = N_c \exp\left(\frac{E_{fn} - E_c}{kT}\right) \tag{25}$$

$$p = N_v \exp\left(\frac{E_v - E_{fp}}{kT}\right) \tag{26}$$

where the effective density of states are defined by [5]

$$N_c = \int_0^\infty g_c(W,\underset{\sim}{r})e^{-W/kT}dW \tag{27}$$

$$N_v = \int_0^\infty g_v(W_h,\underset{\sim}{r})e^{-W_h/kT}dW_h \ . \tag{28}$$

Note that in general N_c and N_v are functions of position and can be evaluated once g_c and g_v are specified.

In the nondegenerage limit it is easy to show that [5]

$$\underset{\sim}{\nabla}\Gamma_n = kT \, \underset{\sim}{\nabla}\ln N_c \tag{29}$$

$$\underset{\sim}{\nabla}\Gamma_p = kT \, \underset{\sim}{\nabla}\ln N_v \ . \tag{30}$$

Equation (29) expresses $N_c = N_c(\Gamma_n)$, provided a reference for Γ_n is stated. For convenience, define Γ_n to be zero for a sample with constant g_c; thus $N_c(0) = N_{co}$. For example, if the sample has parabolic bands, then N_{co} is given by (22) with m_e constant. Likewise, $N_v(0) = N_{vo}$. For this case, the density of states effect is equally well represented by Γ_n and Γ_p as by N_c and N_v.

The current-density equations (12) and (16) can be written as

$$\underset{\sim}{J}_n = eD_n\underset{\sim}{\nabla}n + en\mu_n\underset{\sim}{E} - n\mu_n\underset{\sim}{\nabla}\chi - n\mu_n kT \, \underset{\sim}{\nabla}\ln N_c \tag{31}$$

$$\underset{\sim}{J}_p = -eD_p\underset{\sim}{\nabla}p + ep\mu_p\underset{\sim}{E} - p\mu_p\underset{\sim}{\nabla}(\chi + E_g) + p\mu_p kT \, \underset{\sim}{\nabla}\ln N_v \ . \tag{32}$$

For this case, the Einstein relation is given by $D/\mu = kT/e$.

Rigid-Band Model

The rigid-band model assumes that the density of states for the conduction and valence band is position independent even when the band edges change. For this case, $\Gamma = 0$ and (12) and (16) reduce to

$$\underset{\sim}{J}_n = eD_n \underset{\sim}{\nabla}n + en\mu_n \underset{\sim}{E} - n\mu_n \underset{\sim}{\nabla}\chi \tag{33}$$

$$\underset{\sim}{J}_p = - eD_p \underset{\sim}{\nabla}p + ep\mu_p \underset{\sim}{E} - p\mu_p \underset{\sim}{\nabla}(\chi + E_g) \ . \tag{34}$$

Experiment indicates that highly doped silicon behaves in many respects like a metal. Thus it is reasonable to infer that the rigid-band model is applicable to highly doped silicon [7]. For this case, the band structure is parabolic and the electron density is given by (21) with m_e constant.

4. MINORITY-CARRIER CURRENT

Another important special case is minority-carrier flow under low injection in quasi-neutral material. For concreteness, I assume current injection into n-type material with nonuniform doping bounded by an ohmic contact at $x = 0$ and the edge of a space-charge region at $x = x_E$.

For low injection, $\underset{\sim}{E} \approx \underset{\sim}{E}^o$. The equilibrium field $\underset{\sim}{E}^o$ can be found from (12) with $\underset{\sim}{J}_n = 0$ or from (16) with $\underset{\sim}{J}_p = 0$. For convenience I use the latter method which avoids knowledge of D_n/μ_n; thus

$$e\underset{\sim}{E} = kT \underset{\sim}{\nabla}\ln p^o + \underset{\sim}{\nabla}(\chi + E_g) - \underset{\sim}{\nabla}\Gamma_p^o \ . \tag{35}$$

Here $D_p/\mu_p = kT/e$ was used since the holes are nondegenerate. Using (35) in (16) yields

$$\underset{\sim}{J}_p = - eD_p \underset{\sim}{\nabla}p + p\mu_p [kT \underset{\sim}{\nabla}\ln p^o] \ . \tag{36}$$

In this form, the effects due to degeneracy and nonuniform band structure are embedded in the value of $p^o(\underset{\sim}{r})$. The net built-in field is given by [ ] in (36). Equation (36) coupled with (4) specify the minority density. Similar conclusions follow for electron current and density in p-type material.

For the steady-state case, (4) and (36) can be combined to yield

$$a\nabla^2 u + \underset{\sim}{\nabla}a \cdot \underset{\sim}{\nabla}u - bu = 0 \tag{37}$$

where $a = eD_p p^o$, $b = ep^o/\tau_p$, and $u = p/p^o - 1$. Given a, b, and appropriate boundary conditions, (37) can be solved. See Park et al. [8], and the references therein. When the net recombination is negligible,

$b = 0$. Under this condition, $J_{\sim p}$ is constant and (36) can be solved directly. Recent work by del Alamo and Swanson [9] give measured values of L_p and $D_p p^o$ for phosphorus doped silicon. Thus for this case, a and b are known.

5. SUMMARY AND COMMENTS

An overview of electron and hole motion and density in solids with nonuniform band structure was presented. This includes devices with graded composition, like heterostructures, and devices with highly doped regions, like the emitter region of modern bipolar transistors and solar cells. Expressions for current density, (1) and (2) or (12) and (16), valid for degenerate materials with variable bandgap, electron affinity, and density of states, are given. The relationship between the band edges and the electrostatic potential is given by (8) and (9). These results pertain to device analysis and design. The expressions given can be extended to include thermal diffusion and anisotropy [5].

I conclude with a short list of topics that require additional study.

1. Models to describe the density of states for highly doped material, especially at room temperature, are needed. Also schemes for finding the mobility edges E_c and E_v need to be developed. Recent work in highly doped Si [10] and GaAs [11] at 300 K indicate significant changes in the band structure and density of states.

2. Models for the material parameters like m_e, m_h, ε, χ, and E_g as functions of doping and temperature are also needed. Even limited models can be used to extend understanding and can be employed for first-order analysis.

3. The role of localized states (extended tail states) on minority-carrier mobility needs further examination [12].

4. Although the physics of interfaces has been the subject of many investigations, the energy band discontinuities ΔE_c and ΔE_v at abrupt interfaces are not yet clearly understood [13].

5. Experimental methods need to be developed that can distinguish values of χ and E_g at interfaces from their values in bulk material.

REFERENCES

1. R. P. Mertens, R. J. Van Overstraeten, and H. J. DeMan, "Heavy doping effects in silicon," in *Advances in Electronics and Electron Physics*, L. Marton and C. Marton, Eds., vol. 55. New York: Academic Press, 1981, pp. 77–118.

2. A. H. Marshak and K. M. van Vliet, *Solid-State Electronics*, vol. 21, pp. 417–427, 1978.

3. C. M. Van Vliet, C. G. Van Weert and A. H. Marshak, *Physica*, vol. 134A, pp. 249–264, Dec. 1985.

4. W. Shockley, *Electrons and Holes in Semiconductors*. Princeton, NJ: Van Nostrand, 1950.

5. A. H. Marshak and C. M. Van Vliet, *Proc. IEEE*, vol. 72, pp. 148–164, Feb. 1984.

6. A. H. Marshak and K. M. van Vliet, *Solid-State Electronics*, vol. 21, pp. 429–434, 1978.

7. A. H. Marshak, M. A. Shibib, J. G. Fossum, and F. A. Lindholm, *IEEE Trans. Electron Devices*, vol. ED-28, pp. 293–298, 1981.

8. J. Park, A. Neugroschel, and F. A. Lindholm, *IEEE Trans. Electron Devices*, vol. ED-33, pp. 240–249, Feb. 1986.

9. J. A. del Alamo and R. M. Swanson, *IEEE Trans. Electron Devices*, vol. ED-34, pp. 1580–1589, 1987.

10. H. S. Bennett, *J. Appl. Phys.*, vol. 59, pp. 2837–2844, April 1986.

11. H. S. Bennett and J. R. Lowney, *J. Appl. Phys.*, vol. 62, pp. 521–527, July 1987.

12. A. Neugroschel and F. A. Lindholm, *Appl. Phys. Lett.*, vol. 42, p. 176, Jan. 1983.

13. A. G. Milnes, *Solid-State Electronics*, vol. 29, pp. 99–121, 1986.

248

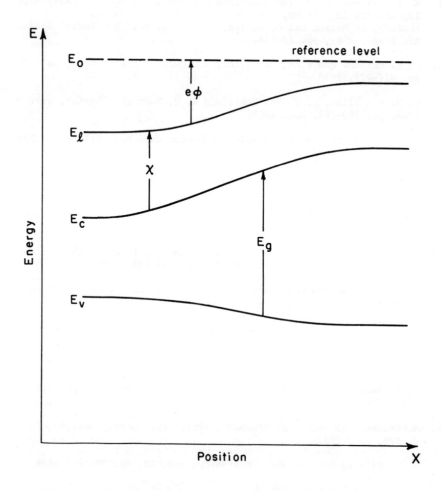

Fig. 1. General one-dimensional energy-band diagram of a nonuniform
semiconductor defining variables of interest; E represents
total electron energy, E_o is the reference level, ϕ is the
electrostatic potential, E_ℓ is the local vacuum level, χ
is the electron affinity, and E_g the bandgap.

OXYGEN-RELATED DEFECTS IN SILICON

G. Pensl, M.Schulz

Institute of Applied Physics
University of Erlangen-Nürnberg
Glückstr. 9, D-8520 Erlangen
F.R.Germany

Cz-grown Si samples containing a high
concentration of oxygen are investigated after
various processing steps by DLTS. Heat
treatments at temperatures around 450 °C
generate a series of discrete double donors
(Thermal Donors) with ionization energies of
approx. 60 meV and 145meV for the neutral and
singly ionized donor species, respectively.
Long time heat treatments at elevated
temperatures (600 to 900°C) generate so-called
New Donors showing a continuous distribution of
trap states with increasing values towards the
conduction band edge. The electronic properties
of the New Donors are explained by the "SiO$_x$-
Interface Model".

INTRODUCTION

Czochralski (Cz)-grown silicon wafers are mainly employed
as starting material for CMOS and bipolar process
technologies. The most abundant impurity in Cz-grown Si
crystals is oxygen which is incorporated into the silicon
lattice during the growth process /1/. The oxygen content
may reach concentrations as high as $2x10^{18}cm^{-3}$
corresponding to the solubility of oxygen in Si at the
melting point (approx. 1430 °C). Consequently, oxygen is
present at a supersaturated concentration at temperatures
where the device manufacturing is conducted. Oxygen may be
either incorporated into the Si lattice as isolated
(monatomic) species or may participate in small
agglomerations and extended precipitates, respectively.
Isolated oxygen has never been observed on a Si lattice
site (Fig.1b). It is usually incorporated on an
interstitial site displaced off the <111> axis (Fig. 1c).
There, it is electrically inactive but gives rise to a
vibrational mode at λ = 9 μm which is employed to determine

250

the interstitial oxygen content by IR absorption /2/. A
further isolated oxygen defect is the A-center which is
identified as a vacancy pairing with an oxygen atom
slightly displaced from the substitutional site in <100>
direction (Fig. 1d). The A-center is a deep acceptor with
ionization energy $E_I = E_C - 0,17eV$ /3,4/.

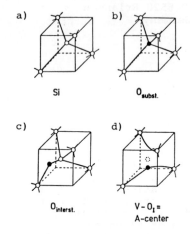

Fig. 1: Possibilities
to incorporate oxygen
into the Si lattice.
a) basic Si lattice,
b) substitutional
oxygen,
c) interstitial
oxygen (Si_2O
molecule).
d) A-center (V-O
complex).

○ = Si

● = O

◌ = Vacancy

At temperatures above 300 °C oxygen becomes mobile;
the supersaturated concentration can be reduced either by
outdiffusion or by forming complexes or precipitates
depending on the heat treatment. The precipitates generate
lattice defects which act as traps for impurities (e.g.
heavy metals). By appropriate thermal cycling /5,6/, a
denuded zone near the surface where the devices are located
may be achieved and oxygen precipitates for internal
gettering may be generated in the bulk of the wafer (Fig.
2).
However, the parameters for this denuded-zone
processing are not universal for a given oxygen content
/7,8/. The oxygen precipitation behavior depends on many
other parameters like the thermal history of the wafer, the
carbon concentration, the microdefect density or the
microinhomogeneity of the oxygen distribution. In addition,
heat treatments during device processing may also cause
oxygen-related defects in the device active region leading
to a degradation of device performance. It is the purpose
of this paper to report briefly on oxygen-related defects
which are generated by various heat treatments.

EXPERIMENTAL

Heat treatments of silicon samples were performed with
a "Thermal Pulse Annealer (TPA)" developed at our institute
/9/, the TPA system is illustrated in Fig. 3. It employs a

251

a)

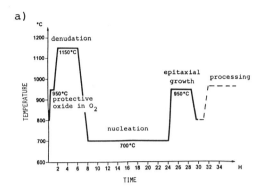

Fig. 2: Thermal
cycling of a CZ-grown
Si wafer to achieve a
denuded zone (after
/5,6/):
a) typical thermal
cycle;
b) oxygen
distribution in the
wafer after denuded-
zone processing.

b)

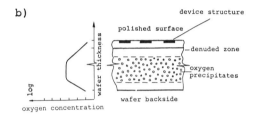

quartz chamber sandwiched between two banks of tungsten
hologen lamps providing a maximum power of 12 kW. The
quartz chamber may be evacuated or filled with a variety of
gases. A W/Re thermocouple covered with silicon serves as a
temperature sensor. Thermocouple feedback combined with
microcomputer control is employed to achieve a multitude of
possible temperature versus time cycles. The TPA system
further offers the advantage of clean processing and of
fast ramp up/ramp down of sample temperatures. It takes
approx. 2 seconds to heat up a Si wafer from room
temperature to 1100 °C.

Oxygen-related defects are investigated by various
experimental techniques like high-resolution electron
microscopy (HREM) /10,11/, small angle neutron scattering
(SANS) /12/, IR absorption /13/, conductivity measurements
/16/, electron spin resonance (ESR) /17,18/, electron
nuclear double resonance (ENDOR) /19/ or various space
charge junction techniques /20/. For the investigations in
this paper, deep level transient spectroscopy (DLTS) has
mainly been employed for analysis. We have used a computer-
controlled system schematically shown in Fig. 4. Data-
acquisation, data-storage, and analysis of stored data are
performed by a HP 9836A computer system; the hardware and
software are described in detail in Ref. 21.

252

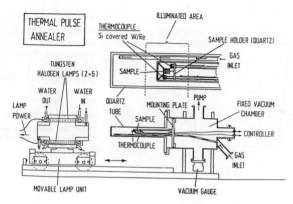

Fig. 3: Schematic of the Thermal Pulse Annealer used for the heat treatments. The light power of 12 kW is provided by two banks of tungsten halogen lamps each consisting of 6 lamps.

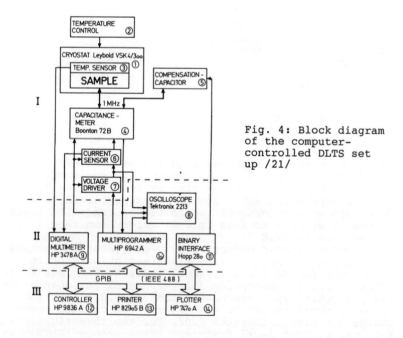

Fig. 4: Block diagram of the computer-controlled DLTS set up /21/

THERMAL DONORS

The Thermal Donors (TD's) are generated by heating at temperatures ranging from 300 to 550 °C for several hours;

253

they are formed independent of the dopants (B, Al, P, Sb).
The TD concentrations may reach values up to 10^{16} cm^{-3}. IR
absorption measurements reveal the existence of neutral and
singly ionized TD's /22/. Based on the similar behavior of
neutral and singly ionized TD's under heat treatments, it
was possible to identify the TD's as double donors. Up to 9
different TD species are observed in IR spectra. The
ionization energies of the different $TD_i°/TD_i^+$ species (i =
1...9) differ slightly (1 to 6 meV); they range from 53 to
69,3 meV and 127,9 to 156,3 meV for neutral and singly
ionized TD's, respectively.

DLTS measurements taken on Schottky barriers (Au/n-
type Si) cannot energetically resolve the different TD_i
species. Fig. 5 shows DLTS spectra of singly ionized TD
species. The solid spectrum originates from an as-grown
sample demonstrating that the TD's may already be generated
by the temperature ramp down after the growth process. It
turns out that the ionization energy determined from an
Arrhenius plot strongly depends on the electric field
applied in the depletion region. Hence, the ionization
energy has to be corrected for the Poole-Frenkel effect
/23/. An average zero-field ionization energy of 155 meV is
obtained for the singly ionized TD_i^+ traps. The dotted
curve is recorded on the same sample after heat treatment
at 500 °C for 26 hrs. The maximum of the DLTS peak shifts
to lower temperatures indicating that the concentration of
TD species with smaller ionization energies ($E_I \approx$ 125 meV)
is increased by the heat treatment.

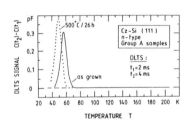

Fig. 5: DLTS signal
versus temperature of
TD_i^+ traps for an as-
grown sample (solid
curve) and for the
same sample after
heat treatment. Group
A samples:
$[O_I]$ =1,7x10^{18} cm^{-3};
$[C_S]$ =3 x10^{16} cm^{-3};
$[P^-]$ =1,9x10^{15} cm^{-3}.

Many TD models have been proposed to explain the
origin and the structure of TD traps (see e.g. /14, 18, 24,
25/ and references therein). The basic idea of the various
TD-models proposed is that oxygen atoms agglomerate
partially in connection with Si atoms and/or vacancies and
form complexes which are electrically active. Initial
growth stages of these agglomerations or rearrangements of
the participating atoms are claimed to be responsible for
the different species of Thermal Donors. For example, the
Ourmazd, Schröter, Bourret (OSB) model /25/ predicts a
complex consisting of one central silicon surrounded by a
number N of oxygen atoms (N $\geqslant$5) where the silicon atom has
two tangling bonds responsible for the donor activity. The
disadvantage of all these cluster models proposed is that
they require an enhanced oxygen diffusivity which is higher

by two orders of magnitude than the experimental data observed /12/.

In order to test the OSB model, hydrogenation and short-time annihilation experiments have been performed. Fig. 6a) shows DLTS spectra taken on the same Si sample(TD concentration: 8×10^{12} cm^{-3}) prior to (solid curve) and after the hydrogenation (dashed curve). The trap level E_0 is related to singly ionized TD's, the trap levels E_1 and E_2 (see Fig. 6b) are generated by the hydrogenation process itself /26, 27/; they are not correlated with TD's as demonstrated in Fig. 6b. The atomic nature of these defects (E_1, E_2) is still not known. The amount of TD passivation is not a drastic effect and is therefore not suited to support the OSB model /25/.

a)

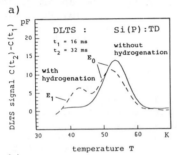

b)

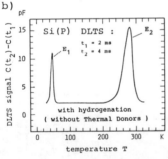

Fig. 6:
a) Comparison of DLTS spectra for TD traps; the solid and the dashed curve were recorded from the reference and hydrogenated (125 °C, 30min) sample, respectively.
[O_I]=$6,4 \times 10^{17}$ cm^{-3}, TD generation: T=470 °C, t=45min.
b) DLTS spectrum of a Si sample containing no TD's.
$E_0 = E_C - 137$ meV, $N_0 = 8 \times 10^{12}$ cm^{-3} (4×10^{12} cm^{-3} after passivation)
$E_1 = E_C - 58$ meV, $N_1 = 2 \times 10^{12}$ cm^{-3} (a), $N_1 = 4 \times 10^{12}$ cm^{-3} (b);
$E_2 = E_C - 498$ meV, $N_2 = 8 \times 10^{12}$ cm^{-3}

In Fig. 7, the short-time annihilation behavior of TD's at T_A=530, 600 and 750 °C, respectively, is investigated (see squares, circles and triangles) /26/. The original set of OSB parameters /25/ was used to perform a fit to the experimental data. No satisfactory agreement between calculated curves and experimental data could be achieved at temperatures above 530 °C as demonstrated in Fig. 7. The TD's can completely be annihilated by heat treatments at temperatures above 600 °C; typically, they are destroyed by an annealing at 700 °C for 10 min /15/.

In summary, we point out that the atomic structure of the TD's is still not solved. The TD complexes extended in <110> direction may be composed of units (or combinations of them) like di-oxygen molecules, self-interstitials or vacancies. The units have to be assembled in such a way that the kinetic features observed are reproduced /14/.

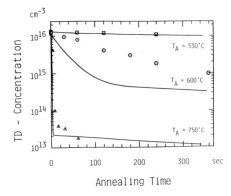

Fig. 7: Short-time
annealing of TD's.
The symbols (squares,
circles, triangles)
are measurement
points, the solid
curves are calculated
after the OSB model
with parameters given
in /25/.

NEW DONORS

Long-time heat treatments of Cz-grown silicon in the temperature range from 600 to 900 °C generate oxygen-induced precipitates and increase simultaneously the electrical activity by forming so-called New Oxygen Donors (ND's)

The formation of precipitates is generaly associated with a decrease of the interstitial oxygen concentration. The growth of SiO_x precipitates results in a large volume increase and causes internal stress fields which may be relaxed by emitting self-interstitials leading to pure interstitial agglomerations. Depending on the heat treatment, various precipitates are observed by transmission electron microscopy (TEM) and by high-resolution electron microscopy (HREM) /28, 29, 10, 11/. The TEM image in Fig. 8 shows one extended dislocation loop and a number of amorphous SiO_x platelets. At temperatures around 1200 °C, the precipitates are resolved again; simultaneously, an increase of the interstitial oxygen concentration is measured by IR absorption.

Heat treatments at temperatures from 600 to 900 °C of Cz-grown Si samples generate New Donors (ND's) independent of the charge type of the intentional doping /15, 30/. The maximum donor concentration observed is about 10^{16} cm^{-3}. It is also reported that preannealing at 450 to 550 °C as well as a high carbon content promote the formation of ND's /15/.

Detailed DLTS studies on the ND's were performed by Hölzlein and coworkers /31, 32, 33, 34/. DLTS spectra were taken on Cz-grown Si samples subsequent to various heat treatments and processing steps. The DLTS measurements were performed on Schottky contacts. The salient feature of all the DLTS spectra measured (DLTS signal over temperature) is

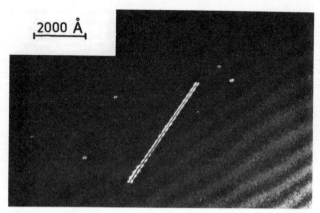

Fig. 8: TEM image of a dislocation loop and of amorphous
(100) platelets; the silicon sample was subjected to the
following heat treatments: 500 °C/25 hrs. + 650 °C/65 hrs.
+ 800 °C/16hrs. The TEM image was taken by W.Bergholz,
Siemens AG, München.

the continuous distribution of the DLTS signal caused by
bulk trap states. It should be noted that such a continuous
distribution is unusual for bulk traps. The energy
distribution of these trap states was calculated from the
DLTS data by an evaluation method described in /32/. The
ND's are assumed to be responsible for the bulk traps
observed.

In Fig. 9, the density of states D_{BT} over E_C-E is
plotted for a set of Si samples annealed at 650 °C;
parameter is the anneal duration. The density of states
monotonically increases towards the conduction band edge
E_C; it also shifts to higher temperatures after prolonged
anneal durations. In addition, comparison of B I and B II
samples demonstrates the effect of a surface layer removal.
The oxygen concentration near the surface is reduced in BI
samples by outdiffusion leading to lower values of the
density of trap states; this outdiffusion effect is avoided
in B II samples by etching off a layer of 30 μm thickness.

Fig. 10 shows the DLTS spectrum of a sample taken
after a three step annealing. The ND traps are generated
during the first two steps; they are annihilated during the
final high temperature annealing at 1000 °C for 2 hrs. as
can be deduced from the disappearance of the continuous
spectrum. The DLTS spectrum now consists of three discrete
peakes originating form discrete bulk traps (E_C-E_T=0,55 eV,
E_C-E_T=0,35eV, E_C-E_T=0,13eV). The chemical nature of these
bulk traps is still not identified.

In Table I, the annealing behaviour of ND's is
summarized and compared with the concentration of amorphous
SiO_x platelets and with the concentration of interstitial

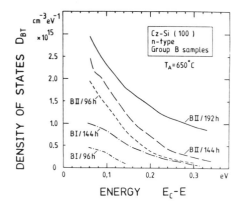

Fig. 9: Density of states D_{BT} versus energy E_C-E; parameter is the anneal duration. BI samples are without and BII samples are with surface layer removal of 30 μm. Group B samples ;
$[O_I] = 7x11^{17}$ cm^{-3};
$[C_S] = 4x10^{16}$ cm^{-3};
$[P] = 1,8x10^{15}$ cm^{-3}.

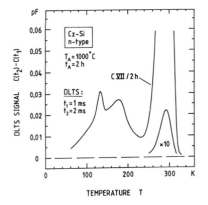

Fig. 10: DLTS spectrum of sample CVII/2h after a three step heat treatment at temperatures: 500°C/26hrs + 650°C/65hrs + 1000°C/2hrs demonstrating the annihilation of ND's. Group C samples;
$[O_I] = 6,4 \times 10^{17}$ cm^{-3};
$[C_S] < 5x10^{15}$ cm^{-3};
$[P] = 8 \times 10^{14}$ cm^{-3}.

oxygen. DLTS, TEM observations, and IR absorption were taken on the same samples. It seems that the formation and annihilation kinetics of ND's and SiO$_x$ platelets are strongly correlated. The properties of ND traps determined by DLTS show striking similarities to the features reported for interface states at a planar MOS structure. Based on these common features, the "SiO$_x$-Interface Model" is proposed to explain the origin of ND traps. The band structure of this model is illustrated in Fig. 11 for the situation that is used in the DLTS measurement. Two different types of defects are proposed to contribute to the trap spectrum observed. The first type of traps are interface states at the surface of SiO$_x$ precipitates which are spatially distributed in the bulk of the Si substrate; these interface states contribute to the entire energy range investigated. The donor-like behavior of the

258

precipitates may be caused by a positive charge associated with the SiO_x precipitates. A fixed positive charge is well known to occur near the $Si-SiO_2$ interface of MOS structures. The fixed positive charge causes a coulombic well around the precipitates as indicated in the figure. As a consequence, it is expected that the second type of traps corresponds to bound states in the coulombic wells.

Table I: Determination of ND traps (DLTS), O_I-concentration (IR absorption), and SiO_x platelets (TEM) for different heat-treated samples. The IR absorption and TEM observations were conducted by W.Bergholz, Siemens AG, München.

annealing	ND traps	O_I $\times 10^{17}(cm^{-3})$	platelets (cm^{-3})
as grown	-	6,4	-
470°C/30 h	-	5,7	-
500°C/26h+650°C/65h	formation	4,1	$2 \cdot 10^{12}$
500°C/26h+650°C/65h 800°C/16h	formation	out of detection efficiency	$\sim 5 \cdot 10^{13}$ $(d \sim 150\text{Å})$
500°C/26h+650°C/65h 1000°C/2h	annihilation	7,3	$< 10^{10}$

SiO_x INTERFACE MODEL

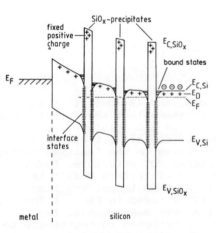

Fig. 11: Schematic of the $Si-SiO_x$ (precipitate) band structure illustrating the "SiO_x-Interface Model" for oxygen-related "New Donor" defects in silicon.

Calculations of ground-state energies in the Coulombic wells of the positive charges and hydrogenation experiments of the ND traps support the SiO_x-Interface Model /31/.

CONCLUSION

It has been demonstrated that oxygen forms isolated
defects, complexes in connection with intrinsic and/or
extrinsic partners, and extended precipitates in silicon
crystals depending on the oxygen content and the processing
steps. Many experimental data are available on all the
different defects; however, the theoretical understanding
of the interacting forces and the atomic structure of the
defects are largely unknown. Most of the oxygen-related
defects like Thermal Donors and New Donors are electrically
active.

ACKNOWLEDGEMENTS

The authors are grateful to Dr.W.Bergholz for giving
permission to present TEM and IR absorption data prior to
publication.

REFERENCES

1. Zulehner, W. and Huber, D., Crystals 8, p.1, Springer
   Verlag, Berlin, 1982
2. ASTM Test Method F 121-80
3. Watkins, G.D. and Corbett, J.W., Phys.Rev. 121, 1001
   (1961)
4. Corbett, J.W, Watkins,G.D., Chrenko, R.M. and
   McDonald, R.S., Phys.Rev. 121, 1015 (1961)
5. Matlock, J.H., Silicon Processing, ASTM STP 804,
   D.C.Gupta, ed., American Society for Testing and
   Materials, 1983, p. 332
6. Borland, J.O., Kuo, M., Shibley, J., Roberts, B.,
   Schindler, R. and Dalrymple, T., Semiconductor
   Processing, ASTM STP 850, D.C.Gupta, ed., American
   Society for Testing and Materials, 1984, p. 49
7. Chion, H.-D., Solid State Technology, 30, 77 (1987)
8. Swaroop, R., Kim, N., Lin, W., Bullis, M., Shive, L.,
   Rice, A., Castel, E. and Christ, M., Solid State
   Technology, 30, 85 (1987)
9. Neumann, J., Diploma thesis, Erlangen (1985)
10. Tempelhoff, K., Hahn, B. and Gleichmann, R.,
    Semiconductor Silicon 1981, p.244, The Electrochem.
    Soc., Inc. Pennington, N.J., Huff, H.R., Kriegler,
    R.J., Takeishi, Y. (eds.)
11. Bourret, A., to be published in Inst. Phys. Conf. Ser.
    (1987)
12. Messoloras, S., Newman, R.C., Stewart, R.J. and
    Tucher, J.H., Semiconductor Science and Technology 2,
    14 (1987)
13. Oeder, R. and Wagner, P., Mat.Res.Soc.Symp.Proc. 14,
    171 (1983)
14. Kaiser, W., Frisch, H.L. and Reiss, H., Phys. Rev.
    112, 1546 (1958)
15. Kanamori, A. and Kanamori, M., J.Appl.Phys. 50, 8095
    (1979)

16. Gaworzewski, P. and Schmalz, K., phys.stat.sol. (a) 55, 699 (1979)
17. Müller, S.H., Siewerts, E.G. and Ammerlaan, C.A.J., Inst.of Physics, Conf.Series 46, 297 (1979)
18. Pajot, B., Compain, H., Leronille, J. and Clerjaud, B., Physica 117B/118B, 110 (1983)
19. Michel, J., Niklas, J.R. and Späth, J.M., Mat.Res.Soc.Symp.Proc. 59, 111 (1986)
20. Benton, J.L., Lee, K.M., Freeland, P.E. and Kimerling, L.C., Proc. of the 13th Int.Conf. on Defects in Semiconductors, p.647, Publ. of the Metallurgical Soc. of AIME, Coronado, 1985, Kimerling, L.C. and Parsey, Jr. J.M. (eds.)
21. Hölzlein, K., Pensl, G., Schulz, M. and Stolz, P., Rev.Sci.Instrum. 57, 1373 (1986)
22. Wagner, P., Holm, C., Sirtl, E., Oeder, R., and Zulehner, W., Advances in Solid State Physics, XXIV, 191 (1984) P.Grosse (ed.), Vieweg Braunschweig.
23. Hartke, J.L., J.Appl.Phys. 39, 4871 (1968)
24. Oehrlein, G.S. and Corbett, J.W., Mat.Res.Soc. Symp.Proc. 14, 107 (1983)
25. Ourmazd, A., Schröter, W. and Bourret, A., J.Appl.Phys. 56, (1984)
26. Fritsch, E., Pensl, G. and Wagner, P., Verhandl. DPG 22, HL 34.5 (1987)
27. Johnson, N.M., Ponce, F.A., Street, R.A. and Nemanich, R.J., Phys.Rev. B 35, 4166 (1987)
28. Bergholz, W., Hutchison, J.L. and Pirouz, P., J.Appl.Phys. 58, 3419 (1985)
29. Bourret, A., Thibault-Desseaux, J. and Seidman, D.N., J.Appl.Phys. 55, 825 (1984)
30. Capper, P., Jones, A.W., Wallhouse, E.J. and Wilkes, J.G., J.Appl.Phys. 48, 1646 (1977)
31. Hölzlein,K., Ph.D. thesis, University of Erlangen, 1985
32. Hölzlein, K., Pensl, G. and Schulz, M., Appl.Phys. A 34, 155 (1984)
33. Hölzlein, K., Pensl, G., Schulz, M and Johnson, N.M., Mat.Res.Soc.Symp.Proc. 59, 481 (1986)
34 Hölzlein, K., Pensl, G., Schulz, M. and Johnson, N.M., Appl.Phys. Lett. 48, 916 (1986)

INTERFACE PROPERTIES OF InP-DIELECTRIC SYSTEMS

R. BLANCHET

Laboratoire d'Electronique, Automatique et Mesures Electriques
de l'Ecole Centrale de Lyon, U.A. (C.N.R.S.) n°848,
36, av. de Collongue, B.P. 163, 69131 Ecully Cedex, France

The development of appropriate passivation schemes for III-V compound semiconductors is of considerable interest for application to high speed, high power and electro-optic devices and integrated circuits. In this paper, special attention is given to InP in connection with insulated gate technology for MISFET's applications.

INTRODUCTION

The potential of III-V compound semiconductors is of considerable interest for application to microwave power generation, high speed logic circuits, and integrated optoelectronics. The demonstration of the potential of InP for the fabrication of inversion mode metal-insulator-semiconductor field effect transistors (MISFET's), first reported by Lile et al. [1], has been confirmed since by many authors and for a wide variety of technological approaches. Despite these promising results, it appears that several outstanding problems remain to be solved [2], the most serious of these being the propensity of InP to decompose readily with a preferential loss of phosphorus ; the dielectric deposition has also been shown to induce surface defects which are deemed to be phosphorus (P) vacancies. The trends which are generally observed are that low temperatures (down to room temperature) of dielectric deposition increase device transconductance and decrease the density of fast interface state [2-4]. However, the gain is often achieved at the expense of an inferior quality of the dielectric [3]. Also, exposure of InP substrates to plasma must be avoided to minimize radiation damages [5]. In most cases, surface damages consist of phosphorus (P) vacancies which result in an increased density of fast surface states near the conduction-band edge. More important is the unresolved problem of long-term stability which has plagued InP metal-insulator-semiconductor (MIS) systems and has coutinued to resist investigators. The origin of drift phenomena has not been elucidated so far although some recent investigations seem to prove they are related to the presence of P vacancies at the interface [6-8]. It has long been recognized that a P overpressure can prevent a disproportionate loss of P in technological processes involving heating of InP substrate at high temperature (for example, LPE growth of InP [9,10], surface preparation prior to MBE growth [11,12], etc.). More recently, it was shown that interfacial properties of Al_2O_3-InP MIS structures prepared in excess organo-phosphorus atmosphere are significantly improved [13] ; Pande et al. reported high channel mobilities in metal-SiO_2-semiconductor field effect transistors using a P-rich interfacial oxide between the SiO_2 gate dielectric and InP substrate [7,14].

We have undertaken a systematic study on InP surface treatments which are based on the use of one of the elements of column V of the periodic table (N, P, As, Sb [15] or Bi) to compensate phosphorus vacancies. In this paper, we present as examples, the recent results obtained with arsenic and phosphorus.

Furthermore, other important problems have been studied by our group with the aim of improving the electronical properties of InP MISFET's : the results of the effect of native oxide and ion beam techniques on the interfacial properties of InP-insulator system are summarized. The ion beam technique has been used for the treatment of the InP surface and for the low energy ion beam enhanced deposition of various insulators.

Finally we give a few informations on the various methods of characterization which have been developed in our group for a tight control of the technological steps.

The devices are processed on n-type (100) oriented, polished InP substrates obtained from MCP, with a carrier density of about 3×10^{15} cm^{-3}. InP substrates are first cleaned in conventional organic solvents and rinsed in deionized water. After a simple etching in a (49%) HF aqueous solution to remove most of the native oxide, the samples are submitted to one of the investigated passivation treatments. The Al$_2$O$_3$ insulator (1000 Å thick) was deposited simultaneously onto the treated samples and on reference InP substrates (without any treatment), using electron gun evaporation of Al$_2$O$_3$. The substrates are held at room temperature during the deposition. The details of the deposition precedure are given elsewhere [16]. We obtained good quality insulators : the resistivity was about 10^{15} Ω cm and the dielectric strength reached 5×10^6 V/cm. Then MIS structures are formed using a mercury probe for the characterization of the insulator-semiconductor interface states via capacitance-voltage (C-V) measurements, performed after annealing the structures in oxygen at 325 °C for the 2 h (this last step allowed for a better reproducibility of the results).

ARSENIC TREATMENT OF InP

It has been demonstrated [17-19] that thermal treatment of InP substrates under an arsenic (As) molecular-beam exposure results in a chemically stable surface up to temperatures as high as 510°C. Although this temperature is much higher than the temperature where noncongruent evaporation of InP takes place, a significant improvement of electronical properties was shown in corresponding MIS devices [19]. Recently, another method for the As treatment of the InP surface has been developed. Consequently, in this work, the InP substrates were submitted to one of the following two treatments :

(i) Introduction in the vacuum chamber (MBE Riber system : base vacuum 10^{-9} Torr) and heating at a temperature between 500 and 600°C for 10 min under an As overpressure varying from 10^{-6} to 10^{-4} Torr. The As overpressure is maintained during the cooling down of the substrate. The excess free As deposited on the surface is subsequently evaporated by heating the substrate in vacuum at various temperatures for 5 min.

(ii) Etch in H$_3$AsO$_4$ solution (arsenic acid 80%) for 5 min in darkness at room temperature [20] followed by a rinse in deionized water.

The surface composition was analyzed by Auger electron spectroscopy (AES) as a function of the annealing temperature of the substrate which was controlled by infrared thermometry. After MBE As treatment, the decrease of the As peak intensity, probably associated with the desorption of As, is accompanied by out-diffusion of P from the bulk towards the surface which results in the observed increase of the P peak intensity. In samples which have not been subjected to the As treatment, a similar increase of the P peak intensity is observed at significantly lower temperatures (about 150°C below). It is followed by a loss of P which appears at temperatures below 450°C in reference samples [21].

In order to optimize the As treatment, we investigated in detail the influence of the As partial pressure and of the substrate temperature during the As treatment : they spanned in the ranges $10^{-6} - 10^{-4}$ Torr and 500-600°C, respectively. Also the temperature of vacuum annealing subsequent to the As treatment was varied from room temperature to 500°C. Interface properties were found not to improve any further with the increase in the temperature and As partial pressure above than 510°C and 2×10^{-6} Torr, respectively. Post-treatment annealing experiments for this last set of substrate temperature and As partial pressure yielded the best results for annealing temperatures of the substrate in the range 300-350°C. This is not surprising since, as can be observed from Auger spectra, most of the unbonded free As is evaporated above 300°C. After annealing in vacuum at 500°C, the electrical properties of the surface remained superior to those of reference samples not annealed in vacuum.

AES spectroscopy shows that As is indeed present on the InP surface after the chemical treatment in H_3AsO_4 solution, but in a smaller amount. Furthermore, the surface chemical composition remains stable up to about 430°C. The desorption of As and the out-diffusion of P from the bulk towards the surface are not observed.

The C-V characteristics plotted at 1,10 and 100 kHz for a reference sample (HF etched) and for a sample chemically or MBE treated with As are shown in Figs 1(a) and 1(b), respectively. It is clearly observed that the As treatment results in a much steeper variation of the capacitance, which indicates that the density of surface states is reduced throughout the band gap. The low carrier density 3×10^{15} cm^{-3} of the substrates allows for the observation of large capacitance swings, which makes the Terman technique more reliable for the determination of the density-of-state distribution in the band gap. Figure 2 (curve b) shows the distribution obtained from C-V plot at 1 MHz for an As treated sample compared with the curve corresponding to a reference sample (curve a). Although the Terman technique must be handled with some care for the determination of the density-of-state distribution in III-V compound semiconductors [22,23], qualitative information is the least it can provide when used in a pertinent way. The As treatment results in a large reduction of the density of surface states (Fig. 2). Pinning of the Fermi-level, due to the steep increase of the density of states toward (from about 1 eV above) the valence-band edge, observed in the reference sample, is considerably reduced after the treatment. A reminiscent peak of states appears to remain at about 1 eV above the valence-band edge. The peak position corresponds to that commonly deemed to be due to phosphorus vacancies [24]. It appears then that, although not fully eliminated, the phosphorus vacancies have been considerably reduced by the As treatment. Similar observations were reported recently by Yamaguchi et al. who fabricated MIS structures on InP with PAs$_x$N$_y$ as the insulator [25]. The drift behavior of the devices was investigated systematically by subjecting them to various bias stresses. The experiment consisted of recording the 100-kHz C-V characteristics at a fast sweeping rate (10 V/s) after stays of different durations (0.1 s, 1 s, 10 s, 100 s) in accumulation regime [19]. The shift of the characteristics is considerably reduced after the As treatment.

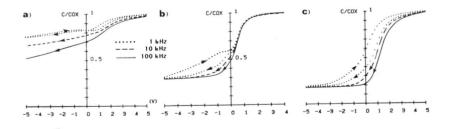

Fig. 1- C-V plots of InP MIS structures : (a) HF treated reference sample ; (b) chemical or MBE As treatment of InP , (c) in situ treatment of InP substrate in phosphorus partial pressure at 480°C before alumina deposition. The hysteresis is only shown on the 1 kHz curve.

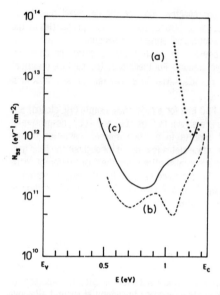

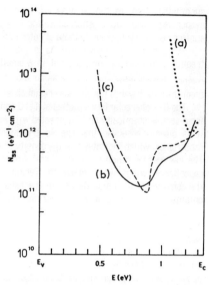

Fig. 2- Approximate density of state distributions obtained from Terman analysis : (a) HF treated reference sample ; (b) chemical or MBE As treatment of InP ; (c) in situ treatment of InP substrate in phosphorus partial pressure at 480°C before alumina deposition.

Fig. 3- Approximate density of state distributions obtained from Terman analysis : (a) HF treated reference sample ; (b) pretreatment of InP substrate in phosphorus partial pressure at 480°C and deposition at 100°C in the same P-overpressure ; (c) deposition of alumina at 100°C in P-overpressure without pretreatment of the InP substrate at 480°C.

PHOSPHORUS TREATMENT OF InP

The aim of the P treatment was the same as that of As, namely, to compensate the P vacancies in the InP-insulator interfacial region.

In this communication we present [26] the effect of (i) the cleaning of InP substrate in P overpressure prior to in situ deposition of the dielectric, (ii) the presence of P overpressure during the first stages of the dielectric deposition.

Experiments are carried out in a conventional vacuum chamber evacuated with a diffusion pump (base vacuum : 2×10^{-7} Torr). Heating of high-purity polycrystalline InP (Cambridge instrument : total impurity content $< 10^{16}$ cm^{-3}) in a modified Knudsen cell is used to generate phosphorus overpressure.

Precleaning of InP surface is done at 480°C in a P partial pressure of 10^{-5} Torr for 10 min. prior to dielectric deposition. This treatment proves to eliminate the poorly controlled native oxide left by the HF chemical etching, without degradation of InP surface. Al_2O_3 is deposited in the same chamber, in P overpressure (10^{-5} Torr) during the beginning of the deposition, at 100°C in order to prevent P condensation on InP substrate. Figure 1(c) shows C-V characteristics versus frequency (1,10 and 100 kHz) of MIS structures. The comparison with C-V curves of a reference sample (Fig. 1(a)) shows that the strong pinning of the surface Fermi level observed in the reference sample is suppressed for the P-treated sample. C-V curve analysis [22,23] leads to significant reductions of interface state density (N_{ss}) near midgap in P-treated samples (Fig. 2). The steep increase of N_{ss} toward the valence band is not any more observable, although the density of states in the vicinity of the conduction-band edge seems to be slightly enhanced. This effect of excess phosphorus in the processing of MIS structures on InP can be considered as general since similar results have been obtained by Kobayashi et al., using a different technique of preparation (Al_2O_3-InP devices prepared in excess of organophosphorus atmosphere) [13].

To distinguish the effect of the precleaning in P overpressure at 480°C from the effect of the presence of P overpressure during the initial stages of the dielectric deposition, we omitted the precleaning step. Figure 3(c) shows the typical density of state distribution of MIS devices whose Al_2O_3 was deposited in P overpressure at 100°C without precleaning. A comparison of this curve (fig. 3(c)) with the curves of Fig. 3(a) and (b) shows that the strength of the Fermi-level pinning in the upper part of the gap in the former is somewhere between that of the reference sample and of the fully P-treated device. It can be concluded that the P overpressure provides an efficient protection of InP surface against degradations induced by the deposition of the dielectric, although it cannot fully suppress Fermi-level pinning.

Precleaning in P overpressure of InP substrates prior to the insulator deposition results in MIS structures with unpinned Fermi level. P overpressure provides also an efficient protection of InP surface during the dielectric deposition, in that it may prevent the formation of P vacancies.

DISCUSSION ON THE As AND P TREATMENTS OF InP

The presence of excess of P during some technological steps is favourable for the realization of InP MIS structures with good electrical properties but the resultats obtained with As, which enhance the stability of the InP surface, seem to be more attractive (cf. Fig. 2). Arsenic treatment is likely to result in the compensation of phosphorus vacancies and/or in the elimination of weakly bonded phosphorus [27]. The effect of InP treatment in an As partial pressure has been recently interpreted as an epitaxial regrowth of an InAs surface on InP [28]. The deposition of alumina on the reference sample may result in the formation of additional phosphorus vacancies, weakly bonded phosphorus being kicked off the surface by impinging alumina molecules. With As atoms being more tightly attached to the InP surface, the deposition of alumina cannot possibly remove them and hardly produces additional phosphorus vacancies. Most of our results have been obtained after electron gun deposition of alumina in a vacuum system different from the As treatment system. This implies that the short intermediate stay of the As treated sample in the outer atmosphere does not lead to any observable deterioration. However, surface arsenic is, in this case, in part oxidized as shown by XPS results [29]. A complete in situ experiment in ultra high vacuum should allow to better understand the effect of arsenic on the electronical properties of MIS structures.

Besides, the InP surface passivation problem with the elements of column V, other important problems have also been studied by our group with the aim of improving the electronical properties of InP MISFET's. We summarize below the essential results obtained on the effects of InP native oxide and of ion beam techniques developed for the treatment of the InP surface, and for the low energy ion beam enhanced deposition of insulators.

EFFECT OF NATIVE OXIDES

The role of native oxides in the insulator-InP interface properties has long been the matter of dispute in the literature. For example, the interface properties of the" Al_2O_3 -native oxide- InP" structures obtained by anodic oxidation of an aluminium layer are clearly improved [30]. It has been shown in this case that the interface oxide is phosphorus rich [31].

Recently, it has been demonstrated in our group that the passivating native oxides can be prepared by a direct anodic oxidation of InP [32-34].

These specific native oxides are identified as condensed phosphates ($In(PO_3)_3$) using XPS chemical depth profiles. The growth and etching of anodic oxides are monitored in situ using spectroscopic ellipsometry. It is suggested that these phosphorus-rich glass-like $In(PO_3)_3$ oxides, which give high quality MIS structures, have better intrinsic passivating properties than $InPO_4$.

DEVELOPMENT OF ION TECHNIQUES

Ion beam techniques have been developed in our group for the treatment of InP surface and for low energy ion beam enhanced deposition of insulators. Large enhancement of photoluminescence intensity emitted by (100) oriented InP surface after exposure to activated hydrogen (low energy (20 eV) hydrogen ion beam), indicating sharp reduction of surface recombination rate was demonstrated in this work [35]. It is shown, using capacitance-voltage measurement performed on metal-insulator-semiconductor structures, that hydrogen treatment results also in a strong pinning of the interface Fermi level. These phenomena are interpreted on the following basis : (i) neutralization by hydrogen species of active recombination centers which are deemed to be phosphorous dangling bonds ; (ii) creation of pinning surface states due to the formation of volatile phosphorus hydride compounds freeing excess indium atoms.

AlN_xO_y alloy films have been deposited on InP by low energy (20 eV) ion beam enhanced reactive evaporation of aluminium in ultra-high vacuum [36]. The resistivity of the films ranges between 10^{15} and 10^{16} Ω cm and the dielectric strength exceeds 10^6 V/cm. The energy of the ions is kept below the threshold energy (found to be in the range 20 eV - 100 eV) beyond which ion bombardment of InP surface results in irreversibly degraded electronic properties. Despite this precaution, further improvement of interface properties (implying lowering of ion energy and appropriate post-annealing schemes of the structures) are required for a full development of this technique, which is compatible with the Molecular Beam Epitaxy technology.

CHARACTERIZATION TECHNIQUES

Several methods of characterization have been developed in our group for a better control of the various fabrication techniques. Besides XPS-UPS photoelectron spectroscopies, which provides a microscopic description of interfaces and dielectrics, use is made of photoluminescence measurements, which give information on recombination processes at surfaces and interfaces and allow for a non invasive assessment of their quality at various stages of their preparation [37,4]. The intensity of the photoluminescence is measured in situ in ultra high vacuum to monitor the formation of interfaces. Also photoluminescence spectra reveal specific features in connection with specific technological steps. Finally scanning photoluminescence microscopy [38] is performed to obtain informations on the uniformity of surfaces and dielectric-semiconductor interfaces ; there is a large dependence of the morphology of the PL images on chemical treatment and annealing of InP samples. Internal photoemission experiments were also carried out in our group for the determination of dielectric-semiconductor barriers and to investigate carrier trapping phenomena in the interface region. Finally, results of such electrical characterization techniques as capacitance and transient current techniques as low noise measurements on MISFET's have been used for the investigation of interface state properties and drift phenomena.

ACKNOWLEDGMENTS

The author is grateful to the "InP-group" members of the "Laboratoire d'Electronique" and the "Laboratoire de Physico - chimie des Interfaces" of the "Ecole Centrale de Lyon" for the stimulating discussions and results. The financial support of the CNRS, the DAII and CNET (French Telecommunications, present contract : 86.1B.156.00.790.92.45.PAB), the "Rhône-Alpes" Region and the French Ministry of Education and Research in conducting this work is gratefully acknowledged.

REFERENCES

[1] D.L. LILE, D.A. COLLINS, L.G. MEINERS and L. MESSICK, Electron. Lett.,**14**, 657 (1978).
[2] D.L. LILE, J. Vac. Sci. Technol. **B2**, 496 (1984).
    P. VIKTOROVITCH, Le Vide, Les Couches Minces (supplément) n°226, 213 (1985).
[3] M.J. TAYLOR, D.L. LILE and A.K. NEDOLUHA, J. Vac. Sci. Technol. **B2**, 522 (1984).
[4] B. SAUTREUIL, P. VIKTOROVITCH and R. BLANCHET, J. Appl. Phys., **57**, 2322 (1985).
[5] L.G. MEINERS, Thin Solid Films, **113**, 85 (1984).
[6] J.A. VAN VECHTEN and J.F. WAGER, J. Appl. Phys., **57**, 1956 (1985).
[7] K.P. PANDE and D. GUTIERREZ, Appl. Phys. Lett., **46**, 416 (1985).
[8] Y. FURUKAWA, Jpn. J. Appl. Phys., **23**, 1157 (1984).
[9] W.Y. LUM and A.R. CLAWSON, J. Appl. Phys., **50**, 5296 (1979).
[10] G.A. ANTIPAS, Appl. Phys. Lett., **37**, 64 (1980).
[11] R.F.C. FARROW, J. Phys., **D7**, 2436 (1974).
[12] J.S. ROBERTS, P. DAWSON and G.B. SCOTT, Appl. Phys. Lett. **38**, 905 (1981).
[13] T. KOBAYASHI, T. ICHIKAWA, K. SAKUTA and K. FUJISAWA, J. Appl. Phys., **55**, 3876 (1984).
[14] K.P. PANDE, M.A. FATHIMULLA, D. GUTIERREZ and L. MESSICK, IEEE Electr. Devices Lett., **EDL 7**, 407 (1986).
[15] S.N. KUMAR, private communication.
[16] R. BLANCHET and C. SANTINELLI, J. Vac. Sci. Technol., **A4**, 1948 (1986).
[17] G.J. DAVIES, R. HECKINGBOTTOM, H. OHNO, C.E.C. WOOD and A.R. CALAWA, Appl. Phys. Lett., **37**, 290 (1980).
[18] K.Y. CHENG, A.Y. CHO, W.R. WAGNER and W.A. BONNER, J. Appl. Phys., **52**, 1015 (1981).
[19] R. BLANCHET, P. VIKTOROVITCH, J. CHAVE and C. SANTINELLI, Appl. Phys. Lett., **46**, 761 (1985).
[20] J. CHAVE, A. CHOUJAA, C. SANTINELLI, R. BLANCHET and P. VIKTOROVITCH, J. Appl. Phys., **61**, 257 (1987).
[21] J. CHAVE, C. SANTINELLI, R. BLANCHET and P. VIKTOROVITCH, in Insulating Films on Semiconductors, edited by J.J. Simonne and J. Buxo (North-Holland, Amsterdam, 1986), p. 89.
[22] B. BAILLY, Thesis, Ecole Centrale de Lyon, France, n°85-02 (1985).
[23] C. SIBRAN, R. BLANCHET, M. GARRIGUES and P. VIKTOROVITCH, Thin Solid Films, **103**, 211 (1983).
[24] W.E. SPICER, P.W. CHYE, P.R. SKEATH, C.Y. SU and I. LINDAU, J. Vac. Sci. Technol. **16**, 1422 (1979).
[25] E. YAMAGUCHI, Y. HIROTA and M. MINAKATA, Thin Solid Films, **103**, 201 (1983).
[26] R. BLANCHET, P. VIKTOROVITCH, A. CHOUJAA, J. CHAVE et C. SANTINELLI, Revue Phys. Appl., **22**, 279 (1987).
[27] C. SANTINELLI, J. CHAVE, R. BLANCHET and P. VIKTOROVITCH, 168th Electrochemical Society Meeting, Las Vegas, October 13-18 (1985).
[28] J.M. MOISON, M. BENSOUSSAN and F. HOUZAY, Phys. Rev., **B34**, 2018 (1986).

268

[29] The XPS analysis carried out after a few hours showed that about 85 % of the arsenic was in the oxide form (G. Hollinger and E. Bergignat, private communication). These results have also been confirmed by the ellipsometry studies (M. Juvin, Thesis, Ecole Centrale de Lyon, France, n°85-08, 1985). In the case of the As chemical treatment, the XPS analysis showed that pratically all of the arsenic was in the oxide form.

[30] T. SAWADA, S. ITAGAKI, H. HASEGAWA, H. OHNO, IEEE Trans. Electr. Devices **ED 31**, 1038 (1984).

[31] E. JALAGUIER, Y. ROBACH, J. JOSEPH, E. BERGIGNAT, G. HOLLINGER, to be published in Revue Phys. Appl.

[32] G. HOLLINGER, E. BERGIGNAT, J. JOSEPH and Y. ROBACH, Journal of Vacuum Science & Technology, **A3**, 6, 2082 (1985).

[33] Y. ROBACH, J. JOSEPH, E. BERGIGNAT, B. COMMERE, G. HOLLINGER and P. VIKTOROVITCH, Appl. Phys. Lett., **49**, 1281 (1986).

[34] G. HOLLINGER, J. JOSEPH, Y. ROBACH, E. BERGIGNAT, B. COMMERE, P. VIKTOROVITCH and M. FROMENT, to be published.

[35] P. VIKTOROVITCH, F. BENYAHIA, C. SANTINELLI, R. BLANCHET, P. LEYRAL and M. GARRIGUES, to be published.

[36] C. SANTINELLI, R. BLANCHET, P. VIKTOROVITCH and F. BENYAHIA, INFOS 87, Leuven, Belgium, 13-15 April 1987, Applied Surface Science, North-Holland, Amsterdam.

[37] S. KRAWCZYK, B. BAILLY, B. SAUTREUIL, R. BLANCHET and P. VIKTOROVITCH, Electron. Lett., **20**, 255 (1984).

[38] S. KRAWCZYK, M. GARRIGUES and H. BOUREDOUCEN, J. appl. Phys., **60**, 392 (1986).

RECENT ANALYTICAL EFFORTS TO MODEL MINORITY CARRIER TRANSPORT IN MODERATELY TO HEAVILY DOPED SEMICONDUCTOR REGIONS

C. R. Selvakumar

Department of Electrical Engineering, University of Waterloo, Ontario, Canada.

Abstract

A brief analysis of minority carrier transport across a region that is moderately to heavily and non-uniformly doped is presented taking into account position-dependent quantities such as mobility, bandgap narrowing and lifetime (SRH and Auger) by invoking the complete continuity equation. The relative influences of net electric field, mobility and their gradients and the lifetime profile on the injected minority carrier profile are evaluated. Based on the conclusions of this study a simple general analytical solution to the minority carrier transport equations in an arbitrarily doped region with and without external generation term (source term) is developed. The usefulness of the present analytical approach is briefly discussed.

1. Introduction

Evaluation of minority carrier current injected into a semiconductor region that is moderately to heavily doped and which has a non-uniform impurity doping profile is an often encountered and one of the most difficult problems to solve. This transport problem is a highly complex one because there are four major effects that strongly influence the flow of minority carriers in the non-uniformly doped region, namely (1) position-dependent bandgap narrowing (BGN), (2) position-dependent Shockley-Read-Hall (SRH) recombination and Auger recombination, (3) varying net electric field [1] and (4) varying mobility [1-2]. In addition, since some of the recently evolving compound semiconductor IC structures have heterojunction bipolar transistors (HBTs) with optically coupled double heterostructure (DH) laser devices suitable for optical integration (for e.g. [10]) the injected minority carrier current into such a non-uniformly doped region should include optically generated current as well. Because of the complexity of this transport problem, only a numerical solution of the complete system of equations typically using a large computer is considered to give sufficiently accurate results. In this paper we present a recently developed analytical approach and certain results obtained using this method. Before we set out to develop an analytical solution to the transport problem we briefly discuss the relative influences of various terms in the complete continuity equation.

2. Importance of Net Electric Field, Mobility and Their Gradients in Non-uniformly Doped Region

For the ease of discussion, let us consider a non-uniformly (with arbitrary doping profile) doped n-type semiconductor region such as an emitter of a bipolar transistor or a solar cell. It is known from simulations, as was recently shown by Adler [3], that the net electric field in a typically doped quasi-neutral emitter region is essentially same as that in thermal equilibrium conditions even when a large current, as large as 10^5 Ampere/cm ^{2}, flows through such a region. This implies that we can essentially ignore Poisson equation in the quasi-neutral region that is moderately to heavily doped with typical non-uniform doping profile. In other words one can assume low level injection conditions in such a region virtually in all practical situations. Therefore we need to solve only the minority carrier continuity equation in a region. The complete continuity equation for excess holes $p(x)$ in an n-type region is given by [1,4].

$$D_p \frac{d^2 p}{dx^2} + \left[\frac{dD_p}{dx} - \mu_p E_p \right] \frac{dp}{dx} - \left[E_p \frac{d\mu_p}{dx} + \mu_p \frac{dE_p}{dx} + \frac{1}{\tau_p} \right] p = 0 \qquad (1)$$

where the various symbols have the usual meaning and E_p is the net field which depends on bandgap narrowing and doping gradients as follows

$$E_p(x) = V_T \left[\frac{1}{n_{ie}^2} \frac{dn_{ie}^2}{dx} - \frac{1}{N} \frac{dN}{dx} \right] \qquad (2)$$

where n_{ie} is the effective intrinsic carrier density and N is donor density. Many authors (for eg. [5],[6]) in the past have made assumptions to the effect that the net field in the emitter is essentially zero and the gradients of mobility or diffusivity can also be assumed to be negligible. This reduces (1) to the following simple equation.

$$\frac{d^2 p}{dx^2} - \frac{p}{D_p \tau_p} = 0 \qquad (3)$$

We have computed the injected minority carrier profiles resulting from the numerical solution of (1) and (3) for a complementary error function (erfc) doped silicon emitter having a thickness of 0.25 μm with a surface doping $N_S = 10^{20}$cm^{-3} and background doping $N_B = 5 \times 10^{17}$cm^{-3}. We have assumed that the emitter surface recombination velocity for holes $S_p = 10^3$cm.s^{-1}. We have used Slotboom's empirical lifetime model which closely matches the experimentally measured lifetimes in silicon [7], and Arora-Hauser-Roulston (AHR) mobility model [17] in solving equations (1) and (3). The results are shown in Fig. 1. As can be seen from Fig. 1, the injected minority carrier charge calculated using (3) is substantially larger than that calculated from (1). The injected current calculated using (3) is about 300% in excess compared with that calculated by solving the complete continuity equation (1). We observe that although we have included position dependent lifetime and diffusivity as in (3), we have obtained unrealistically large injected charge which is mainly due to the neglect of position-dependent net field and its gradients as in (1).

Next, we wish to evaluate the relative magnitudes of various terms in (1) in order to understand which are the factors that influence the injected minority carrier distribution most. Let us rewrite (1) as shown below

$$A(x)\frac{d^2p}{dx^2} + \left[B_1(x) - B_2(x)\right]\frac{dp}{dx} - \left[C_1(x) + C_2(x) + C_3(x)\right]p = 0 \qquad (4)$$

where the newly defined functions are obtained by correspondence with (1). We now specifically compare the relative magnitudes of $C_1(x) = E_p\dfrac{d\mu_p}{dx}$, $C_2(x) = \mu_p dE_p/dx$ and $C_3(x) = 1/\tau_p(x)$ in a typical emitter. Let us consider a silicon emitter with a doping profile $N(x) = N_s erfc\ (x/L)$ where N_S is the surface doping density and L is the characteristic length. Assuming that Slotboom and DeGraaff [8] bandgap narrowing (BGN) model holds good and using Slotboom's empirical lifetime model [7] and AHR mobility model, we have calculated $C_1(x)$, $C_2(x)$, $C_3(x)$ and their sum $C(x)$ are plotted in Fig. 2. We can see that the lifetime term $C_3(x) = 1/\tau_p(x)$ contributes virtually nothing to $C(x)$ and hence does not significantly influence the solution of (1). We have found that the lifetime profile $\tau_p(x)$ **does not** influence the injected minority carrier profile $p(x)$ in any significant way for real emitters as thick as 2 μm. The total injected charge in a 10 μm gaussian emitter is affected by less than 15% when infinite lifetime is used instead of realistic lifetime profiles [9]. Although lifetime does not significantly affect the injected minority carrier profile, it **does** strongly influence the injected minority carrier current because the injected current depends on the integral of the ratio $p(x)/\tau_p(x)$. To illustrate that $p(x)$ is not significantly influenced by the choice of $\tau_p(x)$ model, we compute $p(x)$ by solving (1) in a relatively thick emitter of 2

μm using two different lifetime models, namely a) $\tau_p(x) = \infty$ and b) $1/\tau_p(x) = 2.25 \times 10^{-19} N^{1.36}$. The results are shown in Fig. 3. As can be seen from Fig. 3 the excess minority carrier profile in the emitter is affected very little by this vastly different lifetime models even in such a thick emitter. We conclude from the above exercises that 1) the net field and its gradient have a strong influence on the excess minority carrier profile and hence can not be neglected while solving the continuity eqution and 2) the lifetime profile does not significantly influence the excess carrier profile although the excess minority carrier **current** is strongly infuenced by lifetime. In the following sections we derive analytical solutions to the minority carrier continuity equations without and with external generation of carriers that takes into account position-dependent quantities such as lifetime (SRH and Auger), bandgap narrowing (BGN) and mobility.

3. Analytical Solution to the Continuity Equation Without External Generation Rate Term.

Let us consider an n-type region having an **arbitrary** doping profile $N(x)$. We wish to write the current and continuity equation in terms of a normalised variable. Let us normalise the excess minority carrier density $p(x)$ by its thermal equilibrium density $p_o(x)$ and call it $u(x)$

$$u(x) = p(x)/p_o(x) \qquad (5)$$

In terms of $u(x)$ the transport equations assume the following simpler forms [11]

$$J_p(x) = -qD_p p_o \frac{du}{dx} \qquad (6)$$

$$\frac{1}{q} \frac{dJ_p}{dx} = \frac{-p_o u(x)}{\tau_p(x)} \qquad (7)$$

The above two equations can be solved analytically if there is a particular mutual relationship between the transport parameters and **indeed there is one**. Minority carrier lifetime τ_p is found to be related to diffusivity and thermal equilibrium minority carrier density as shown Ref. 11. Recently Del Alamo and Swanson have experimentally measured [12] the quantity $(D_p p_o)$ and diffusion length L_p as functions of doping density. Using their experimental values we have computed [13] the ratio $D_p p_o / L_p$ as a function of doping density, N and the results are shown in Fig. 4. One can observe that this ratio is essentially a constant ($= C_S$). Since the measurements of [12] were done at 292° K, we have calculated the value of C_S at 292° K merely from interpolations of values of C_S at 300°K and 290°K given in [1], and we find that it fits the results calculated from [12] quite well. Assuming

$$\frac{D_p p_o}{L_p} = C_s \tag{8}$$

we can obtain the continuity equation as

$$\frac{d^2 u}{dx^2} + \frac{1}{L_p}\frac{dL_p}{dx}\frac{du}{dx} - \frac{u}{L_p^2} = 0 . \tag{9}$$

One can show that the above equation has a simple general solution as

$$u(x) = A\sinh Ki(x) + B\cosh Ki(x) \tag{10}$$

where the dimensionless quantity $Ki(x)$ is give by

$$Ki(x) = \int\limits_0^x \frac{d\lambda}{L_p(\lambda)} \tag{11}$$

The arbitrary constants A and B in (10) can be easily evaluated from known boundary conditions. The most important feature of (10) is that it is entirely similar to the simple text-book solutions obtained for a uniformly doped region except that x/L_p is replaced by the integral in (11). Thus one can show that the excess electron density distribution $n(x)$ in an arbitrarily graded p-type base is given by

$$n(x) = n_o(x)\frac{\sinh[Ki(W_B) - Ki(x)]}{\sinh Ki(W_B)} \tag{12}$$

where n_o is thermal equilibrium electron density in the p-type base, W_B is the width of the quasi-neutral base; similarly the base transport factor α_T is given by

$$\alpha_T = [\cosh Ki(W_B)]^{-1} \tag{13}$$

and if $Ki(W_B)$ is small one can write

$$\alpha_T = 1 + \frac{1}{2}[Ki(W_B)]^2 \tag{14}$$

As can be readily verified, for the case of a uniformly doped region equations (10) - (14) reduce to the corresponding text-book expressions. Various results calculated from the present analytical solution have been recently compared with a more rigorous computer simulation using BIPOLE program and have been found to be remarkably accurate [14] in most of the cases.

4. Solution in the Presence of a Non-thermal Source Term

The complete continuity equation in terms of the normalised excess minority carrier density $u(x)$ including the non-thermal generation rate $G(x)$ can be shown to be

$$\frac{d^2u}{dx^2} + \frac{1}{L_m}\frac{dL_m}{dx}\frac{du}{dx} - \frac{u}{L_m^2} = -\frac{G(x)}{C_S L_m} \tag{15}$$

where the subscript m assumes p for holes and n for holes and as well in the expression $u = m/m_o$. In order to evaluate the accuracy of the analytical solution, let us set out to solve (15) in the emitter of a solar cell and let the emitter be assumed to have an arbitrary doping profile. The general solution of (15) is then given by

$$u(x) = [A_0(x) + A_1] + \sinh Ki(x) + [B_0(x) + B_1]\cosh Ki(x) \tag{16}$$

and the minority carrier current is given by

$$|J_m(x)| = qC_S[\{A_0(x) + A_1\}\cosh Ki(x) + \{B_0(x) + B_1\}\sinh Ki(x)] \tag{17}$$

where the arbitrary constants A_1 and B_1 can be evaluated from the standard boundary conditions:

$$u(W) = 0 \tag{18}$$

$$J_m(0) = qS_m m_0(0)u(0) \tag{19}$$

where S_m is the recombination velocity of the minority carriers at the emitter contact $(x = 0)$ and the junction is assumed to be at $x = W$. The functions $A_0(x)$, $B_0(x)$ and the constants A_1 and B_1 are given by

$$A_0(x) = -\frac{1}{C_S}\int_0^x G(\lambda)\cosh Ki(\lambda)d\lambda \tag{20}$$

$$B_0(x) = \frac{1}{C_S}\int_0^x G(\lambda)\sinh Ki(\lambda)d\lambda \tag{21}$$

$$A_1 = -\frac{A(W)\sinh Ki(W) + B(W)\cosh Ki(W)}{\sinh Ki(W) + S_F\cosh Ki(W)} \tag{22}$$

$$B_1 = S_F A_1 \tag{23}$$

and the two **dimensionless** parameters, Ki number and the Surface Factor S_F, are given by (11) and (24) respectively

$$S_F = \frac{D_m(0)}{S_m L_m(0)} \tag{24}$$

Now we make some specific calculations using the expression in (17) pertaining to silicon solar cells to show the accuracy of the present analytical approach. Assuming gaussian doping in the n-type emitters we calculate the internal quantum efficiencies in emitters of different thicknesses for AM1 illumination. Our calculations are compared with 'exact' computer calculations of Del Alamo and Swanson [15] and are presented in Fig. 5. We have used AM1 illumination rates given in [16]. We have assumed that the temperature $T = 300\,^\circ$K and $C_S = 3.16 \times 10^6$ cm^{-1}s^{-1}. For the analytical computations we have used the Slotboom-DeGraaff BGN model [8] and the Arora-Hauser-Roulston mobility model [17]. As can be seen from Fig. 5 our simple analytical calculations give remarkably close results to the more elaborate 'exact' calculations.

5. Conclusion

By simple numerical solution of the complete continuity equation, the influences of net field and its gradients are shown to be important and hence was seen that they can not be neglected without introducing large errors in the calculation of minority carrier profiles in non-uniformly doped quasi-neutral regions. Similarly the interesting role played by the lifetime in not influencing the injected minority carrier profile and yet strongly influencing the injected current was recognised. Next, by recognising an empirical relationship between transport parameters, the transport equations were solved **analytically** that properly takes into consideration the various position-dependent effects in an arbitrarily doped semiconductor region (Si, GaAs, and InGaAsP.). The analytical solution was also remarkably similar in form to the simple uniformly doped region solution. Finally expressions for excess minority carrier charge and current in the presence of an arbitrary external generation source in an arbitrarily doped region were analytically developed. Its usefulness was demonstrated by calculating the internal quantum efficiency in a gaussian doped silicon emitter under AM1 illumination and comparing it with 'exact' computer solutions.

Although all the numerical results presented in this paper are for silicon devices the analytical results are more general and thus hold good for **any** semiconductor for which the special relationship as given in (8) holds. Equation (8) has been found to be valid for GaAs and InGaAsP [18] but for other semiconductors the needed transport parameters data over a range of doping densities are not readily available and hence (8) has not been verified.

References

1. C.R. Selvakumar, *Analysis of Minority Carrier Transport in Heavily Doped Semiconductors*, Ph.D thesis, Indian Institute of Technology, Madras, India (1984).

2. R. Amantea, "A new solution for minority carrier injection into the emitter of bipolar transistor." *IEEE Trans. Electron. Devices*, ED-27, 1231-1238 (1980).

3. M. S. Adler, "An operational method to model carrier degeneracy and bandgap narrowing" *Solid-St. Electron.*, 26, 387-393 (1983).

4. C.R. Selvakumar, "New insights into minority carrier transport phenomenon in heavily doped emitters by a simple semi-analytical approach," 2nd International Workshop on Physics of Semiconductor Devices, New Delhi, p. 182 (1983).

5. W. W. Sheng, "The effect of Auger recombination on the emitter injection Efficiency of Bipolar Transistors", *IEEE Trans. Electron Devices, ED-22*, 25-27 (1975).

6. J. G. Fossum and M. A. Shibib, "An analytical model for minority-carrier transport in heavily doped regions of silicon devices", *IEEE Trans. Electron Devices, ED-28*, 1018-1025, (1981).

7. P.A.H. Hart, "Bipolar transistors and integrated circuits" in Handbook on Semiconductors (Ed. T. S. Moss) p. 97 (1981).

8. J. W. Slotboom and H.C. DeGraaff, "Measurements of bandgap narrowing in Si bipolar transistors", *Solid-St. Electron.*, 19, 857-862 (1976).

9. C. R. Selvakumar, BIPOLE simulation, unpublished work (1986).

10. T.R. Chen, Y.H. Zhuang, B. Chang, M.B. Yi and A. Yariv, "A Stripe-Geometry InGaAsP/InP heterojunction bipolar transistor suitable for optical integration," *IEEE Electron. Device Lett. EDL-8*, 191-193 (1987).

11. C. R. Selvakumar, "A simple general analytical solution to minority carrier transport in heavily doped semiconductors" *J. Appl. Phys.*, 56, 3476-3477 (1984).

12. J. A. Del Alamo and R. M. Swanson, Record of 18th IEEE Photovoltaic Spec. Conf. (1985).

13. C. R. Selvakumar, private communication to R.M. Swanson dated Dec. 13, 1985.

14. C. R. Selvakumar and D. J. Roulston, "A new simple analytical emitter model for bipolar transistors", *Solid-St. Electon.*, *30*, 723-728 (1987).

15. J. A. Del Alamo and R. M. Swanson, "The physics and modeling of heavily doped emitters" *IEEE Trans. Electron Devices*, ED-31, 1878-1888 (1984).

16. J. Furlan and S. Amon, "Approximation of the carrier generation rate in illuminated silicon", *Solid-St. Electron*, *28*, 1241-1243 (1985).

17. N. D. Arora, J.R. Hauser and D. J. Roulston, "Electron and hole mobilities in silicon as a function of concentration and temperature" *IEEE Trans. Electron Devices*, ED-29, 292-295 (1982).

18. C. R. Selvakumar, "A New Minority Carrier Lifetime Model for Heavily Doped GaAs and InGaAsP to obtain Analaytical Solutions", Solid-St. Electron., Vol. 30, pp. 773-774 (1987).

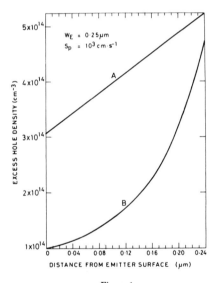

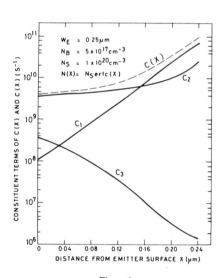

Figure 1

Calculated excess hole density distributions in a 0.25 μm emitter at a forward bias of 0.7V in two situations. Curve A is the result of neglecting net electric field, its gradient and mobility gradients, i.e. solution of (3). Curve B is the result of the solution of the complete continuity equation (1).

Figure 2

A plot of the constituent terms of $C(x)$ and their sum for a typical emitter with a width $W_E = 0.25$ μm. Note $C_3 = 1/\tau_p(x)$.

278

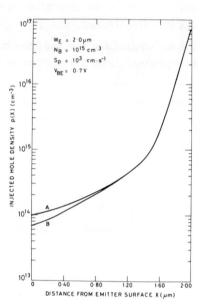

Figure 3
Injected hole density distributions in a thick emitter $(W_E = 2.0\ \mu m)$ resulting from the use of two vastly different lifetime models. Curve A is the result of using an infinite hole lifetime and Curve B is the result of using a realistic position-dependent lifetime model due to Slotboom. Emitter doping is assumed to be $N(x) = N_S\ erfc(x/L)$ and background doping is N_B.

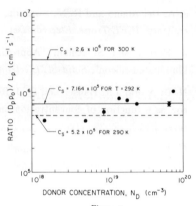

DONOR CONCENTRATION, N_D (cm^{-3})

Figure 4
The newly defined quantity $D_p p_o / L_p$ plotted as a function of donor density N_D in silicon. The product $D_p p_o$ and diffusion length L_p are experimentally determined by Del Alamo and Swanson [12]. The horizontal lines are from Ref. 1.

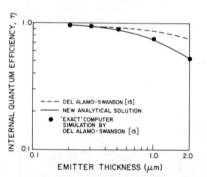

EMITTER THICKNESS (μm)

Figure 5
Comparison of the new analytical evaluation of internal quantum efficiency with the 'exact' computer solutions of Ref. 15. The dashed curve is due to the analytical method of Del Alamo and Swanson in Ref. 15. Silicon gaussian emitters with surface doping density of $N_S = 10^{20}$ cm^{-3} are assumed. Surface recombination and background p-type doping $N_B = 10^{16}$ cm^{-3} velocity for holes $S_p = 10^4$ cm.s^{-1}.

DEPOSITION AND DIRECT WRITING OF Al ON Si
BY LIGHT-ASSISTED CHEMICAL PROCESSING

J.E.BOUREE
CNRS, Laboratoire de Physique des Solides
F-92195 Meudon Cedex, France

ABSTRACT - A brief description is presented on the mechanisms of the photon-induced chemical vapor deposition. Aluminum deposition on silicon is studied as an example where pure pyrolytic decomposition of the trimethylaluminum (TMA) molecules with an Ar ion laser (514 nm) results in a high C contaminated film. On the contrary, UV heterogeneous photolysis of TMA by mercury lamp at 254 nm is shown to be effective at room temperature for controlled pre-nucleation, prior to subsequent Al deposition. Therefore, using a light-stimulated process with the two sources, Al direct writing on Si is performed from a flow of TMA diluted in hydrogen. The Al contamination free lines (50 μm wide) are obtained by varying the power of the Ar+ laser at a scanning speed of 41 μm/s. Thus an autocatalytic controlled pyrolysis is taking place, allowing a growth rate of 10 μm/s nearby the room temperature (50°C). Two point probe electrical measurements give ohmic type resistivity around 5 mΩ.cm. An alternatively improved direct writing process was obtained at room temperature with a slower scanning speed of 5 μm/s by using a high power UV laser (257 nm), resulting in a high photolytic deposition rate (700 Å/s).

1. INTRODUCTION

Lasers or more generally monochromatic optical sources have been extensively used in recent years to induce or enhance chemical interactions at gas-solid or liquid-solid interfaces. The rapid expansion of the field, as evidenced by specialist conferences (1 - 9), is driven by both the basic scientific interest and the potential technological applications in material processing and fabrication of new materials. From a fundamental point of view, the chemical reactions occurring at or near the surface can be extremely complex and a good understanding of the underlying mechanisms is desirable if photon-induced processes are to be optimised. The microelectronics industry investigates on photo-assisted processes as the numerous applications based on low temperature processing or on highly localized processing look promising : substrate cleaning, doping, oxidation, etching, deposition of thin films and recrystallization of silicon on insulators. More precisely a lot of effort has been devoted to the production of thin layers and discrete patterns of dielectrics, metals and semiconductors on a variety of substrates using coherent or incoherent optical sources whose wavelength ranges from ultraviolet (UV) to infrared (IR).

In this paper, we will concentrate solely on the formation of thin metallic films on semiconductor substrates and more specifically on the localized deposition of aluminum on silicon as well as on direct writing (process based on rastering with a focused laser beam) of Al on Si, which is a very promising technique to perform new interconnection networks in very large scale integrated

(VLSI) circuits without using mask. In the following section 2, some brief description is presented on the reaction mechanisms of the photon-induced chemical vapor deposition. The experimental arrangement as well as the characterization techniques will be described in section 3. Finally, experimental results based on different light sources will be discussed in section 4, as an illustration of the various mechanisms introduced in the second section of the paper.

2. MECHANISMS OF THE PHOTON-INDUCED CHEMICAL VAPOR DEPOSITION

Monochromatic lamp or laser irradiation of a substrate surface placed in a reaction chamber containing reactant gases can induce surface phase or gas phase reactions leading to the more or less localized growth of structure.

The two techniques for deposition can be, in first approximation, classified into thermochemical or pyrolytic deposition (10) and photochemical or photolytic deposition (11). In the former, the substrate is heated to decompose the gases above it. The mechanism is identical to that corresponding to classical CVD process, i.e. thermal activation of chemical reaction. In the latter, the gas or a weakly bound film is directly dissociated near or on the substrate following one electronic transition or many vibrational transitions in the parent molecule.

In the case of pyrolytic deposition on a semiconductor substrate, optical radiation in the visible range is strongly absorbed by the semiconductor inducing generally interband carrier excitation, whereupon the absorbed energy is transferred to the lattice by collisional de-excitation of the carriers by phonon interaction. The phonon population and lattice temperature increase according to the rate of supply of incident photons (i.e the fluence and intensity of the laser beam), the absorption coefficient of the material for the wavelength of radiation used and the various scattering times of the excited carriers as well as the thermal characteristics of the semiconductor.

In the case of photolytic deposition, both UV and IR photo-assisted processes can be used. In UV processes, the molecular decomposition can be achieved either through direct single-photon excitation of the reactant molecules or indirect photo-sensitized decomposition by collisional energy transfer from an absorbing sensitizer to the donor molecules. The basic principle of IR photochemistry is the resonant absorption of photons by IR active vibrational modes of molecules. IR photochemistry relies on non - linear excitation processes because many IR photons must be piled up in the electronic ground state of a molecule to reach the dissociation threshold.

Let us consider the case of localized deposition of aluminum on silicon or on silica, which is quite interesting both because of obvious technological applications and because photolytic and pyrolytic deposition processes may be sequentially or simultaneously involved. Several authors have demontrated that a two-step process consisting of surface nucleation followed by growth was needed to deposit Al thin film on Si and SiO2 substrates. Tsao et al. (12) used a focused cw UV irradiation (frequency-doubled Ar ion laser) followed by a cw IR CO2 laser irradiation, triisobutylaluminum (TIBA) being the gaseous source. Following the same idea, Bouree et al. (13) first illuminated a flow of trimethyaluminum (TMA) molecules with UV light (mercury lamp) and then submitted the sample to a focused cw visible Ar ion laser. In these two examples, the two-step process was performed by using two different light sources. Higashi et al. (14) used a similar approach with only one light source, namely a pulsed UV (KrF excimer) laser while heating the substrate up to 250°C. Roughly speaking, for each of these experiments, UV exposure is used to photolytically decompose organometallic surface adsorbates, leaving Al sites which then serve to activate the subsequent thermal decomposition of TIBA or TMA to grow Al films. In the following sections, we will discuss in detail the work based on the two-step process and will show the possibility of Al direct writing on Si in a single-step process (surface nucleation not disconnected with growth stage) by using a single light source : high power UV laser.

3. EXPERIMENTAL

3.1. Sample preparation
The substrates are (100) oriented p-type silicon. The sample surface is chemically polished prior to the cleaning procedure. A careful degreasing in organic solvents followed by an etch in buffered HF, a rinse in deionized water and a drying with N2 are conducted before introduction into the cell.

3.2. Experimental apparatus
The experimental light CVD apparatus is schematically shown in Fig. 1. It is composed of a gas feeder connected to the cell and of three light sources : a visible laser, a UV lamp (Fig. 1.a) and a UV laser (Fig. 1.b).

The stainless steel gas feeder was described in detail in a previous work (13).

The laser used as the source of visible radiation is a cw multimode (488 nm and 514.5 nm lines) Ar ion laser with a maximum available output power of 25 W. The laser beam is focused on the Si surface by means of a 80 mm focal length lens. A typical gaussian radius ω , as measured with a CdTe Schottky diode, is about 25 μm (15).

The low pressure mercury lamp used in this work irradiates preferentially at 254 nm. The measured power density on the Si surface is about 5 mW/cm^2 .

The cw UV laser beam employed is generated by frequency doubling an Ar ion laser beam (514.5 nm) in intracavity mode. The maximum available output power at 257 nm is 50 mW, as measured with a calorimeter. The beam is focused to about 15 μm spot diameter through a lens assembly onto the Si surface (see Fig. 1.b).

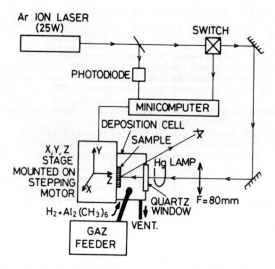

Fig. 1.a. Experimental set-up of the light (visible laser and/or UV lamp) CVD system.

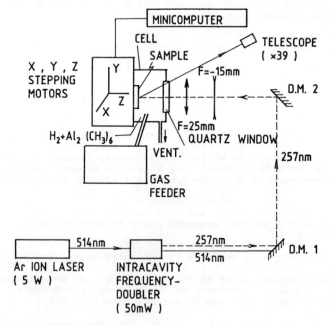

Fig. 1.b. Experimental set-up of the UV light CVD system dedicated to Al direct writing via UV laser. (D.M. = DICHROIC MIRROR)

3.3. Experimental procedure
    In order to remove residual gas impurities and passivate the
different pipes before beginning each experiment, the cell and the
pipes are repeatedly pumped out using a sorption pump to 10^{-3} Torr
and flushed with pure H_2. The system is then degassed at about 100°C
during three hours in flowing H_2 (DH_2 = 1l/min). Typical
experimental conditions are listed below :

Total cell pressure   : 1 atm
Partial pressure of TMA   : 9.5 Torr
Hydrogen flow in TMA container   : 0.6 l/h
Substrate temperature   : 293 K
Water vapor concentration   : 0.7 ppm vol
Average scanning speed along the X axis   : 5 to 100 μm/s.

3.4. Characterization techniques
    The deposited layers are visualized in situ with a telescope
and extensively analysed ex situ by the following techniques : dark
field optical microscopy, scanning electron microscopy (SEM), energy
dispersive X-ray analysis (EDAX) dedicated to light elements, Auger
electron microspectroscopy (AES) as well as electrical characterization techniques.

4. RESULTS AND DISCUSSION

4.1. Pyrolytic deposition of Al by cw visible laser
    In previous work (15), a Si sample placed in a flow of TMA was
irradiated with the cw visible laser. SEM with EDAX as well as Auger
microanalysis indicated in the region corresponding to the laser
spot the mixed presence of a large amount of C and of lower values
of Al, Al_xO_y and SiO_z. Thus the pyrolytic decomposition of the
adsorbed TMA molecules was shown to lead to an uncontrolled
deposition process where surface nucleation barrier of Al on Si was
not overcome. Consequently another approach was attempted, which was
a pure photolytic process related to the use of a UV light source.

4.2. Photolytic deposition of Al induced by UV lamp
    It was shown previously (16) that when a Si substrate placed
in flowing TMA is exposed for a few minutes to the UV lamp, its
surface is subject to a minute change which, as analysed by SEM,
results in the nucleation of an Al fine grain structure. As the
photolysis is going on (UV exposure of 30 minutes), these grains are
shown to be made up of Al domed crystallites and some spheroids of
about 50 nm size (see Fig. 2). Dome type grains are most probably
the sign of heterogeneous nucleation resulting from photo-
dissociation of adsorbed TMA molecules whereas spheroid type grains
are most probably associated with homogeneous nucleation (17) in the
gas phase. It must be noted that all these grains are nucleated at
room temperature because low fluence UV light does not induce any
temperature rise. When UV exposure time increases, the grains
coalesce and finally after three hours a continuous thin film is
observed (16). An ex situ Auger microanalysis performed on the
photodeposited film indicates the presence of aluminum, oxygen and
traces of carbon (see Fig. 3). Al metal and Al_xO_y peaks can be

284

distinguished from their energy position (64 and 52 eV respectively). Sputtered erosion profiles combined with point and scanning
analysis suggest that the chemical nature arises form initial oxygen
contamination of the Si surface and subsequent Al atmospheric
oxidation during the transfer of the sample to the Auger
spectrometer.

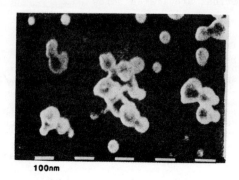

100nm

Fig. 2. SEM micrograph of a UV lamp pre-nucleated Si substrate after
an exposure of 30 minutes.

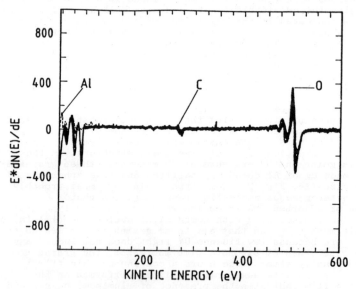

Fig. 3. Sputtered Auger spectroscopy profile of an Al thin film
deposited by UV lamp.

Thus it may be said that the photolytic deposition of Al induced by the low fluence UV lamp is the key for the understanding of the nucleation preceding the growth. However the deposition rate obtained with this UV source is too slow for a possible technological application (16).

4.3. Photolytic deposition by UV lamp followed by pyrolytic deposition by cw visible laser

In order to increase the deposition rate, a combination of UV lamp (pre-nucleation step) and visible laser (continuous growth step) was used. The Si sample is exposed to the UV lamp for a time τ and then irradiated with the Ar+ laser for different laser powers. For a given laser power, the time to necessary for the nucleation of Al to occur (onset of nucleation) is measured in situ by using a laser reflectometry technique, as described in a previous work (13). Fig. 4 shows the influence of the pre-exposure time τ on the time to for the onset of nucleation. This figure reveals the high sensitivity of to at low laser power for rather short pre-exposure time (τ < 30 min.), result consistent with the fact that the time to necessary to reach the onset of nucleation and growth can be interpreted as the time necessary to overcome a physical energy barrier in order to grow a cluster from a chemisorbed phase (17,18). At this stage, it is interesting to estimate, for a given power, the

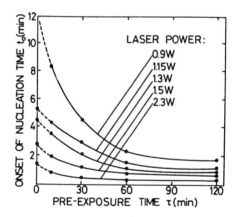

Fig. 4. Variation of the time to for the onset of nucleation as a function of UV pre-exposure time τ for a given laser power.

temperature induced by the laser on the silicon substrate : e.g. 50°C is calculated at the spot center corresponding to 1 W visible laser power. Thus in this two-step two sources process, Al deposition is observed far below the threshold temperature of pyrolytic decomposition of the TMA molecule (350°C) : the UV exposure is shown to subsequently induce an enhanced surface pyrolysis of TMA (pyrolytic autocatalysis). Moreover three main features are remarkable. The resulted Al deposit width is completely

determined by the laser spot size impinging on the Si surface.
Deposits contain traces of oxygen, but they are carbon contamination
free, as determined from SEM + EDAX analysis (19).

4.4. UV lamp + visible laser-assisted direct writing of Al

The two-step two sources process is used for direct writing
of Al on Si. One additional parameter is needed to obtain uniform
lines : the average scanning speed along the X axis. Optimized
results were obtained (13) for a scanning speed of 41 µm/s, an incident visible laser power of 2 W and with TMA diluted in hydrogen :
1.2 l/h hydrogen is sent in TMA container and further mixed with 1.8
l/h hydrogen flow. A few millimiter-long and 50 µm wide Al line
drawn in that conditions is shown in the SEM micrograph of Fig. 5.

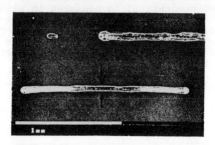

Fig. 5. SEM micrograph of a direct writing Al line using UV lamp +
visible laser (P = 2W, V = 41 µm/s).

This line has a small variation (typically 4%) of width definition.
Big round-shaped grains up to 10 µm diameter are formed between
nodular edges. The thickness and the shape of the lines obtained
with this procedure are determined by using a mechanically moving
stylus (Alphastep). Fig. 6 shows that a crater like shape line is
obtained for a laser power of 2.3 W while for 2W the shape of the
line follows the laser intensity distribution. Taking account in

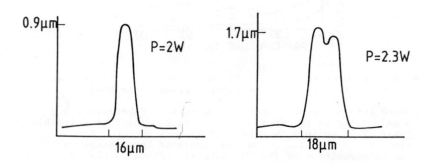

Fig. 6 . Kinetic parameters of direct writing Al lines.

the latter example the effective laser beam dwell time, an optimized growth rate of 10 μm/s is determined. This confirms indeed that the UV lamp gives the basis for the further enhanced pyrolytic decomposition of TMA molecules by visible laser. The electrical measurements are performed on as-grown Al lines connecting Ti/Au/Ti plots deposited on Si substrate. Two point probe measurements give ohmic resistivity around 5 mΩ.cm (19), as compared with 2.6 μΩ.cm for the bulk. Surface contamination as related to grain morphologies (14) and deposition rate appear to play a major role in the increase of resistivity. So, more work is necessary to improve the electrical characteristics.

4.5. UV laser-assisted direct writing of Al

It has been possible to get direct writing of Al on Si substrate by using the two-step two sources process. The low level of C contamination obtained after dilution of TMA seemed to indicate that the out-diffusion time of the generated carbon species is the limiting step. So a different approach comes up naturally which makes use of a single source, namely a high power UV laser, to photo-lyse the adsorbed TMA molecules. The direct writing has been obtained in a one-step process at a low scanning speed of 5 μm/s for an incident laser power of 8 mW and a spot diameter of 16 μm. Surprisingly enough, for such a line, a high growth rate of about 700 Å/s is registred. Furthermore, the optical micrograph of Fig. 7 typically displays a remarkable increase in size, along the scanning direction, of the deposited Al grains (up to 30 μm length). Both of these growth effects are contrasting with the variable line definition showed up. So, every time, a very small grain insures the electrical continuity between two big grains. To optimize this unfavourable microstructure, experiments are under completion to improve the line definition.

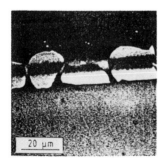

UV laser
power = 8mW
(257nm)

Spot laser
diameter =
16μm

Scanning
speed =
5μm/s

× 1000

20 μm

Fig. 7. Optical micrograph of a direct writing Al line using UV laser.

5. CONCLUSIONS

It has been shown that it is possible to overcome the nucleation barrier of Al on Si single crystal, using TMA heterogeneous photolysis at very low UV fluence and room temperature. The pre-nucleation technique can be generalized to other CVD systems having large physical nucleation barriers. Moreover the surface photoreaction induced by the lamp gives the basis for the further enhanced surface pyrolysis of TMA molecules (pyrolytic autocatalysis) observed at near room temperature and due to an Ar ion laser. The laser direct writing of Al has been obtained after TMA dilution in hydrogen. Electrical measurements show ohmic resistivity higher than in the bulk.

Finally using a high power UV source, the results show the interest of UV continuous cold direct writing in a one-step rapid process.

ACKNOWLEDGEMENTS

This work, supported by a contract from CNET-Meylan, was jointly done with Dr. J. FLICSTEIN and Dr. Y.I. NISSIM (CNET-Bagneux). The author would like to acknowledge Dr. M. RODOT (CNRS-Meudon),Dr. G. AUVERT (CNET - Meylan) for their interest in this study and Mr. R. DRUILHE (CNRS - Meudon) for his support on the experimental part.

REFERENCES

1. Surface Studies with Lasers, ed. F.R. AUSSENEGG, A. LEITNER, M.E. LIPPITSCH (Springer-Verlag, Berlin) 1983.
2. Laser Diagnostics and Photochemical Processing for Semiconductor Devices, ed. R.M. OSGOOD, S.R.J. BRUECK, H.R. SCHLOSSBERG (North-Holland, New York) 1983.
3. Laser-Controlled Chemical Processing of Surfaces, ed. A.W. JOHNSON, D.J. EHRLICH, H.R. SCHLOSSBERG (North-Holland, New York) 1984.
4. Laser Assisted Deposition, Etching and Doping, ed. S.D. ALLEN, SPIE, 459 (1984).
5. Laser Processing and Diagnostics, ed. D. BAUERLE, (Springer-Verlag, Berlin) 1984.
6. Laser Chemical Processing of Semiconductor Devices, Extended Abstracts, ed. F.A. HOULE, T.F. DEUTSCH, R.M. OSGOOD (MRS, Pittsburgh) 1984.
7. Beam Induced Chemical Processes, Extended Abstracts, ed. R.J. VON GUTFELD, J.E. GREENE, H. SCHLOSSBERG (MRS, Pittsburgh) 1985.
8. Laser Processing and Diagnostics, ed. D. BAUERLE, K.L. KOMPA, L. LAUDE (Ed. Physique, Les Ulis) 1986.
9. Photon, Beam and Plasma Stimulated Chemical Processes at Surfaces, ed. V.M. DONNELLY, I.P. HERMAN, M. HIROSE (MRS, Pittsburgh) 1987.
10. D. BAUERLE, in reference 5, p. 166.
11. D.J. EHRLICH, R.M. OSGOOD and T.F. DEUTSCH, J. Vac Sci. Technol. 21, 23 (1982).
12. J.Y. TSAO and D.J. EHRLICH, Appl. Phys. Lett. 45, 617 (1984) ; in reference 6, p. 84.
13. J.E. BOUREE, J. FLICSTEIN and Y.I. NISSIM, in reference 9.
14. G.S. HIGASHI and C.G. FLEMING, Appl. Phys. Lett. 48, 1051 (1986) ; G.S. HIGASHI, G.E. BLONDER and C.G. FLEMING, in reference 9.
15. J.E. BOUREE, Y.I. NISSIM, J. FLICSTEIN, C. LICOPPE and R. DRUILHE in Energy Beam-Solid Interactions and Transient Thermal Processing, ed. V.T. NGUYEN and A.G. CULLIS (Ed. Physique, les Ulis) 1985, p. 119.
16. J.E. BOUREE, J. FLICSTEIN, Y.I. NISSIM and C. LICOPPE, in reference 7, p. 71.
17. B.K. CHAKRAVERTY, in Crystal Growth : an introduction, ed. P. HARTMAN (North-Holland, Amsterdam) 1973, p. 50.
18. J.Y. TSAO and D.J. EHRLICH, J. Cryst. Growth 68, 176 (1984).
19. J.E. BOUREE and J. FLICSTEIN, to be published in NATO ASI Series, 1987.

1/f NOISE IN SEMICONDUCTOR DEVICES: THEORY AND APPLICATIONS

B. Pellegrini

Istituto di Elettronica e Telecomunicazioni
Università di Pisa, Via Diotisalvi 2, 56100 Pisa, Italy

On the basis of a recent theorem of electrokinematics, which in particular generalizes the Ramo-Shockley theorem to any conduction medium and boundary condition, a new island model of 1/f noise is proposed which attributes its origin to the defect centers and unifies approaches ascribing it to fluctuations of carrier number and mobility. Unlike preceding models, it also shows that the weight of the defects in generating noise at low frequency is squarely dependent on their relaxation time and that this fact makes it easier to account for the ubiquity of 1/f noise down to however low measurable frequency. Finally, some applications of 1/f noise apt to determine properties of semiconductor devices are described.

1. INTRODUCTION

As is well known, flicker noise is a general phenomenon which occurs in virtually all devices crossed by a current and has a power spectrum proportional to $1/f^\gamma$ with an exponent γ of frequency f near one down to the lowest possible measurable frequency, for instance down to 5×10^{-7} Hz in MOS transistors [1].

Unlike the other two general types of noise, the thermal and shot ones whose cause has been well known ever since they were disovered experimentally, the origin $1/f^\gamma$ noise, though it was discovered in the 1924 and, since then, much theoretical and experimental research, numerous conferences and a lot of models have been devoted to it, is still the subject of lively debate.

At present three main standpoints are held and compared. One, based on the Hooge empirical formula [2] or on the Handel quantum model [3], ascribes the origin of the flicker noise to fluctuations in carrier mobility, whereas a second set of models, including the McWorter [4] and island ones [5], is founded on the number fluctuations of the carriers.

A third, more recent, approach [6] removes such a division by showing that, due to their capture, storage and release from the defects, or islands, the number fluctuation of the free carriers determines their hemimicroscopic mobility fluctuations too.

On the other hand, Handel's quantum 1/f noise model, based on the self-interference of the wave packet of a carrier upon an its scattering, has been shown by Kiss and Heszeller not to be valid on the basis of rigorous quantum electrodynamics [7], so that the island approach, which attributes the origin of $1/f^\gamma$ noise to the defects of the conducting medium and unifies the standpoints of the fluctuations in the number and mobility of the carriers, seems to remain the only model that, at present, is still valid.

Therefore such a model and some its applications will be described here.

2. ISLAND MODEL

The island model of $1/f^\gamma$ noise will be presented here in a new way on the basis of a recent theorem of electrokinematics [8] which, in particular, generalizes the Ramo-Shockley theorem to any medium and boundary condition and, unlike this theorem, allows us to write the current i of any cylindrical bipole of lenght L in the form

$$ i = - \frac{q}{L} \sum_{j=1}^{N} v_j \qquad , \qquad (2.1) $$

where q is the electron charge, v_j is the carrier velocity along the cylinder axis of jth of the N electrons contained in the sample at time t.

292

The equation (2.1) holds true when the terminal potentials are kept constant (in this case $N=\bar{N}$ is independent of t) and the displacement current across the lateral surface is negligible [8].

In particular, it holds good for a macroscopically homogeneous sample to which an electric field F along the cylinder axis is applied. In this case, which is the one we are dealing with, we have

$$v_j(t) = \mu_j(t)F + u_j(t) \rho_j(t) \qquad . \qquad (2.2)$$

where u_j is the fluctuation of v_j due to any scattering process during the free path of the electron between its emission from one island and the ensuing capture in another, μ_n is its microscopic mobility during such a path. $\mu_j(t)=\mu_n \rho_j(t)$, rather, is the hemimicroscopic mobility which takes into account the drift velocity modulation produced by the electron capture, storage and release due to the defects. We take into account such a modulation by means of the stochastic telegraph function $\rho_j(t)$ that, at random, goes from 1 to 0 and viceversa.

By indicating with $\alpha=\bar{\rho}_j$ the time average value of ρ_j, we have $\mu_j=\bar{\mu}+\Delta\mu_j$, with $\bar{\mu}=\alpha\mu_n$ and $\Delta\mu_j(t)=\mu_n[\rho_j(t)-\alpha]$, and then, from (2.1) and (2.2), $\bar{i}=-q\bar{\mu}FN/L$.

Then the fluctuations of i, according to (2.1) and (2.2), have two contributions. One, $\Delta i_t=-(q/L) \sum_{j=1}^{N} u_j(t)\rho_j(t)$, is due to the scattering and capture processes and it generates the thermal noise [6, 8].

The other

$$\Delta i_I = \frac{\bar{i}}{\bar{\mu}N} \sum_{j=1}^{N} \Delta\mu_j \qquad , \qquad (2.3)$$

is generated by the hemimicroscopic mobility fluctuation $\Delta\mu_j$ due to island modulation of the drift velocity and it can be considered to be the origin of $1/f^\gamma$ noise [6].

From a symmetric standpoint Δi_I and, accordingly, $1/f^\gamma$ noise may be ascribed to the fluctuation $\Delta N'$ of the number $N'(t)=\sum_{j=1}^{N} \rho_j(t)$ of the free carriers, i.e. which at t are not stored by islands.

Indeed, according to previous definitions of $\mu_j(t)$ and $N'(t)$ and from (2.1) and (2.2), we also have $i=-(q/L)\mu_n FN'(t)+\Delta i_t$ so that, since $\overline{N'}=\alpha N$, we again obtain the previous value of $\overline{i}$ whereas for Δi_I we now get

$$\Delta i_I = \frac{\overline{i}}{N'} \Delta N' \qquad . \qquad (2.4)$$

On the other hand, if in the sample there are N_I islands with a single-energy-level E_k and an occupation factor $\phi_k(E_k,t)$, we have $N=N'+\sum_{k=1}^{N_I}\phi_k$ so that, since $N=\overline{N}$, (2.4) becomes

$$\Delta i_I = -\frac{\overline{i}}{N'} \sum_{k=1}^{N_I} \Delta\phi_k \qquad , \qquad (2.5)$$

that is, Δi_I is also proportional to the sum of the occupation-factor fluctuations of the single islands.

Therefore, if these are independent of each other, the power spectral density S_I of Δi_I, according to (2.5), directly becomes

$$S_I = \frac{4\ \overline{i}^2}{\overline{N'}^2} \sum_{k=1}^{N_I} \frac{\tau_k\overline{\phi_k}\ (1-\overline{\phi_k})}{1 + \tau_k^2\omega^2} \qquad , \qquad (2.6)$$

where $\overline{\phi}_k(E_K)=\{1+\exp[(E_k-E_F)/KT]\}^{-1}$ is the mean value of the occupation factor of E_k, E_F is the Fermi level, K is the Boltzmann constant and T is the temperature. In its turn the island relaxation time $\tau_k(E_k)$ is given by

$$\tau_k = \frac{\overline{\phi_k}}{c_k\ \overline{n'}} \qquad , \qquad (2.7)$$

where c_k is the capture coefficient, $\overline{n'}=\overline{N'}/\Omega$ is the mean density of the free electrons, Ω is the sample volume and $n_I=N_I/\Omega$ is the island density.

By eliminating $\overline{\phi}_k$ between (2.6) and (2.7), after the substitution $\tau_k=\tau$, $c_k=c$ and after putting an integral in the place of the sum of (2.6), this becomes

$$S_I = 4 \frac{\overline{i}^2}{N'} n_I \iint \frac{\tau^2 (1 - \tau c \overline{n'}) c}{1 + \tau^2 \omega^2} D(c,\tau) \, dc d\tau \qquad , \qquad (2.8)$$

where $D(c,\tau)$ is the distribution of c and τ.

According to other previous models, (2.8) may be also written in the form

$$S_I = \frac{\overline{i}^2}{N'} \int \frac{\tau \, Y(\tau)}{1+\tau^2\omega^2} D_\tau(\tau) d\tau = \frac{\overline{i}^2 \, G}{N'} \int \frac{\tau \, D_{\tau e}(\tau)}{1+\tau^2\omega^2} d\tau \qquad , \qquad (2.9)$$

in which we have put $Y(\tau)D_\tau = GD_{\tau e}(\tau) = 4n_I\tau\int(1-\tau c\overline{n'})cD(c,\tau)dc$, where $D_\tau(\tau)$ and $D_{\tau e}(\tau)$ are the distribution and an equivalent distribution of τ, respectively, and G is the normalization factor.

For instance, when $E=E_k > E_F+3kT$, so that (2.7) gives $c\overline{n'}\tau \ll 1$, and $D(c,\tau)=D_c(\tau)D_\tau(\tau)$, we have

$$G = 4n_I < c > < \tau > \qquad , \quad D_{\tau e} = \tau \, D_\tau / < \tau > \qquad . \qquad (2.10)$$

Now, if in a range τ_1, τ_2 of τ we have $D_{\tau e} \propto 1/\tau^{2-\gamma}$ with $0 < \gamma < 2$, from (2.9) we obtain $S_I \propto 1/f^\gamma$ in almost all the frequency band $f_1=1/2\pi\tau_2$, $f_2=1/2\pi\tau_1$.

In general, however, such an assumption does not prove to be true and we meet two difficulties in determining the frequency dependence of S_I through (2.8) or (2.9); this is due to the fact that D or $D_{\tau e}$ is often unknown and the integrals of (2.8) and (2.9) themselves, except in the special case mentioned above, are not easy to compute in closed analitycal form.

However, some general features of $S_I(f)$ can be inferred from (2.9) using the following procedure.

For this let us indicate a reference spectrum and frequency with S_r and $f_r=1/2\pi\tau_r$, respectively, and let us make the Taylor-series expansion

$\ln(S_I/S_r) = \ln[S_I(f_0)/S_r] - \gamma(f_0)[\ln(f/f_r) - \ln(f_0/f_r)]$ of $\ln(S_I/S_r)$ with respect to $\theta_\omega = \ln(f/f_r)$ about $\theta_0 = \ln(f_0/f_r)$, f_0 being a given frequency and

$$\gamma(f) = - \frac{\partial \ln(S_I/S_r)}{\partial \ln(f/f_r)} = 1 + \delta \qquad . \quad (2.11)$$

Therefore, in a proper band around f_0 [5], the spectrum S_I can be put in the power form

$$S_j(f) = S(f_0) \, (\frac{f_0}{f})^{\gamma(f_0)} \qquad . \quad (2.12)$$

The problem now is to compute the frequency exponent γ. For this, let us make the variable changes $\tau = \tau_r \, \exp(\theta)$ and $\omega = 2\pi f = \tau_r^{-1} \, \exp(\theta_\omega)$ so that (2.9) becomes

$$S_I = \frac{\overline{i}^2 \, G\tau_r}{2N'} \, \exp(-\theta_\omega) \int \frac{D_{\theta e}(\theta)}{\cosh(\theta + \theta_\omega)} \, d\theta \qquad , \quad (2.13)$$

where $D_{\theta e} = \tau_r \, \exp(\theta) \, D_{\tau e}[\tau(\theta)]$ is now the equivalent distribution of θ. From (2.11) and (2.13) we get

$$\delta = \int \tanh(\theta + \theta_\omega) \, H(\theta) d\theta \qquad , \quad (2.14)$$

where $H(\theta)$ is a distribution-like function given by

$$H(\theta) = \frac{D_{\theta e}(\theta)}{\cosh(\theta + \theta_\omega)} \, / \int \frac{D_{\theta e}(\theta)}{\cosh(\theta + \theta_\omega)} \, d\theta \qquad . \quad (2.15)$$

Therefore, according to (2.14) and (2.15), δ is the average value of $\tanh(\theta + \theta_\omega)$ so that we have $|\delta| \leq 1$ and, from (2.11), $0 \leq \gamma \leq 2$, as must happen as a result of the superimposition of Lorentzian spectra which, according to (2.6), (2.8) and (2.9), gives S_I.

When all the islands are equal and, accordingly, have the same $\tau = \tau_b = \tau_r$

$\exp(\theta_b)$, that is $D_{\theta e} \propto \delta(\theta-\theta_b)$, from (2.14) and (2.15) we obtain the value $\delta=\tanh(\theta_b+\theta_\omega)=(\tau_b^2\omega^2-1)/(\tau_b^2\omega^2+1)$ of the single Lorentzian which we have in this case.

When, on the other hand, the values of τ are dispersed, but in such a way that $D_{\theta e}\neq0$ only for $|\theta+\theta_\omega|>1.5$, from (2.14) and (2.15) we get $|\delta|=|\tanh(\theta+\theta_\omega)|=1$ with a negligible error.

In all the other cases of dispersed τ's, owing to $\cosh(\theta+\theta_\omega)$ of (2.15), $H(\theta)$ tends to become essentially different from zero and an even function of θ around θ_ω, around which, on the contrary, $\tanh(\theta+\theta_\omega)$ is an odd function, so that from (2.14) we have $|\delta|\rightarrow0$ and $\gamma\approx1$.

Therefore, in order to have a $1/f^\gamma$ noise with $\gamma\approx1$ in a band $f_1=1/2\pi\tau_2$, $f_2=1/2\pi\tau_1$, around a frequency $f_0=1/2\pi\tau_0$, the defect relaxation times τ must have dispersed values in the range τ_1 and τ_2 and broad maxima or a slow variation of $D_{\tau e}$ in it. For instance, for a band of 10^{-3}, 10^3 Hz, the sample has to have relaxation times in the range of 10^{-3}, 2×10^2 sec. However, unlike what happens in previous models which in (2.9) assume $D_{\tau e}=D_\tau$, this fact does not mean that the distribution D_τ of τ has to be prevalently allocated and/or to have its maxima in such a range. It is, rather, the equivalent distribution $D_{\tau e}=\tau D_\tau/\langle\tau\rangle$, which indeed, in relation to D_τ, greatly increases the weight of the greater τ's and shifts the maxima of D_τ twards themselves, that has to have such an allocation.

Furthermore, from another standpoint, the noise between f_1 and f_2 cannot be ascribed to the defect density $n_{I12}=n_I \int_{\tau_1}^{\tau_2} D_\tau d\tau$, but to the equivalent one $n_{I12e}=n_I\langle\tau\rangle^{-1} \int_{\tau_1}^{\tau_2} \tau D_\tau d\tau$ which, for $\tau_1> \langle\tau\rangle$, may become several orders of magnitude greater than n_{I12} [6].

Such a shift of $D_{\tau e}$ and an increase of n_{I12e} in relation to D_τ and n_{I12}, respectively, seem to account satisfactorily for the ubiquity of the flicker noise down to the lowest possible frequency. Finally, its spectrum, according to (2.8), (2.9) and (2.12), can be written in the form

$$S_I = \beta\, f_0^\delta\, \frac{\overline{i}^2}{N'f^\gamma} \qquad\qquad , \qquad (2.16)$$

where $\beta(f_0)$ is a coefficient which, usually, slowly depends on f_0 and is proportional to the island density n_I, and, even for a very small fraction of dispersed relaxation times around τ_0, we have $\gamma \approx 1$.

Equation (2.13) is verified by almost all the unipolar bipoles, including the semiconductor ones.

3. APPLICATIONS

The $1/f^\gamma$ noise models and measurements may be used to study and characterize electron materials, devices, technologies and circuits.

When $D_{\theta e}$ is a slowly varying function in relation to $\cosh(\theta+\theta_\omega)$, from (2.13), and then from (2.8) and (2.9), we get

$$S_I = \frac{\overline{i}^2 G \pi}{2\omega N'} D_{\theta e}(-\theta_\omega) = \frac{\overline{i}^2 n_I}{fN'} \tau^2 \int c(1 - \tau c \overline{n}')D(c,\tau)dc \quad , \quad (3.1)$$

where now $\theta=-\theta_\omega=-\ln(f/f_r)$, i.e. $\tau=1/2\pi f$.

Therefore by measuring, for instance by means of a signal analyser, $fS_I(f)$ versus $\ln(f/f_r)$, we obtain directly the distribution $D_{\theta e}$ of θ, that is, since from (2.7) we have $\theta=-E/KT+\lambda$, λ being a proper variable, we can evaluate the distribution of the energy levels of the defects and, hence, their mean activation energy and variance.

More directly, in the case of (2.10), from (3.1) we have $n_I D_\tau(\tau)=$ $(2\pi)^2 f^3 S_I(f)\overline{N}'/<c>\overline{i}^2$ with $\tau=1/2\pi f$, that is we can determine distribution of τ of the defects.

This technique has been applied to thick film and metal film resistors in order to explain their conduction and noise mechanisms [5,9,10]. The same technique with similar objectives may be applied to semiconductor resistors.

If two defect groups are characterized by relaxation time τ_a and τ_b and density n_a and n_b, respectively, from (2.8) or (2.9) we have $S_{Ia}(0)/S_{Ib}(0) \approx n_a \tau_a^2/n_b \tau_b^2$ so that slow defects can be screened and detected by means of noise measurements with a much greater sensitivity than with

298

any other method [11].

If defects are allocated on the surface of the device and they have a Fermi level which is independent of the bulk level, in (2.7) the density $\bar{n}'$ relative to surface is also independent of that of the conduction zone, so that from (2.6) and (2.16) we have two contributions to S_I in the form

$$S_I = \frac{\bar{i}^2 f_0^{\delta}}{f^{\gamma}} \left(\frac{WL\sigma_s}{\bar{N}'^2} + \frac{\beta_b}{\bar{N}'}\right)$$
, (3.2)

where coefficients σ_s and β_b are relative to the surface and to bulk, respectively, and WL is the lateral surface, for instance the gate surface in MOS transistors.

In these latter $\bar{N}' \propto WL(v_G-v_T)/d$, where v_G and v_T are the gate and threshold voltage, respectively, and d is the oxide thickness, so that (3.2) can be easily verified, as has been done [12].

In the n channel MOS's the term depending on β_b is negligible and the parameter σ_s thus obtainable may be usefully employed to qualify the oxide properties and the relative technologies, especially in the VLSI field.

The previous model, rather, cannot directly applied to bipolar devices because the island-charge fluctuations produce modulation of the generation-recombination current which has not been taken into account in the present approach and which, as found experimentally, may lead to an exponent of the current in (2.16) of less than two.

The $1/f^{\gamma}$ noise, with $\gamma \leq 2$, may also be usefully utilized to study the electromigration phenomena, the mean time of failure and the reliability of thin metal films employed as interconnections in the IC's [13]. In particular, it makes it possible one to determine the activation energy of the electromigration process.

As shown by the wide literature, the $1/f^{\gamma}$ noise may be employed to study, from several standpoints, heterojunctions, photoelectron devices, semiconductor detectors, bipolar and unipolar transistors, and any conducting material and devices.

An important objective of reasearch on the $1/f^\gamma$ noise, which, unlike thermal and shot noises, is not a fundamental phenomenon but an excess noise due to conducting medium defects, is of course that of eliminating or, at least, of minimizing it in order to improve properties, sensitivity and performance of materials, devices, circuits and instruments.

AKNOWLEDGMENT

The author wishes to thank the Committee of the 4th International Workshop on the Physics of Semiconductor Devices for its invitation to present this talk. The work is supported by the Ministry of Education, by the National Reaserch Council (CNR) of Italy and by the Finalized Research Project "Material and Devices for Solid State Electronics" of the CNR".

---

1 - M.A. Caloyannides, J. Appl. Phys. <u>45</u>, 307 (1974).

2 - F.N. Hooge, Physica <u>60</u>, 130 (1972).

3 - P.H. Handel, Phys. Rev. <u>A22</u>, 745 (1980).

4 - A.L. McWhorter, MIT Lincoln Lab. Rep. 80 (1955).

5 - B. Pellegrini, Phy. Rev. <u>B24</u>, 7071; ib. <u>335</u>, 57 (1987).

6 - B. Pellegrini, Phy. Rev. <u>B26</u>, 1791 (1982); Solid-State Electronics <u>29</u>, 1279 (1986).

7 - L.B. Kiss and P. Heszeler, J. Phys. <u>C19</u>, 1631 (1986).

8 - B. Pellegrini, Phy. Rev. <u>B34</u>, 5921 (1986).

9 - B. Pellegrini, R. Seletti, P. Terreni and M. Prudenziati, Phys. Rev. <u>B27</u>, 1233 (1983).

10 - P. Dutta, P. Dimon and P.M. Horn, Phys. Rev. Lett. <u>43</u>, 646 (1979).

11 - B. Pellegrini, B. Neri, R. Saletti, Alta Frequenza <u>55</u> (English Issue), 245 (1986).

12 - A.A. Abidi, J. Chang, C.R. Viswanathan, J.A. Wikstron, and J.J.M. Wu, Abstrats of 9th Int. Conf. on Noise in Physical Systems, Montreal, (May 1987).

13 - A. Diligenti, B. Neri, P.E. Bagnoli, A. Barsanti, and M. Rizzo, IEEE Electron Dev. Lett. <u>ED-6</u>, 605 (1985).

Recent Developments of Low Temperature Epitaxial Growth
of Semiconductors

T. Hariu
Department of Electronic Engineering, Tohoku University
Sendai 980, Japan

Basic requirements for low temperature epitaxy are
briefly reviewed in view of the recent developments ôf
ion beam epitaxy, photo-assisted epitaxy and plasma-
assisted epitaxy. The experimental results of plasma
assisted epitaxy are described in some detail on in-
situ surface cleaning, the effects of applied rf
power, substrate temperature, discharging gas, supply
rate of constituent atoms etc. on the crystallographic,
electrical and optical properties of grown layers.

1. Introduction

Low temperature epitaxial growth of semiconductor crystal
layers is an important process technology which is under
development to fabricate controlled device structures with
minute dimensions in order to achieve the highest device
performance and new device functions. Considering that it is
only the surface of the substrate that is required to be
activated for such processes as etching, crystal growth,
oxidation and deposition of passivating insulator films, some
additional energy other than thermal energy should be supplied
to the surface with the rest of the substrate kept at as low
temperatures as possible.

This additional energy can be supplied as photons, accelerated
ions and/or activated radicals. The purpose of this
contribution is firstly to briefly review the recent
developments of low temperature epitaxial growth technologies
and then to describe in some detail the plasma-assisted
epitaxy (PAE) which has been developed in our group for
several years[1] and which is considered to be the most
promising technology for this purpose in near future before
some other new technology is developed.

2. Basic Considerations on Low Temperature Epitaxial Growth

Two kinds of additional energies should be given to the
growing surface in order to successfully reduce the temperature
for epitaxial growth. Firstly constituent atoms with kinetic
energy sufficient enough for them to migrate to the lowest-
energy site, should be supplied to the growing surface. This
is the right site for perfect crystal where it is difficult
for them to reach with thermal energy at low temperatures.
These atoms with only thermal energy request a longer time for
epitaxial growth as the substrate temperature is reduced. The
dependence of this required time on temperature should be

exponential because an atom supplied and adsorbed at an
arbitrary site on the growing surface should overcome a
potential barrier before it sits on a right site. Therefore
the growth rate for epitaxial growth should be exponentially
decreased as the growth temperature is reduced. It may be
worthwhile to mention that the epitaxial growth temperature,
therefore, is not uniquely difined but is given as a function
of growth rate.

Another problem of contamination is a trouble at very low
growth rate at low temperatures because the probability of
incorporating unrequired atoms is increasingly raised, as is
recognized in MBE.

Kinetic energy for sufficient migration can be added by
accelerating ions or ionized clusters[2] or supplying atoms
from high temperature sources such as plasmas. This kinetic
energy should not be excessive, otherwise it can cause crystal
damages and even sputtering instead of growth.

Second energy which should be added is the internal energy of
atoms which is required to activate chemical reaction or
bonding. This is particularly important to grow compound
crystals which are usually grown at high temperatures by
thermal processes. Although the photo-chemistry, in which
photons are employed to activate chemical reaction, is rather
classical, it is only recently that photo-activated processes
began to attract wide attention for use in semiconductor device
technologies, particularly in film deposition or crystal
growth.[3] It is to be noted here that the momentum of a photon
with a given energy is very small compared with electrons or
atoms, then it can not directly increase the kinetic energy of
atoms to migrate over the growing surface.

The principal advantages of plasma processes consists in the
capability of increasing the above both energies simultaneously.
Although the temperature of neutral atoms or ions is not so
high as that of electrons at several tens thousands degree
centigrade in low temperature, non-equilibrium plasmas, the
internal energy of them can be raised, as simply confirmed by
the light emission from the plasma, so that chemical reaction
are activated.

Although the average fraction of ionized particles is of the
order of 10^{-3}, the fraction of ions supplied to the growing
surface is increased by the presence of ion sheath in front of
the substrate and the kinetic energy can be increased by the
spontaneous accelerating electric field across the sheath or by
applying additional dc voltage to the substrate.

Other advantages of the plasma-assisted epitaxy will be
described below: (1) cleaning effect of the substrate and the
growing surface[4], (2) high efficiency in impurity doping[5],
(3) high growth rate, if necessary and (4) good surface
morphology across a large area.

As described above, the mechanism of plasma-assisted epitaxy
is complicated, including the effects of excited radicals,
their kinetic energy, higher kinetic energy of ions, high
temperature electrons, photons etc. On the contrary, the
kinetic energy of supplied particles can precisely controlled
with small spreading in ion beam epitaxy (IBE) and ionized-
cluster beam epitaxy (ICBE), and the supplied photon energy can
be purely selected in photo-activated growth process, depending
upon the required reaction. These processes are then suitable
for the investigation of each mechanism occuring in the
processes and they will also find their own application when
some mechanisms to transform kinetic energy to internal energy
or vice-versa, are included appropriately.

3. Experimental Apparatus for PAE

Two experimental apparatuses and typical deposition conditions
are shown in Fig.1 and Table 1, respectively. After evacuating
the reaction vessel, hydrogen or argon gas is flown in up to a
desired pressure. Discharging plasma is excited by rf power at
13.56MHz in this study through inductive coupling (Fig.1(a)) or
capacitive coupling (Fig.1(b)). A dc coil is also provided
outside the vacuum chamber in Fig.1(a) to apply magnetic field
which can focus and stabilize the discharging plasma. Metallic
Ga or trimethylgallium and metallic As for the growth of GaAs,
metallic Ga and metallic Sb for GaSb, metallic Zn and metallic
Se for ZnSe, were used as source materials. Pande and Seabough
used trimethylgallium and arsine in dc-excited plasma, as shown
in Fig.2.[6]

Electron density, electron temperature and plasma potential

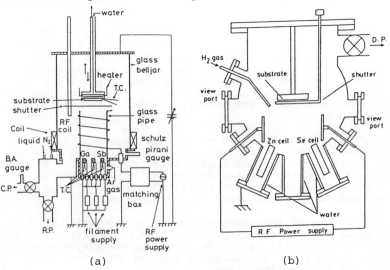

(a)   (b)

Fig.1   Two different experimetal PAE apparatuses.

with respect to the earthed substrate are shown in Fig.3(a) for argon plasma as a function of applied rf power and their distribution across the reaction vessel are shown in Fig.3(b). In the case of hydrogen plasma, electron density is lower by a factor of about 5.

4. Results on Plasma Assisted Epitaxy

4.1 Cleaning of the substrate surface

It is essential to remove native oxide from the substrate surface in order to make the successive layer grow epitxially. It was found from depth-composition profiles by AES that the surface of substrates was cleaned by plasma. As shown in Fig.4 even when oxygen was not incorporated in the deposited films, oxygen was detected by AES at the substrate-GaSb interface when the GaSb layer was deposited without plasma but not when it was produced by PAE. We consider that the gettering action also occurs during sebsequent growth of the layer beyond the initial stage and can prevent the incorporation of oxygen.

In plasma CVD epitaxial growth of Si, Townsend and Uddin considered that the predeposition clean up in hydrogen plasma is effective to get epitaxial layers of high surface quality.[7] Suzuki and Ito has reported that the addition of GeH_4 to SiH_4 plasma at the beginning of deposition is effective to remove native oxide.[8]

Table I Typical deposition condition of PAE

H_2 Pressure	7×10^{-2} Torr
RF Power (13.56 MHz)	50 W
Substrate Temperature	550 °C
Growth Rate	200 - 700 Å/min
Ga Temperature	1000 °C
As Temperature	390 °C

(a) for GaAs

H_2 Pressure	5×10^{-2} Torr
Ar Pressure	8×10^{-3} Torr
RF Power (13.56 MHz)	0 - 100 W
Substrate Temperature	340 - 470 °C
Growth Rate	100 - 2000 Å/min
Ga Temperature	920 - 970 °C
Sb Temperature	500 °C
Supply Ratio of Ga to Sb	1:3 - 1:6

(b) for GaSb

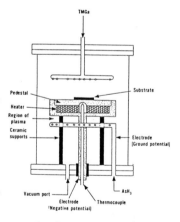

Fig.2 An experimental PAE apparatus using trimetylgallium and arsine in dc-excited plasma by Pande and Seabough.[6]

Donahue et al. also have found that the in-situ cleaning by
sputtering in an argon plasma at 775°C with a dc bias applied
to the susceptor is essential for achieving epitaxial growth
of Si at 775°C by plasma CVD.[9]

It is now clear that one of the essential features of PAE is
the removal of native oxide on the substrate with the help of
plasma.

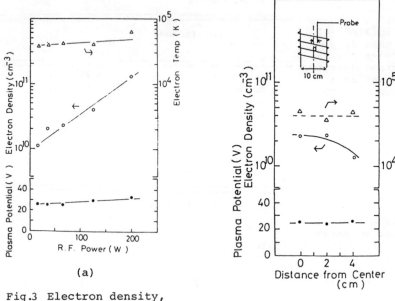

(a)

(b)

Fig.3 Electron density,
electron temperature
and plasma potential
with respect to the earthed substrate in the present
PAE system (a) as a function of applied rf power and
(b) their distributions across the reaction chamber.

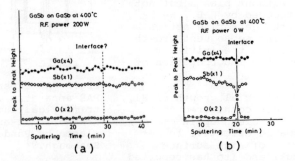

(a)   (b)

Fig.4 Depth profiles of GaSb films deposited on
GaSb substrate, (a) by PAE and (b) without plasma.

4.2 Growth rate and reduction of epitaxial temperature

Fig.5 shows the increase of growth rate of ZnSe and ZnS by PAE, compared with the deposition without plasma, by a factor of about 2, particularly at higher substrate temperature where the growth rate is reduced. The effect of plasma is also remarkable, as shown in Fig.6, in reducing the temperature for epitaxial growth and in assisting the crystallization, as observed by the decrease of half-widths of X-ray diffraction lines. It is to be noted that even the growth without plasma was performed in hydrogen gas flow, different from MBE, after the substrate surface had been cleaned by hydrogen plasma.

Epitaxial layers of GaAs and GaSb were grown on (100)GaAs substrate at $350°C$10,11)in the plasma of applied power, say, 100W, while only polycrystalline layers were grown without plasma. However, it was found, as shown in Fig.7, that the

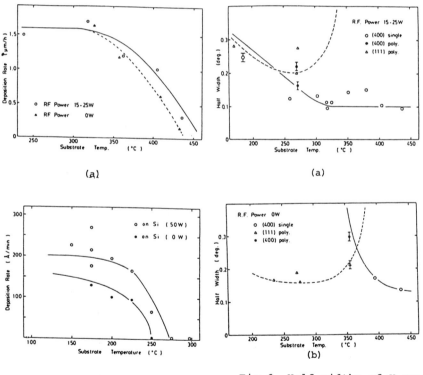

(a)

(a)

(b)

(b)

Fig.5 Growth rate of (a)PAE ZnSe and (b)PAE ZnS, compared with the deposition without plasma.

Fig.6 Half-widths of X-ray diffraction lines of PAE ZnSe grown (a)with and (b) without plasma. Solid lines are of epitaxial single crystals and dotted line of polycrystals.

epitaxial layers were deposited only within a limited range of applied plasma power and that polycrystalline layers were produced at excess powers. It was also confirmed that the epitaxial growth is achieved up to higher applied power in hydrogen plasma than argon, probably due to weaker bombardment in the former.

4.3 The effects of plasma power and substrate temperature on electronic properties

Electrical and optical measurements on grown layers also indicated that the applied rf power should be optimized to improve these properties. Fig.8 shows the variation of mobility in undoped p-type GaSb layers grown in hydrogen plasma as a function of rf power at various substrate temperatures. Optimum plasma power exists at each substrate temperature, and it shifts to a higher value as the substrate temperature is reduced.

Fig.9 shows the mobility of undoped GaSb films deposited by PAE in comparison with those by other methods like MBE, LPE, VPE and MOCVD as a function of substrate temperature. GaSb films by PAE have comparable mobilities in spite of lower substrate temperatures.

Photoluminescence spectra also indicated that the

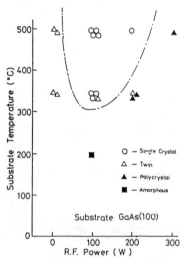

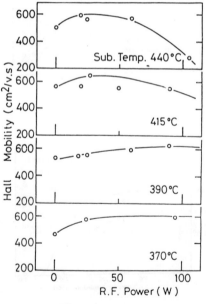

Fig.7 Structural change of GaAs layers deposited on (100)GaAs as a function of the substrate temperature and the plasma power.

Fig.8 Hall mobilities of undoped p-type GaSb films deposited on semi-insulating (100) GaAs as a function of plasma power at various substrate temperatures.

applied rf power should be optimized to maximize the
luminescence intensity due to bound excitons in relation to
donor-acceptor or band acceptor emission. Fig. 10 shows the
variation of luminescent intensity at maximum peak near the
band edge as a function of applied rf power. The maximum
intensity is obtained for a layer deposited with plasma power
of 20-30W at a substrate temperature of 440°C in the present
deposition system. It is also shown that the luminescent
intensity is always higher for a layer grown on (100)GaSb
substrate than (100)GaAs, probably due to the exact lattice
matching in the former.

4.4 Discharging gas

In applying PAE to the growth of some particular semiconductor
crystal, proper discharging gas should be employed by considering its chemical reactivity and its physical effect like ion or
atomic bombardment as well as its doping as an impurity.
Hydrogen plasma was found to be better than argon plasma in the
present case in terms of higher carrier mobility, lower carrier
density and higher intensity of photoluminescence. Fig.11
shows the effect of growth rate on free hole density and Hall
mobility in undoped GaSb layers deposited in hydrogen and
argon plasma at a substrate temperature of 415°C. The

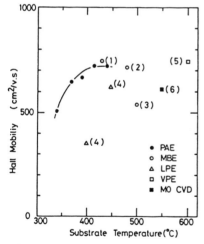

Fig.9 Hall mobilities of
undoped p-type GaSb films
deposited by PAE in comparison
with those by other methods as
a function of substrate
temperature. (1) Naganuma et
al.[12], (2)Goto et al.[13], (3)
Yano et al.[14], (5)Kakei et
al.[15], (6)Manasevit et al.[16]

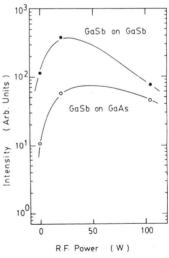

Fig.10 Maximum lumi-
nescent intensity in GaSb
layers deposited on (100)
GaSb and (100)GaAs as a
function of applied rf
power in PAE.

deposition in hydrogen plasma gives lower carrier density and higher mobility than argon plasma.

The effect of mixed plasma of hydrogen and argon in PAE GaSb, and of hydrogen and nitrogen in PAE ZnSe on the photoluminescent properties are shown in Figs.12 and 13, respectively. The photoluminescence emission due to bound excitons is very much enhanced by optimizing the applied rf power in hydrogen plasma and is reduced by mixing Ar-gas into the plasma. However, donor-acceptor pair emission is enhanced by mixing Ar-gas by about 40%, probably due to the increase of acceptor density. The similar tendency as GaSb was observed in PAE ZnSe grown in mixed plasma of hydrogen and nitrogen, where nitrogen should be an acceptor in ZnSe. The intensity of donor-acceptor pair emission in the layers grown in the mixed plasma with about 30% nitrogen, is enhanced by more than two orders of magnitude compared with the emission intensity due to bound excitons in layers grown in pure hydrogen plasma. Instead the luminescence due to bound excitons in the former is reduced.

4.5 Relative supply rate of constituent atoms

Fig.14 shows carrier densities and Hall mobilities of undoped n-type GaAs layers

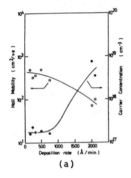
(a)

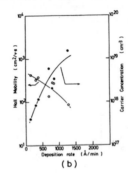
(b)

Fig.11 Effect of growth rate on free hole density and Hall mobility in undoped GaSb layers grown at substrate temperature of 415°C in (a)hydrogen,(b)argon plasma

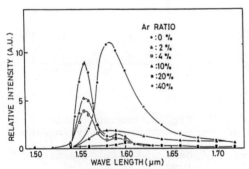

Fig.12 Change of photoluminescence spectra of undoped PAE GaSb grown in mixed plasma of hydrogen and argon.

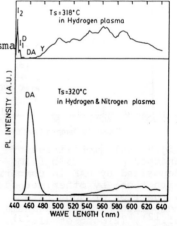

Fig.13 Photoluminescence spectra of PAE ZnSe grown in pure hydrogen and mixed plasma with 30% N_2.

grown on semi-insulating (100)GaAs as a function of the supply ratio of As to Ga (not the flux ratio As_4/Ga as in MBE). An optimum supply ratio was around 2.5, much less than in MBE. arsenic vapours produced by resistive heating are considered to be primarily composed of tetramic As_4 molecules and contain a dimer As_2 component, however, they are likely to be dissociated into lower molecular weight fragments such as As_3, As_2 or As atoms by frequent collision with a mean free path of the order of several millimeters in the present plasma. This dissociation will be favourable in improving crystal quality, as observed in MBE.[17]

5. Conclusions
Several favourable effects of PAE have been shown: high chemical reactivity, reduction of epitaxial temperature, cleaning effect of the substrate and growing surface, high efficiency in impurity doping, good surface morphology etc. Further optimization is expected to improve the quality of PAE layers at lower temperatures with respect, particularly to residual impurities and defects.

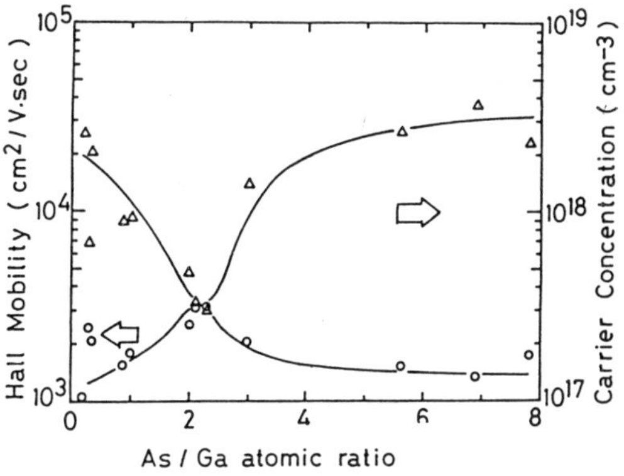

References
1) T. Hariu et al, Inst. Phys. Conf. Ser. No.74, p.193(1984)
2) T. Takagi et al, J. Vac. Sci. Technol. 12,1128(1975).
3) M. Kumagawa et al, Jpn. J. Appl. Phys. 7,1332(1968) and J. Nishizawa et al, Ext. Abstracts, 16th Conf. Solid State Devices and Materials p.1, Kobe(1984)
4) T. Hariu et al, Thin Solid Films 80,235 (1981)
5) K. Matsushita et al, Jpn. J. Appl. Phys. 22, L602(1983)

Fig.14 Carrier density and Hall mobility in n-type undoped PAE GaAs layers as a function of the supply rate As/Ga.

6) K.P. Pande et al, J. Electrochem. Soc. 131,1357(1984)
7) W.G. Townsend et al, Solid State Electron. 16,39(1973)
8) S. Suzuki et al, J. Appl. Phys. 54,1466(1983)
9) T.J. Donahue et al, Appl. Phys. Lett. 44,346(1984)
10) K. Takenaka et al, Jpn. J. Appl. Phys. Suppl.19-2,183(1980)
11) Y. Sato et al, Appl. Phys. Lett. 44, 592(1984)
12) M. Naganuma et al, Shinkuu 23, 326(1980)(in Japanese)
13) H. Goto et al, Jpn. J. Appl. Phys. 20, L893(1981)
14) M. Yano et al, Jpn. J. Appl. Phys. 17,2091(1978)
15) M. Kakei et al, Jpn. J. Appl. Phys. 9,1039(1970)
16) H.M. Manasevit et al, J. Electrochem. Soc. 126,2031(1979)
17) H. Kuenzel et al, Appl. Phys. Lett. 37,416(1980)

ELECTRONIC PROPERTIES OF CVD AMORPHOUS SILICON

Stanislaw M.PIETRUSZKO

Warsaw University of Technology
Institute of Microelectronics and Optoelectronics
IMO PW, Koszykowa 75, 00-662 Warszawa, Poland

1.INTRODUCTION

The chemical vapour deposition (CVD) technique is widely used in the electronic industry. The epitaxial layers and poly-crystalline films for gate electrodes, load resistors, inter-connects and other applications in silicon integrated circuits technology are produced. Recently, there is a growing interest in the use of amorphous and polycrystalline films as an active la-yers in thin film devices for large area applications: flat panel displays, image sensors, solar absorbers, etc.

In this paper the properties of amorphous silicon obtained by thermal decomposition of silane or higher silanes are revie-wed. A systematic studies of the properties of these layers has started only few years ago. Some of the motivation for studing such a high temperature deposited materials comes from the expec-tation that it may be well reconstructed and constitute a proto-type of "perfect" amorphous silicon (a-Si). But this material has high density of dangling bonds due to a low hydrogen content. On the other hand, by post-hydrogenation hydrogen can be introduced to films after deposition and it allows to make independent stu-dies of the effect of hydrogen which have implications for other methods of preparation of amorphous silicon. The hydrogenated amorphous silicon (a-Si:H) of a quality comparable to that pro-duced by much more common glow discharge technique was obtained. Several extended studies on the CVD material have already been published [1-3].

2. DEPOSITION TECHNIQUES AND PARAMETERS

There are mainly three configurations of deposition reactors used in CVD technique. They are shown schematicaly in Fig.1. The first one, reactor A is a "cold wall" atmospheric pressure type (APCVD) used for classical deposition of epitaxial layers at at-mospheric pressure. Typically, pure silane diluted with hydrogen (argon, nitrogen or helium can also be used) is admitted into quartz bell jar with silicon carbide susceptor heated by induc-tion. The epitaxial layers on silicon substrates are grown at a temperature around 1050°C. In the case of polycrystalline films the deposition temperature is usually 650-800°C. Amorphous films are obtained at the temperatures below 650°C. These films can be subsequently annealed.

Reactor B is a "hot wall", low pressure type (LPCVD). In this case the reactor chamber is a quartz tube inserted in a resistively heated furnace. Substrates are held parallel to each other. In our case pure silane (70 ml/min) with carier gas nitrogen (4-20 ml/min) is introduced to the tube. The pressure is typically 0,5 Tr. Deposition occurs both on the substrates and on the walls. Operating at low pressures enhances gas diffusion rates permitting uniform deposition on the stacked substrates. This type of reactor can also be used in CVD deposition from higher silanes.

Reactor C is used for the HOMOCVD (homogeneous CVD). Pure silane is introduced to a quartz tube where the water or nitrogen cooled substrate holder is placed. The crucial difference between B and C is that the gas temperature is higher (typically 625°C) than the substrate temperature (20-400°C). The low substrate temperature allows incorporation of hydrogen.

In all these reactors doping can be achieved by mixing phosphine or diborane with silane in a proportion which ranges from 10^{-7} to 10^{-2}.

In Fig.2 the deposition rates of undoped silicon films obtained by these different methods are compared [4]. In APCVD reactor Beers and Bloem [5] observed two activation energies, one of 2,3 eV at low temperature - low silane dilution and the other of 1,4 - 1,65 eV at high temperature - high silane dilution. Various workers obtained activation energies in the first [3] or second range [6], due to different reactor configurations and different carrier gases.

In the LPCVD reactor we have obtained activation energy of 1,5 - 1,6 eV in the temperature range of 550 < T_D < 650°C and at 0,5 Tr of pressure [7]. A similar activation energy of 1,4-1,7 eV has been reported by other authors under similar deposition conditions [8]. Higher deposition rates are obtained from higher silanes CVD films [9]. These deposition rates are similar to APCVD rates but at a deposition temperatures 150°C lower.

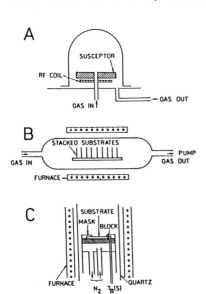

Fig.1. CVD reactors

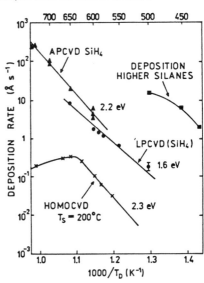

Fig.2. Deposition rate versus temperature.

In HOMOCVD reactor the deposition rate depends on the combination of gas and substrate temperatures. In Fig.2 the rates obtained at $T_s = 200°C$ are shown [10]. At lower gas temperatures ($T_g < 650°C$) the homogeneous decomposition of SiH_4 occurs with an activation energy 2,3 eV, at higher gas temperatures the growth rates are limited by silane depletion at the reactor wall and homogeneous nucleation.

The main chain of reaction in the gas phase most probably is to be the following:

$$SiH_4 \rightleftharpoons SiH_2 + H_2 \tag{1}$$

$$SiH_2 + SiH_4 \rightleftharpoons Si_2H_6 \tag{2}$$

$$SiH_6 + SiH_2 \rightleftharpoons Si_3H_8, \text{ etc.} \tag{3}$$

The first of these reactions has been well characterized thermodynamically [11]. The chain can lead eventually to homogeneous nucleation of solid polymers which manifests itself by the appearence of "snow" in the the reactor. In order to prevent homogeneous nucleation, one has to lower the silane partial pressure, either by working at low total pressure (LPCVD) or by dilution in the case of APCVD.

Since only hybrides are produced in the homogeneous reaction it is generally considered that heterogeneous reactions are needed to form the silicon. There have been studies [12,13] attempting to understand the complete deposition process, but the matter is still unclear, especially in the low range of temperature. For example, it is not unambiguously known whether silane is directly decomposited at the surface, or whether it is some other precursor formed in the gas phase. Some information on the latter type of process may be obtained from HOMOCVD experiments [14].

3. STRUCTURE OF CVD FILMS NEAR CRYSTALLIZATION

The common point of thermal CVD techniques is that the deposition temperature is the main parameter which controls the thin film structure and properties. Several authors [2,5,7,16] investigated transition from amorphous to polycrystalline structure as the deposition temperature increased.

The first question one has to answer is if the CVD films are truly amorphous. Janai et al. [2] and Thompson-CSF group [4] presented an extensive studies on films prepared by APCVD. They have found signs of crystallisation around 670°C. According to them in many cases a gradual transition with partially crystallized films which are inhomogeneous in depth occurs. The inhomogeneous distribution of structure can give erroneous results in the interpretation of the electrical and optical measurements in films near crystallization.

We have investigated structure of LPCVD films. The results are shown in Fig.3. We have found that films deposited below 575°C are amorphous and above this temperature are polycrystalline. It has been [16] found that there exists a discontinuity of the coherence length on going from polycrystalline to amorphous material. In the case of pure, as well as hydrogenated silicon, the lowest limit of the crystalline size for polycrystalline material with the diamond lattice appears to be about 30 Å [16 and literature cited therein]. Below this limit, the X-ray diffraction pattern drastically changes. It means that instead of the 220 and 311 diffraction peaks of the diamond lattice, a second, broad maximum of the diffuse scattering from the amorphous mater-

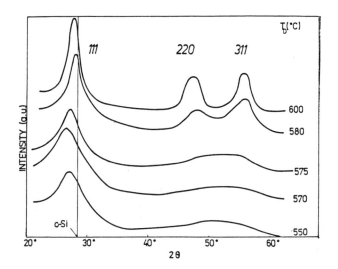

Fig.3.Intensity of X-ray diffraction from films deposited at
different temperatures [7].

ial appears at a position between the 220 and 311 peaks, as in
our case. On the contrary, Thompson-CSF group has reported [4]
some inhomogeneous structure in films deposited near 580°C.

4. HYDROGEN CONTENT AND SPIN DENSITY

Intrinsic CVD amorphous Si is composed of a pure amorphous
Si network with a density close to the crystalline value and a
spin density of about 10^{19}/cm^3 [15,17-20]. The high growth temperature of CVD a-Si produces the compact network and ensures that
the film hydrogen concentration is neglgible (this explains the
high spin density).
Several workers have studied the dependence of N(s) on CVD
a-Si growth temperature [17-20] and the following values are
typical:
Spin density   3,0   8,8   14   20   (x 10^{18}/cm^3)
Growth temp.   500   550   600   650   (°C)
The increase in N(s) with temperature is ascribed to a decreasing
film hydrogen concentration which is typically below 0,1 at %
above 550°C. Gaczi and Booth [21] note N(s) to halve when the
growth temperature is increased from 650 to 700°C and attribute
this to the onset of polycrystallinity.
The spin density of CVD a-Si may be dramatically reduced by
post-growth hydrogen ion implantation or by exposure to hydrogen
plasma [1,3,22]. For example, Hirose [1] notes N(s) to fall from
10^{19} to below 10^{17}/cm^3 after the introduction of less than 0,5 at
% H accompanied by sharp increses in the doping efficiency and
photoluminescence. A feature of CVD a-Si:H versus conventional GD
a-Si:H is that much less hydrogen is required to achieve a similar reduction in spin density; this is related to the higher
growth temperature of CVD a-Si which ensures a nearly ideal amorphous network. Nakashita et al. [23] found that the incorporated

H concentration profile after plasma exposure was well described by a complementary error function.

CVD of higher silanes at low temperatures ($400 - 500°C$) is a good candidate for producing amorphous films with appreciable hydrogen content. Ashida et al. [24] have produced films by CVD decomposition of disilane with N(s) about $10^{16}/cm^3$.

Scott and co-workers [10,14,25,26] have grown a-Si films by HOMOCVD of silane held at $650°C$ and the spin density of such films has a remarkable dependence on the substrate temperature. HOMOCVD produces relatively defect-free a-Si:H containing up to 40 at % H at $T_s = 30°C$ with N(s)= 10^{16} cm^{-3}. The hydrogen content and spin density decreases when T_s increases. At $T_s = 250°C$ hydrogen content is 8 at % and N_s is 2×10^{15}cm^{-3}. These films exhibit efficient visible photoluminescence at the room temperature. The spin density of HOMOCVD films grown at low substrate temperature is much lower than N(s) of comparable GD amorphous Si:H films and this is ascribed to lack of scouring at the film growth surface.

5. TRANSPORT PROPERTIES

Conductivity of undoped films

The room temperature conductivity, CE, of amorphous silicon produced by thermal CVD or LPCVD of silane is in the range 10^{-7} to 10^{-6} S/cm. Hey and co-workers [27] note CE to fall from about 4×10^{-7} to 5×10^{-9} S/cm as the growth temperature falls from 650 to $525°C$ (CVD films) and ascribe this to a reduction in the defect density. This decrease in CE is very marked in view of the small hydrogen concentration (0,8 to 0,2 at %) and the data add weight to the view that defect passivation by hydrogen is far more efficient in CVD than in conventional GD a-Si:H.

Many workers [1,3,20,24,28-31] have hydrogenated CVD a-Si by subsequent annealing (at $350 - 400°C$) in a hydrogen plasma. This leads to a drastic reduction in defect density (by at least two orders of magnitude) and in CE by about one order of magnitude. The other hydrogenation technique employs hydrogen ion implantation and in this case, CE is also reduced by one order of magnitude [32].

Films grown by CVD of higher silanes with the growth rate one or two orders of magnitude higher, have CE in the range of 10^{-8} to 10^{-7} S/cm [16,33]. HOMOCVD a-Si with appreciable hydrogen content has activated CE (2×10^{-7} S/cm) in films deposited at the substrate temperature of $275°C$ [14].

Conductivity of doped films

The conductivity of CVD a-Si can be controlled over a wide range by phosphorus or boron doping as shown in Fig.4. The figure also shows the doping dependence of the spin density and the activation energy. Since this material contains a big quantity of native defects P and B dopant atoms interact with the defects to reduce both the gap states and the spin density, before any substitutional doping can occur. The conductivity of as-deposited films shows a large increase above the threshold dopant concentration in both types of doped samples.

For P doping, CE decreases by about one order of magnitude [1,15,20,34] or remains constant [28,35] at low P doping (up to PH_3/SiH_4 plasma doping ratio around 3×10^{-4}). At higher doping

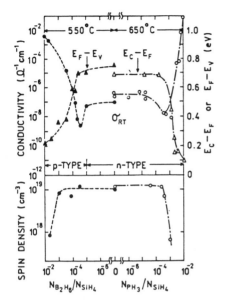

Fig.4. Room temperature conductivity, activation energy and spin density plotted as a function of doping ratio [1].

there is a sudden increase followed by saturation at a conductivity of 10^{-1} S/cm for $PH_3/SiH_4 = 10^{-2}$. This behaviour of the conductivity is attributable to a reduction of dangling bonds and defects due to the compensating action. This conclusion is in agreement with ESR measurements [1,15,28,35].

In the case of B doping CE initially decreases mainly because the Fermi level shifts towards the valence band and p-type conductivity appears at B_2H_6/SiH_4 ratios of 5×10^{-6} to 10^{-5} and than steeply rises to a value of about 10^{-1} S/cm. Nakashita et al. [40] have measured lightly-boron-doped films and found at a doping ratio of 4×10^{-6} intrinsic amorphous Si, of which CE is of the order of 10^{-10} S/cm.

Post-hydrogenation of P doped films leads to a dramatic improvement in doping efficiency, particularly for films with low doping level [28,35], CE increases by five orders of magnitude towards smaller dopant concentrations once hydrogen is diffused into the films. The ESR signal decreases largely what is consistent with an increase in CE.

Values of CE resulting from post-hydrogenation of B doped films differ from those obtained for P doping. The most striking result is the decrease in doping efficiency for strong hydrogenation observed in highly B doped films [28,35]. It is suggested that hydrogenation leads to reduction in the number of electrically active B atoms.

Liao et al. [29] have obtained fairly good doping efficiency by P or B ion implantation of LPCVD amorphous Si films without hydrogenation. The maximum CE obtained in B doped films is 0,3 S/cm and in P doped films is about 10^{-2} S/cm. The hydrogenation did not improve the doping efficiency except at very low implanted impurity concentration. (NB The implantation doping efficiency is about two orders of magnitude higher then that for GD films under the same implantation conditions.)

Doping of the films produced by CVD of higher silanes was investigated [9,37-39] and it was found that electrical properties of disilane CVD films are improved compared with those monosilane films. Disilane CVD films do not exhibit hopping conduction possibly due to bonded hydrogen which reduces number of dangling bonds. As a consequence, CE increases monotonically with P doping up to more than 10^{-2} S/cm for $PH_3/Si_2H_6 = 10^{-1}$. For B doping, CE slightly decreases and p-type conductivity appears at B_2H_6/Si_2H_6 ratios above 7×10^{-6} and than rises to a value of about 10^{-2} S/cm for ratio 10^{-2}. Low B doping efficiency in disilane CVD compared with that in silane CVD is basically attributable to diborane-induced gas phase nucleation of silicon particles in disilane.

Šcott and co-workers [41] have found the CE of HOMOCVD films to saturate at 1 - 2 orders of magnitude greater than the CE of GD films. It is suggested that the lower operating energy of the HOMOCVD process (0,1 eV as opposed to 10 - 100 eV in plasma processes) gives more efficient doping and results in a lower surface mobility of adsorbed dopant atoms and prevents diffusing to non-active sites.

Temperature dependence of conductivity

Measurement of the conductivity temperature dependence, CE(T), give information about the transport in a-Si. Typical temperature characteristic is shown in Fig.5 [7].

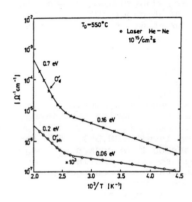

Fig.5. The conductivity and photoconductivity versus reciprocal temperature [7].

The main feature of CE(T) of a-Si produced by the thermal CVD is that the conductivity occurs mainly via hopping in deep traps near the Fermi level, at temperatures below 370 - 420 K, while at higher temperatures conductivity via extended states is dominant. The activation energy at low temperatures is typically 0,14 to 0,17 eV and at higher temperatures is 0,6 to 0,8 eV. The pre-factor of the high temperature conduction is in the range 10^2 to 10^3 S/cm.

Hey et al. [20,27] noted a sharp drop in the high temperature conductivity pre-factor by four orders of magnitude and in the activation energy by 0,4 eV as the CVD a-Si growth temperature is raised from 525 to 650°C. They attribute these results to a change in the dominant conduction mechanism from extended state conductivity to hopping caused by changes in the very small H concentration in CVD a-Si films.

The increased H concentration due to post-hydrogenation led to a reduction in the hopping conductivity and to a single activated conductivity over a wide temperature range. Suzuki et al. [32] implanted hydrogen ions into CVD films and found that hopping conduction is eliminated after post-implantation annealing at 400°C.

Results for doped CVD amorphous Si films show that at low doping levels CE(T) is non-linear and similar to that of undoped samples [1,20,23,28-29,34-35]]. In the higher doping range this dependence shows a single activation energy which decreases when the doping increases. For the most heavily P and B doped films the activation energy is above 0,1 eV. The post-hydrogenation reduces the hoping conductivity and CE(T) shows a single activation for temperatures between 200 and 420 K for the whole doping range with the pre-factor in the range 10 to 10^4 S/cm. Hasegawa et al. [35] found that hydrogenation largely decreases the activation energy for films with doping ratio below 10^{-3}.

Beyer and Overhof [36] reported that the "kink" at temperatures about 400 K seen in the CE(T) of many GD a-Si:H films (connected with the formation of a density of states peak near midgap influencing transport and reducing the doping efficiency) is absent in CVD a-Si. The temperature shift of the Fermi level is almost linear over wide temperature range. It is suggested that CVD a-Si has improved microstructural homogeneity.

A-Si grown by CVD of higher silanes has a single activation energy between 0,7 and 0,8 eV [37-39]. The use of higher silanes, mainly disilane, leads to reduction in the hoping conduction, as the disilane CVD films incorporate bonded hydrogen much more than the monosilane CVD films. Some workers have investigated doped material [9,37-39]. Hirose [39] has found that the single activation energy for P doped films does not change at low P doping (up to PH_3/Si_2H_6 doping ratio around 5 x 10^{-4}). At higher doping there is a sudden decrease to 0,2 eV for $PH_3/Si_2H_6 = 10^{-2}$. For B doping, the activation energy slightly increases for $B_2H_6/Si_2H_6 = 7 \times 10^{-6}$ and than decreases to 0,4 eV for this ratio equal to 10^{-2}. Akhtar et al. [9] have observed a single activation energy to change from 0,61 eV (undoped) to 0,12 (P doped).

Scott and co-workers [14] have obtained amorphous Si by HOMOCVD and found CE(T) to be linear with activation energy 0,7 to 0,6 eV and pre-factor 10^{-3} to 10^{-4} S/cm in films deposited at substrate temperatures between 250 and 350°C. They have also achieved very efficient doping with an activation energy of 0,15 eV at highest dopant concentrations.

6. PHOTOCONDUCTIVITY

The room temperature photoconductivity of CVD amorphous Si is in the range 10^{-10} to 10^{-9} S/cm [15,34]. The temperature dependence of photoconductivity is nonlinear on an Arrhenius plot (Fig.5). Films deposited at lower temperature (490°C) have photoconductivity by more than an order of magnitude higher and the temperature dependence is modified [34].

The exposure to hydrogen plasma after deposition changes the photoconductivity drastically and increases the H content to a few percent. Hirose [1] has reported that the AM1 photoconductivity exceeds 10^{-9} S/cm. Hey and Seraphin [20] have obtained a photoconductivity 2 x 10^{-5} S/cm. Berman et al. [31] have reported that films possessing photoconductivity of the order of 10^{-4} S/cm under AM1 can readily be prepared. Hasegawa and co-workers [22] have maximised the photoconductivity at 4 x 10^{-6} S/cm for a 350°C anneal for about 30 min. Nakashita et al. [30] found that the photoconductivity at room temperature is increased by more than four orders of magnitude to 10^{-6} S/cm by post-hydrogenation.

CVD of higher silanes, mainly disilane at low temperature (450 - 550°C) gives a-Si with higher hydrogen content and hence with higher photoconductivity than CVD of monosilane. The reported photoconductivity falls in the range 5 x 10^{-7}- 5 x 10^{-5} S/cm, depending on the substrate temperature and deposition parameters [9,24,33,38,39]. Scott et al. [14] have grown a-Si:H films by HOMOCVD and obtained AM1 photoconductivity as high as 10^{-4} S/cm for films grown at 275°C.

The general trend of increasing photoconductivity with decreasing deposition temperatures should be related to the increase in hydrogen content and decrease in dangling bond density; resulting in a lifetime increase. This agrees with increased photoconductivity after post-hydrogenation.

It is important to note that amorphous Si produced by CVD or HOMOCVD exhibits no significant Staebler-Wronski effect.

318

7. APPLICATIONS OF CVD AMORPHOUS FILMS

The first proposed application for CVD a-Si has been for photothermal convertion [42]. A proposed structure was Si_3N_4/ a-Si/Ag/substrate. Replacement of crystalline silicon by amorphous silicon which has a higher absorption coefficient allows the use of thinner layers. It has been shown [43] that suitable dopant (eg. carbon) can prevent high temperature recrystallization and stabilize amorphous structure while maintaining adequate optical properties.

Janai and Moser [44] note that the band-edge optical absorption of CVD a-Si is reduced by about an order of magnitude by subsequent ruby laser anneal and suggest CVD a-Si a suitable material for optical storage devices.

CVD a-Si as-deposited and post-hydrogenated has been used to obtain Schottky barrier [45] and p-i-n [46] devices for photovoltaic applications as well as for fast photoconductors [47]. The effords are made to produce matrix of thin film transistors for flat panel displays and image sensors.

8. CONCLUSIONS

The density of defects and general properties of CVD a-Si depend mainly on deposition temperature, in contrast to glow discharge a-Si:H where the properties of films depend on a large number of deposition parameters. However material obtained by this technique has been given less attention due to large density of states which make it less suitable for electronic applications. But appearance of new low temperature techniques like HOMOCVD, CVD of higher silanes or a post-hydrogenation have stressed the advantages of the CVD technique: low density of gap states due to high hydrogen content, reproducibility and absence of Staebler-Wronski effect.

Also from a fundamental point of view CVD a-Si can be considered as a reference material, good for the investigation of basic questions concerning amorphous silicon.

ACKNOWLEDGMENTS

This work was carried out under Polish Central Program for Fundamental Research CPBP 01.08 coordinated by University of Lodz

REFERENCES

1. M.Hirose, J.Phys.Colloq. (France) 42 C-4(1981)705
2. M.Janai, D.D.Allred, D.C.Booth, B.O.Seraphin, Solar Energy Mat. 1(1979)11
3. N.Sol, D.Kaplan, D.Dieumegard, D.Dubriel, J.Non-Cryst.Solids 35 & 36 (1980)291
4. J.Magariño, in Poly-micro-crystalline and amorphous semiconductors, Ed. by P.Pincard and S.Kalbitzer, Editions de Physique (1984)651
5. A.M.Beers and J.Bloem, Appl.Phys.Lett. 41(1982)153
6. M.Hirose, M.Taniguchi, Y.Osaka, J.Appl.Phys. 50(1979)377
7. S.M.Pietruszko, in Semiconductor and Integrated Circuit Technology, Ed. by X.Y.Wang and B.X.Mo, World Sci.Publ. (1986)318
8. G.Harbecke, L.Krasbauer, E.Steigmeir, A.E.Widmer, H.F.Kapport,

G.Neugebauer, RCA Rev, 44(1983)287
9. M.Akhtar, V.L.Dalal, K.R.Ramaprasad, S.Gau, J.A.Cambridge,
    Appl.Phys.Lett. 41(1982)1146
10.B.A.Scott, J.A.Reimer, P.A.Longeway, J.Appl.Phys. 54(1983)6853
11.J.H.Purnell, R.Walsh, Proc.R.Soc. A 293(1966)543
12.M.L.Hitchman, J.Kane, A.E.Widmer, Thin Solid Films 59(1979)231
13.W.A.Bryant, Thin Solid Films 60(1979)19
14.B.A.Scott et al., Appl.Phys.Lett. 40(1982)973
15.G.Harbeke, A.E.Widmer, J.Stuke, J.Phys.Soc.Jpn. 49(1980)
    Suppl.A 1229
16.S.Veprek, Z.Iqbal, F.A.Sarott, Phil.Mag. B 45(1982)137
17.S.Hasegawa, T.Kasajima, I.Shimizu, Solid State Commun.
    29(1979)13
18.D.Kaplan, Phys.Scr. (Sweden) 124 2(1981)396
19.T.Nakashita, M.Hirose, Y.Osaka, Jpn.J.Appl.Phys. 20(1981)471
20.P.Hey, B.O.Seraphin, Solar Energy Mater. 8(1982)215
21.P.J.Gaczi, D.C.Booth, Solar Energy Mater. 4(1981)279
22.S.Hasegawa, D.Ando, Y.Kurata, T.Shimizu, Phil.Mag. B
    47(1983)139
23.T.Nakashita, Y.Osaka, M.Hirose, T.Imura, A.Hiraki,
    Jpn.J.Appl.Phys. 22(1983)1766
24.Y.Ashida, Y.Mishima, M.Hirose, Y.Osaka, K.Kojima,
    Jpn.J.Appl.Phys. 23(1984)L129
25.D.J.Wolford, J.A.Reimer, B.A.Scott, Appl.Phys.Lett.
    42(1983)369
26.B.A.Scott, W.L.Olbricht, B.A.Meyerson, J.A.Reimer,
    D.J.Wolford, J.Vac.Sci. & Tech. A 2(1982)450
27.P.Hey, N.Raouf, D.C.Booth, B.O.Seraphin, AIP Conf. 73(1981)58
28.J.Magariño, D.Kaplan, A.Friedrich, A.Deneuville, Phil.Mag. B
    45(1982)285
29.X.B.Liao et al., Solar Energy Mater. 6(1982)147
30.T.Nakashita, M.Hirose, Y.Osaka, Jpn.J.Appl.Phys. 23(1984)146
31.A.Berman et al., J.Non-Cryst.Solids 59&60(1983)751
32.T.Suzuki, M.Hirose, Y.Osaka, Jpn.J.Appl.Phys. 19-2(1980)91
33.S.C.Gau et al., Appl.Phys.Lett. 39(1981)436
34.M.Taniguchi, Y.Osaka, M.Hirose, J.Elect.Mater. 87(1979)689
35.S.Hasegawa, D.Ando, Y.Kurata, T.Shimizu, Jpn.J.Appl.Phys.
    22(II)(1983)L 815
36.W.Beyer, H.Overhof, J.Non-Cryst.Solids 59&60(1983)301
37.F.B.Ellis, R.G.Gordon, W.Paul, B.G.Yacobi, J.Non-Cryst.Solids
    59&60(1983)719
38.R.E.Rochelean, S.S.Hegedus, B.N.Baron, MRS Proc. 49(1985)15
39.M.Hirose, in Amorphous Semiconductor Technology and Devices
    Ed. by Y.Hamakawa (1984)67
40.T.Nakashita, K.Kohno, T.Imura, Y.Osaka, Jpn.J.Appl.Phys. 23
    (I)(1984)1547
41.B.A.Scott, in Semicond. & Semimetals 21A(1984)123
42.B.O.Seraphin, J.Vac.Sci. & Tech. 16(1979)193
43.D.C.Booth, D.D.Alfred, B.O.Seraphin, Sol.Ener.Mat. 2(1979)107
44.M.Janai, F.Moser, J.Appl.Phys. 53(1982)1385
45.Y.Mishima, M.Hirose, Y.Osaka, Jpn.J.Appl.Phys. 20(1981)593
46.N.Szydlo, E.Chartier, N.Proust, J.Magariño, D.Kaplan, Appl.
    Phys.Lett. 40(1982)988
47.D.H.Auston, P.Lavallard, N.Sol, D.Kaplan, Appl.Phys.Lett. 36
    (1980)66

THERMAL CONSIDERATIONS IN THE DESIGN OF MILLIMETER WAVE DEVICES

ISHWAR CHANDRA
Solidstate Physics Laboratory
Lucknow Road, Delhi-110007,
INDIA

Introduction

There have been many technological advances in the development of semi-conductor devices for microwave and millimeter waves. These have been possible because of better device design, growth of material of required specifications, fabrication procedure and discovering newer structures. During the operation of active devices like Gunn and IMPATT diodes the temperature of the active layer or the junction increases considerably. Thermal considerations, therefore, form an important design aspect to determine the device geometry etc. Taking the example of these devices the role of skin effect is brought out in this paper.

Heat Sink Considerations

The key to proper thermal design of a Gunn diode or an IMPATT diode is to achieve high power and low active layer or junction temperature. Consider the Gunn diode device structure as given in fig. 1. The active layer temperature for a particular device geometry can be calculated by a conventional heat flow model [1]. The contribution of the substrate is neglected here and the heat flow assumed only towards the sink. The maximum temperature is expected to occur at the active layer substrate interface.

$$T_{max} = \left\{ T_0 + Q \left(\frac{d}{2K} + \frac{l_c}{K} \right) \right\} \exp \left\{ Q \left(\frac{l_n}{120} + \frac{l_n}{300} \right) \right\}$$

Where T_0 is the ambient temperature, Q is the input heat flow.

For safe operating active layer temperature between 450-500 K in the case of copper heat sink, the device diameter works out to 50-70 microns, with the low field resistances 1.10 - 0.55 ohms. Similar calculations can be made for IMPATT diodes. In the above analysis losses from the substrate are neglected.

In practical devices, however, the substrate has a finite thickness which contributes to the D.C. resistance and therefore to the device heating. When the device becomes operational, the current distribution no more remains homogeneous. The skin effect causes the current displacement in the contact region towards the edge of the bond patch [2,3]. Since the current is confined towards the periphery of the substrate the effective resistance considerably increases. Effective normalized ohmic resistance calculations made by Lay and Kuchne [3] for 100 GHz Si IMPATT device are shown in fig. 2. Similar calculations made by us for 47 GHz GaAs device are shown in fig. 3.

According to these calculations if the doubling of the substrate resistance is acceptable, the dimensioning rule $\delta/a \rangle 0.2$ and $L/a\langle 2/3$ result to avoid additional losses. Accordingly the maximum permissible thickness could be about 9 microns in the case of very high conductivity Si substrates for 100 GHz IMPATTs, 12 microns in the case of InP devices with 3×10^{18} cm^{-3} doping level for W-band CW operation [3], while our results show maximum permissible 10 microns substrate of high conductivity GaAs for 47 GHz Gunn diodes.

Thinning of substrate to such small dimensions is a serious technological problem. A bare sample of GaAs of this thickness cannot be handled and so a gold plated support has to be provided. This support leads to curling of the wafers. (Photo-1) shows curling of a 15 micron thin GaAs with 30 microns thick gold support. This has been avoided by us by using an etch-stop pattern provided on the epitaxial side of the wafer, plating it to provide the support structure, and lapping it from the substrate side to reveal an array of GaAs device mesas interspersed in the plated support matrix (Photo-2).

Skin Effect Consideration in Bonding Wire

The contribution of skin effect in the bonding ribbon/wire in the device performance is not available in literature and some approximate calculations have been made by us [4].

Considering the high frequency and DC resistances of wire/ribbon as given by $R_{rf} = \dfrac{l}{\pi \sigma \, \delta (D-\delta)}$ and $R_{dc} = \dfrac{4l}{\pi \sigma D^2}$ where δ is the skin depth at frequency of operation, l & D the length and diameter of the bonding wire. Taking the skin depth for Gold at 47 GHz as 0.36 u dc and rf resistances are calculated. Assuming that all the heat generated flows longitudinally, the temperature will assume a parabolic profile with the maxima in the centre when the two ends are at equal temperature. The rise in temperature ΔT will be given by

$$\Delta T = \frac{1}{8} \frac{\sigma}{K} i^2 r^2 \simeq 2 \times 10^4 \times i^2 r^2$$

where i is the current flowing r the resistance, σ & k the electrical & thermal conductivity of the wire. Rise in temperature of the wire/ribbon in the dc and rf cases for 50 and 100 GHz has been calculated for various geometries of wire and ribbon for different current values. The results are shown in the table.

These preliminary calculations are approximate and the results indicative of the effect. The rise in temperature is quite significant and explains the experimental observations. It has been noted that in experimental Gunn devices whereas the current-voltage characteristics are absolutely steady below threshold voltage with almost the same current flowing beyond this voltage, a 25 micron bonding wire fuses at a spot almost midway between the device chip and the collar at which its other end is bonded (Photo-3). The need to provide bonding straps of large overall periphery is evident, specially in 100-150 GHz devices, where the skin depth in Gold ~ 0.1 micron. This also indicates that apart from device design the bonding lead design is equally important.

Conclusions

It has been shown that in MMW devices the skin effect considerations are important not only in determining the maximum substrate thickness but also for determining the bonding strap dimensions. These results indicate that skin effect may have to be more closely looked into while designing the mm wave MMIC's also.

Acknowledgement

The author is grateful to Director SSPL for his encouragement and inviting him to present this work at the conference. Contribution of all the members of GaAs device processing group in the laboratory is gratefully acknowledged.

References

1.   Narayan S.Y., RCA Review 33 (4) p 752 (1972).

2.   De Loach B.C., IEEE Trans MTT. 18, p 72 (1970).

3.   Luy J.F. and Kuchne, Solid State Electronics
      29 (4) pp 471-476 (1986).

4.   Agarwal A.K., Gulati Mrs. R. and Chandra I., to be published.

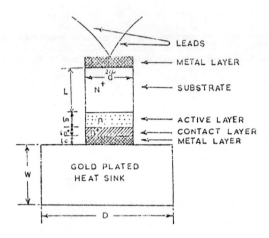

FIG.-1   GUNN DIODE STRUCTURE

325

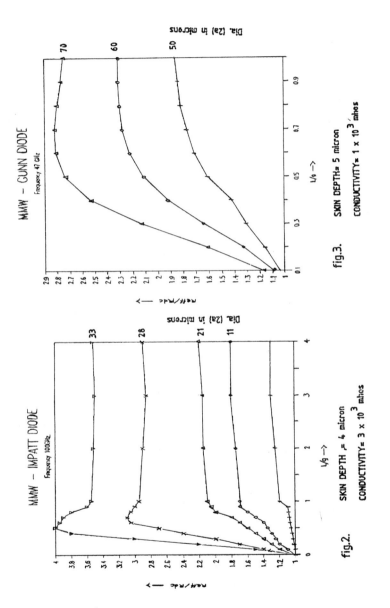

MMW – GUNN DIODE
Frequency 47 GHz

Dia. (2a) in microns

Reff/Rdc →

fig.3.

SKIN DEPTH= 5 micron
CONDUCTIVITY= 1 x 10³ mhos

MMW – IMPATT DIODE
Frequency 100GHz

Dia. (2a) in microns

Reff/Rdc →

fig.2.

SKIN DEPTH = 4 micron
CONDUCTIVITY= 3 x 10³ mhos

TABLE 1

Current Amp	25 micron dia wire			50x25 micron ribbon		150x25 micron ribbon	
	ΔT_{dc}	ΔT_{50GHz}	ΔT_{100GHz}	ΔT_{50GHz}	ΔT_{100GHz}	ΔT_{50GHz}	ΔT_{100GHz}
0.25	0.5	237	3801	74	1195	16	253
0.50	2.5	951	-	298	-	65	1012
0.75	5.4	-	-	675	-	146	-
1.00	9.7	-	-	1195	-	259	-
1.50	21.8	-	-	-	-	406	-
2.00	38.7	-	-	-	-	564	-
2.50	60.5	-	-	-	-	796	-
3.00	87.1	-	-	-	-	1036	-

Table showing rise in temperature (°C) of various geometry 450 micron gold strape at 50 and 100 GHz

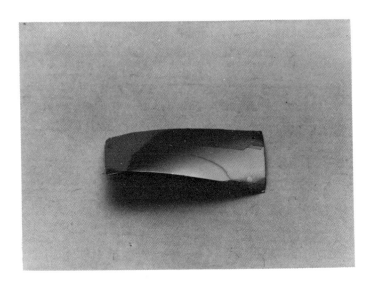

Photo 1

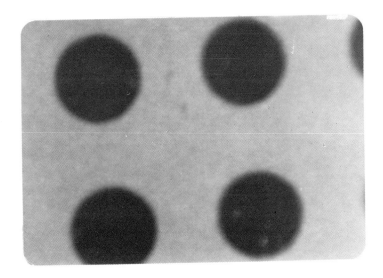

Photo 2

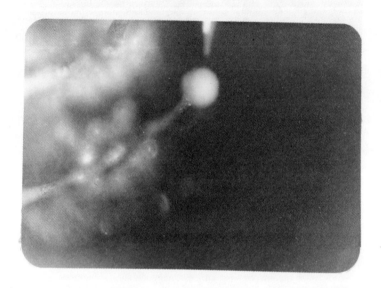

Photo 3

OVERVIEW OF POWER SEMICONDUCTOR DEVICES

Harshad Mehta*
Electric Power Research Institute
3412 Hillview Avenue
Palo Alto, CA 94303

Victor A.K. Temple
Corp. R&D Center
General Electric Company
One River Road
Schenectady, NY 12301

Abstract

This paper presents an overview of power semiconductor device technologies. Recent developments and future trends are discussed. The content of paper is based mainly on technical information available in published papers. Conventional discrete devices such as the thyristor, the bipolar transistor, unipolar transistor and BIMOS devices are referred to by dividing them into groups depending on whether their on-state current flows through 0, odd or even P-N junctions. There has been steady progress with each of these power devices in current carrying and voltage blocking capabilities as well as switching performance.

1. Introduction

Since the development of the junction transistor in the early 1950s, there has been a steady growth in the ratings of silicon devices. The ever increasing power handling capability of these devices at higher and higher frequencies has opened up new markets for their application. Today the market for silicon power devices already exceeds one billion dollars and a steady growth has been projected for the future. The applications of power devices extends from low power levels where the devices are used in commercial and military applications, to very high power levels for industrial applications. These devices also cover a very broad range of frequencies from the line frequency (60 Hz) on up to frequencies as high as 100 KHz. Some of the most recent devices introduced into the market in the last two years have extended the application of silicon power devices beyond 1 MHz. In addition, silicon power devices being developed in industrial laboratories have already proven their performance up to frequencies as high as 1 GHz.

The current status of the application of these devices spans a wide range in both the consumer and the industrial sectors. Silicon power devices have already made a place in home appliances such as toasters, blenders, refrigerators, ovens, washing machines, television and radio receivers, as well as in lighting control. A revolution in

*At Present in Washington, D.C. on special assignment.

**MCT development is sponsored jointly by EPRI and GE.

automotive electronics in the coming years is expected to extend their
state controls on a car such as windshield wipers, fuel injection
systems, and antiskid controls. In the industrial sector, these
devices have made a tremendous impact on power generation and
distribution, transportation systems, and industrial processing
equipment. Their application have extended from motor controls for
machine tools, hoists, and forklifts, to transporation systems, and
industrial processing equipment. Their applications have extended
from motor controls for machine tools, hoists, and forklifts, to
transportation systems which include railroads and electric cars. In
addition, solid state uninterruptible power supplies are currently
being widely used in hospitals, air traffic control systems, and radio
and television stations.

2.Device Technologies

Power semiconductor devices are based on either Bipolar or MOS (Metal
oxide semiconductor) technology. Table 1 briefly outlines the major
features of these technologies. Further power semiconductor devices
can be classifed by number of junctions the on-site current flows
through – 0 for most FETs, 1 for diodes, 2 for transistors or 3 for
thyristors. Even junction devices usually have lower forward drop but
also generally requires one or more central regions to be charged from
an external source which injects the carrier type that can not be
injected from the outer most semiconductor layers. Odd junction
devices have the advantage of eliminating third terminal (for charge
injection) but have a forward drop that includes at least a diode
drop.

The internally and externally charged subclasses have, as prime
example, the MOSFET and the BJT. The advantage of the internally
charged device is its high ratio of controlled energy which vastly
amplifies device control and allows for single chip control and
protection. Naturally all MOS gated devices fall into this
category. However, at least in their latch condition, so do diodes,
thyristors and, In fact, all of the odd junction class of devices.
This will be expound– ed in more detail later. It may also be
surprising that the SIT (Static Induction Transistor) could be
considered with the FET as a O junction device but, unlike the FET, is
an externally charged device.

CHARACTERISTICS	BIPOLAR (THYRISTOR)	METAL-OXIDE SEMICONDUCTOR
CONDUCTION	•MAJORITY & MINORITY CARRIERS	•MAJORITY CARRIERS ONLY
ADVANTAGES	•HIGH POWER CAPABILITY	•LOW TURN-OFF ENERGY
	•LOW FORWARD DROP	•HIGHER THERMAL LIMIT (250^0c)
	•LOW COST	•INEXPENSIVE TO DRIVE
DISADVANTAGES	•NO TURN-OFF CAPABILITY	•LOW VOLTAGE
	•HIGH TURN-OFF ENERGY	•LOW CURRENT
	•THERMAL LIMIT	•HIGH-ON RESISTANCE
	•LIMITED SWITCHING SPEED	

Table 1. - Power semiconductor device technologies

3. ZERO JUNCTION DEVICES

In power device design there are 3 main tradeoffs associated with power handling - voltage capability, forward drop and switching speed. All zero junction devices have two advantages, a resistive forward drop that allows lowest VF, at least at low current densities, and high switching speed since turn-off requries no charge recombination phase. Breakdown voltage is sometimes achieved by inserting a junction that is bypassed by a voltage controlled channel or by inserting a grid by which current flow can be pinched off.

The disadvantage of the zero junction device is that current flow is resistive, making them high in forward drop at high current density. What constitutes high current density goes down very rapidly as device breakdown voltage increases, making FETs poor power devices for voltages above several hundred volts.

Table 2 lists some of the better known 0 junction devices. These fall into the internal and external charged sub-classes depending on whether they are junction gated as in the J-FET and SIT, or whether they are MOS gated, as in the MOSFET. We have recognized that latching devices may be confusing in that the charge needed to keep current flowing is internally derived but that the charge that needs to be inserted or removed for turn-on or turn-off is generally externally derived. Accordingly Tables 2-5 subdivide the control charge column into on-state and switching. Note that "I" and "E" stand for internal and external, respectively. Also note that all of the J-FET devices can be run in the bipolar mode if desired in which cases a certain fraction (nearly all) of the on-state current flow would relay through the gate region turning the device into a 2

332

junction device.

To be competitive most zero junction devices must be built with very
dense gate repeat distances which often compromise cost and yield.
This is especially so in J-FET sturcture which have the added
difficulty of needing a low gate resistance. For this reason, and
because of the advantages the internally charged subclss has in
gating, the MOSFET is the clear winner except in those applications
where one must have FET forward drop at low current and bipolar
forward drop at high current. One exception to the dense gate
requirements is the Schottkey diode, although some new concepts in
Schottkey's utilize a grid structure.

DEVICE	CHARGED - ON-STATE/ SWITCHING	COMMENTS
RESISTOR	I / I	NO VOLTAGE CAPABILITY
MOSFET	I / I	SOME MAY HAVE 2 JUNCTION PARASITICS
J-FET	I / E	OPERATED IN VOLTAGE CONTROL MODE
SIT	I / E	OPERATED IN VOLTAGE CONTROL MODE
SCHOTTKEY DIODE	I / I	SOME MAY HAVE 1 JUNCTION PARASITIONS

Table 2. - Zero Junction Devices

4. ONE JUNCTION DEVICES

Junction diodes might be thought to be the only devices in this
category. However, the IGBT (Insulated Gate Bipolar Transistor also
termed the IGT or COMFET) also satisfies the definition since, in the
on-state the current flows through a diode in series with a FET.
These devices have the advantage of good modulation and internal
charging at the expense of a diode minimum forward drop and a
recombination limited switching speed. The IGBT's transistor name is,
however, apt in the it is used as a transistor and turns off like an
open-base wide-base transistor. Its FET limited injection capability
and extra drop means that the IGBT drop greatly exceeds that of the
diode.

DEVICE	CHARGED ON-STATE/ SWITCHING	COMMENTS
P-N DIODE	I / I	"OPTIMUM" BIPOLAR DEVICE
IGBT (IGT, COMFET,ETC)	I / I	HAS PARASITIC 2 AND 3 JUNCTION MODES
SITH (FTD, FCT, ETC)	I / E	AT LOW CURRENT LEVEL ONLY. BECOMES 3-JUNCTION AT HIGHER CURRENT

Table 3 - One Junction Devices

The SITH (Static Induction Thyristor, also termed the FCT or Field Controlled Diode) would also seem to be diode--like in the on-state. This is true, however, only for a limited current range. This explains the comment next to SITH in Table 3.

5. Two Junction Devices

Although the comments next may specifically refer to bipolar junction transistors they qualitatively apply to other 2*n junction devices (n>0). All of these devices need charge flow into the center region to allow current flow. If that flow is large enough device will have a minimum forward drop (sometimes called saturation voltage) that is less than a diode drop. Essentially, once the mid-zone (base) background doping is overwhelmed, the two junctions no longer exist as barriers. The actual high current drop will depend on how well both base and the off-state voltage supporting part of the collector are modulated. This will not be as modulated as in a diode because minority carrier injection will be present only at one point in the device, whereas, in the diode, it is from both sides. This makes it possible for a diode to have the same modulation profile at the same voltage rating as a transistor yet need only one-fourth the carrier lifetime.

DEVICE	CHARGED ON STATE/ SWITCHING	COMMENTS
BJT	E / E	MAIN VARIATIONS ARE WIDE AND NARROW BASE
J-FET	E / E	OPERATED IN CURRENT CONTROL MODE
SIT (ALSO FTD, ETC.)	E / E	OPERATED IN CURRENT CONTROL MODE
MOS-DAR- LINGTON	I / I	MAY HAVE 3-JUNCTION PARASITICS

Table 4. Two Junction Devices

One also has to distinguish between wide and narrow base transistors which support voltage largely in their base and collector, respectively, in the externally charged widebase transistor the higher the turn-off base drive the faster the turn-off. This is taken advantage of in low voltage bipolar switching devices that operate in the gigahertz range. High voltage, wide-base, externally charged transistors do not yet exist owing to the need for the high voltage base zone to be externally contacted. Finally, there are numerous MOS/BJT combinations in which on-state current crosses 2 p-n junctions. One example is the MOS Darlington in which an FET is integrally connected between the collector and the base. This gives the important advantage of FET gate control and puts the device into the internally charged category if the FET is MOS gated.

6. Three Junction Devices

Forward drop, voltage and speeds of 3-junction devices are qualitatively similar to those of the 1-junction device. What we have added with the extra junctions is reverse blocking capability needed for ac circuits and control capability which is exersized by inserting or removing base charge, usually through a high current gate terminal. Devices include the standard thyristor, the GTO (Gate Turn-Off thyristor, an SCR with a highly interdigitated gate) and the SITH (also called Controlled Diode). The later device can be viewed as a diode with a grid inserted for voltage controlled turn-off. Unfortunately, this grid must also serve to remove minority current and a good deal of the modulated device's stored charge. Optimal SITHs must have a much higher interdigitation level than a GTO which in turn is much more highly interdigitated level than a GTO which in turn is much more highly interdigitated than the SCR. Where it suffers is in ease of control - from no control in the SCR to fairly difficult and high current in the GTO and SITH. Note that the SITH should be classified as a 1-junction device at low current densities.

DEVICE	CHARGED ON-STATE/ SWITCHING	COMMENTS
SITH (FTD, FCT, ETC)	I / E	3-JUNCTION AT NORMAL TO HIGH DEVICE CURRENT.
GTO	I / E	OPERATES LATCHED.
SCR, THRISTOR	I / E	OPERATES LATCHED UNTIL CURRENT REVERSAL.
MOS-THYRIS-TOR	I / I	MOS GATED TURN-ON BUT CAN'T BE TURNED OFF.
MCT (ALSO MOS-GTO	I / I	OPTIMUM HIGH POWER DEVICE.

Table 5.   Three Junction Devices

MOS Controlled Thyristors (MCTs) are a new class of devices which
allow MOS control of the thyristor.   This device class adds the ease
of MOS gate control to devices with already the best high voltage,
high current capability.   Because this class of devices is relatively
recent an additional section of this paper describes some of the first
MCTs in more detail.   Table 5 lists some 3-junction devices including
the MOS-SCR, not to be confused with the MCT, in which an integral FET
connects layers 1 to 3 or 2 to 4 and is activated to turn the
thyristor on.

7.   MOS CONTROLLED THYRISTORS/MOS GTO**

There are several ways to combine MOS and thyristor elements to form
advanced devices.   Recently, both GE and Siemens have described a new
device termed the "MOS GTO" by Siemens and the MCT, a more inclusive
name, by GE.   The key element of the device is the MOS device located
across the upper emitter-base junction of the vertical thyristor
element which, when turned on, shorts out the upper emitter and
quenches the thytristor internal positive feedback, thus allowing it
to turn off.   Looking at the equivalent circuit it is clear that, when
on, the FET will be able to divert an upper base current given by the
on-state emitter-base forward drop divided by the FET path
resistance.   Because the base is normally doped to the order of
1e17/cm3 and, in addition, is well modulated we find that the off-FET
path is dominated by its channel resistance alone.   As in the GTO
there is some turn-off gain with the result that the approximately
current that can be turned off is given by:

$$I\ off = 2\ VJ/(AI \ast Rfet)$$

where VJ is the on-state emitter-base junction drop, AI is the current
gain of the lower transistor contained in the thyristor structure and
Rfet is the FET resistance.   Typically 1000 to 2000 A/cm2 can be
turned off at room temperature with 20 volts applied to the MOS
gate.   The largest turn-off density reported has been 6000A/cm2 while
the largest reported current turned off has been 200A at room
temperature with a resistive load.

Device turn on has been accomplished in 4 different ways.   We have
used light incident on a perforated upper emitter contact, we have
used noise (This alternative is possible because the standard emitter
shorts can be omitted, their function more than adequately taken over
by the FET controlled emitter shorts.), we have used a standard gate
(ie current directly into the upper base) and we have used a built-in
"on-FET" which would be connected between the emitter and collector of
the upper transistor.   This last choice is much the preferred one and
results in a single gate - electrode which is pulsed on polarity for
turn-on and opposite & polarity for turn-off.

336

Conclusions:

Power devices play an important role in system applications. Various devices are described and their major advantages and disadvantages are outlined. It is very likely that no one device exists or will exist that answers all needs and there always be the need to match devices with applications appropriate to their special features and requirements. Power devices will certainly be a key to increase both performance and reliability of power electronic systems.

References

1. V.A.K. Temple, "Power Device Evolution," SATECH Conference, September (1987)

2. V.A.K. Temple, "MOS Controlled Thyristors," IEDM Technical Digest (1984)

3. A Herlet, "The Forward Characteristics of Silicon Power Rectifiers at High Current Densities," Solid State Electronics, Vll. P. 717 (1968)

4. B.J. Baliga, "Silicon Power Field Controlled Devices and Integrated Circuits," Applied Solid State Science, Supplement 2B, P. 109

5. S.M. Sze, "Physics of Semiconductor Devices" Wiley (Interscience), New York 1969

6. S.K. Ghandhi, "Semiconductor Power Devices" Wiley, New York 1977

Thin Film Polycrystalline Devices using Ternary Semiconductors

R. W. Miles, M. Carter, A. Knowles, S. Arshed, H. Oumous and R. Hill
(Newcastle Photovoltaics Applications Centre, Newcastle, UK)

Abstract

The methods used to fabricate the ternary semiconductor copper indium diselemide are discussed with reference to the utilization of this material in solar cells and the procedure followed to produce high efficiency cells is also discussed. The potential of the mercury zinc telluride alloy system for use in solar cell structures is also briefly considered.

Introduction

Although amorphous silicon is the most developed thin film material for fabricating commercial solar cells there is considerable interest in other materials particularly for use in tandem solar cell structures [1]. Of these materials the ternary semiconductor $CuInSe_2$ and CdTe (and its ternary alloys with HgTe and ZnTe) are of considerable interest.

As discussed by [2] and [3] the interest in the use of $CuInSe_2$ arises because it is one of the few materials to satisfy all of the following criteria:

1. Its energy gap ($Eg = 0.96 - 1.04$ eV) is near the optimum value for use in homojunction or heterojunction solar cells particularly as the absorber in the lower cell of a tandem structure.

2. It has a direct band gap and a very high absorption coefficient ($> 10^5$ cm^{-1}) over the solar spectrum which means that light is absorbed within 1 μm at most which minimises the need for a long minority diffusion length (and hence expensive materials processing) and also minimises the quantity of material used (and hence minimises material costs).

3. Both n-type and p-type material can be produced thus allowing the formation of homojunctions and a variety of heterojunction types.

4. The lattice mismatch with the established n-type window materials CdS and CdZnS is low (eg 1.16% with CdS) which minimises the density of interface states.

5. Its electron affinity is compatible with CdS, CdZnS and indium tin oxide such that conduction band spikes are avoided in heterojunction formation.

6. The material has proved to be exceptionally stable eg Boeing has exposed unencapsulated cells to artificial illumination for periods of over 9000 hours and so far no significant degradation of cell parameters have been observed.

7. A wide range of low cost deposition technologies may be used to fabricate the material.

Interest in the use of HgCdTe (with $Eg \approx 1$ eV) as the absorber in the lower cell of a tandem structure and CdZnTe (with $Eg > 1.5$ eV) for use as the absorber in the top cell of a tandem structure arises mainly because of the highly developed state of CdTe technology coupled with the fact that the bandgap of CdTe is not optinum for use in tandem structures.

We are currently investigating a new alloy, mercury zinc telluride, HgZnTe, as it could potentially be used as the top or bottom absorber in a tandem cell and it may even be possible to fabricate a stacked structure out of this one material.

This short review will concentrate on the methods used to produce $CuInSe_2$ solar cells and will also discuss the HgZnTe alloy system.

$CuInSe_2$

A wide variety of techniques have been used to deposit thin films of $CuInSe_2$. These include single source evaporation, dual source evaporation, three source evaporation, sputtering, electrodeposition, spray pyrolysis and laser induced synthesis.

Detailed reviews of the electrical, optical, structural and morphological properties of the $CuInSe_2$ produced by these techniques are given by [2] and [4].

Of these techniques only three source evaporation, usually, with the use of an E.I.E.S. (electron impact excitation spectroscopy) feedback system to control the evaporation rates of the copper and indium has resulted in films of good enough quality to fabricate solar cells with efficiencies > 10%. (The selenium evaporation rate is usually monitored with a quartz crystal monitor and controlled to give a supersaturation ratio of two to ensure the deposition of p-type films. A selenium deficiency results in n-type films). The electrical, optical, structural and morphological properties of the films deposited depend critically on the stoichiometry of the layers deposited, particularly on the ratio of Cu:In.

Representative results of how the resistivity varies with molecular non-stoichiometry is given in Figure 1 (from [6] to [8]). A positive value of non-stoichiometry (an excess of indium) results in high resistivity films and a negative value (an excess of copper) results in low resistivity films. R.E.D. and X-ray studies of the film structure ([7] and [9]) indicate that copper rich $CuInSe_2$ consists of approximatly stoichiometric $CuInSe_2$ mixed with a distinct phase of Cu_{2-x} Se. Indium rich $CuInSe_2$ appears to consist of an alloy of $CuInSe_2$ with In_2Se_3. For films deposited onto·glass the grain size increases with substrate temperature as shown in Figure 2. The influence of crystallite orientation angle is also shown in Figure 3.

(NB The grain size of the thicker films used to make solar cells are greater than those shown in Figure 2, typically ≥ 1 μm).

Mickelson and Chen [5] were the first workers to use the three source method for producing $CuInSe_2$ and also to use this material to fabricate solar cells. The highest reported efficiencies are shown in Table 1 and it is evident that the cells produced by Boeing using a similar cell structure to that developed by [5] have achieved the highest efficiency.

Two layers of $CuInSe_2$ are used. The bottom layer is deposited slightly copper rich (the low resistivity layer) and a layer which is slightly indium rich (the high resistivity layer) is deposited onto this layer. The use of the low resistivity layer is essential to minimise the total series resistance of the device. The use of this layer alone in cell structures has been found to result in unstable devices because of the formation of copper cones which seriously degrade the device performance [5]. It is highly probable that these are due to the electrochemical reduction of $Cu_{2-x}Se$ to Cu in a similar manner to that observed in Cu_2S/CdS solar cells ([7] and [11]). The inclusion of the high resistivity (indium rich) layer thus preserves junction integrity.

Added advantages of including the copper rich layer are that the low resistivity layer favours the formation of an ohmic contact to the base metallisation, possesses a larger grain size, is highly adherent, and creates the possibility of back surface field effects in the cell structure [5].

An essential step to produce high efficiency cells involves heating the entire cell in an atmosphere containing oxygen typically at a temperature in the range 150-200°C for a period of one hour [5].

The role of oxygen is not fully understood and is the subject of recent studies [12] to [15]. It appears that dipping a device into a chemical solution containing an oxidising agent has an equivalent effect to annealing in an oxygen containing environment [13]. Heating a device in a reducing environment (eg hydrogen) [15] or dipping in a chemical solution containing a reducing agent [13] degrades device performance. Moreover reduction followed by oxidation or vice versa appears to be reversible [15] providing the annealing temperature is < 500 K [12]. The influence of a reducing environment in degrading device performance is consistent with an increase in $CuInSe_2$ resistivity [15]. The photoluminescence studies of [12] indicate that annealing in oxygen reduces the density of native defects whilst there is a corresponding increase in acceptor concentration implying that annealing in oxygen reduces the degree of compensation. A comparison of the photoluminescence from a free $CuInSe_2$ surface with that covered with CdS shows that the defect density on the former is lower than that on the latter implying that the CdS may act as an encapsulent during annealing.

EBIC studies of the $CuInSe_2/CdS$ junction have been made by [13] and [14] but their data and interpretations are controversial and are disputed by [16].

Spray pyrolysis, sputtering, electrodeposition and laser induced synthesis are techniques which could potentially produce larger areas of $CuInSe_2$ at lower cost than the three source method.

As reviewed by [17] to [19] spray pyrolysis consists of spraying an ionic solution containing the elements onto a heated substrate. An aqueous solution of $CuCl_2$ is used as the copper source, $InCl_3$ as the In source and N,N dimethyl senourea as the Se source. The $CuInSe_2$ concentrations used are in the range 0.005-0.1 mol ℓ^{-1} with different ratios of Cu:In:Se and the substrate temperature is varied from 150-450°C. As with the other techniques for a given substrate temperature and a supersaturation of Se the resistivity varies with molecular non-stoichiometry in a similar manner to that given in Figure 1. For a given ratio of Cu:In:Se the resistivity is found to increase with substrate temperature. Although the grain size of the layers produced are small, typically < 0.5 μm annealing in an inert atmosphere is found to increase the grain size to 1 μm [20]. The efficiencies of the cells made from the material produced using this technique are however rather disappointing ≈ 2%. [17] attribute the low efficiencies to shorting paths which are probably due to pinholes or islands of degenerate second phase materials.

The use of sputtering to produce thin films of $CuInSe_2$ is discussed by [2], [19], [21] and [22].

R.f. sputtering consists of preparing a target of CuInSe$_2$ with a given ratio of Cu:In:Se and sputtering this material onto a heated substrate in an argon atmosphere. The target composition, the substrate temperature, the r.f. sputtering power, the argon pressure in the deposition chamber, the deposition rate and the bias applied to the substrate all influence the properties of the films deposited.

[4] indicate that the Brown University group have managed to prepare stoichiometric CuInSe$_2$ with grain sizes of the order of several microns and with a minority carrier diffusion length $\approx$ 1 μm and they have used this material to produce cells with efficiencies $\approx$ 5%.

The d.c. magnetron sputtering method of [21] consists of co-sputtering from two planar magnetrons with Cu and In targets onto a heated substrate in a H$_2$Se/Ar gas mixture. The best cells they have produced have an efficiency $\approx$ 4% (with Jsc = 33 mA cm^{-2} , Voc = 0.25 V and F.F. = 50%).

The electrodeposition of thin films of CuInSe$_2$ has been studied by [23] to [26]. Kapur et al [23] have used the sequential deposition of copper and indium followed by selenisation (by annealing in hydrogen selenide) to produce CuInSe$_2$ of sufficiently good quality to produce solar cells with efficiencies > 7%.

[24], [25] and [26] have electroplated CuInSe$_2$ directly from a single electrolyte solution and [26] report devices with efficiencies $\approx$ 4%. [25] have considered the electrodeposition of a Cu/In alloy followed by selenisation to produce thin films. Of the alternatives to the three source technique the results of Kapur [23] indicate that this technique is the most promising and it is highly likely that it will be used commercially to produce low cost solar cells.

The laser induced synthesis of CuInSe$_2$ from the elements as reported by [27] and [28] is a relatively new development. Both sets of workers sequentially deposit layers of Cu, In and Se (using thermal evaporation) with thicknesses corresponding to an atomic ratio 1:1:2 and then anneal the sandwich (or sandwiches) with an argon ion laser.

[27] have obtained single phase CuInSe$_2$ with grain sizes of the order of 20 μm for crystallites obtained on free standing films. The transmittance versus wavelength curves of the CuInSe$_2$ obtained by [28] indicate the presence of a second phase material which they suggest is indium selenide.

This technique may also be used to produce CuInSe$_2$ from electroplated layers of Cu, In and Se or layers of Cu, In and Se produced by any of the other techniques and so it has the potential to form the basis of a low cost method for producing CuInSe$_2$. [28] believe that larger areas of usable CuInSe$_2$ may be produced using this technique than by the three source method.

Mercury Zinc Telluride

Mercury zinc telluride is an alloy of mercury telluride and zinc telluride. It is a novel material which has been fabricated in bulk form ([29] to [31]), using liquid phase epitaxy [32] to [34] and using molecular beam epitaxy [35]. Recent interest in this alloy system arises from its potential use as an infrared detector material as an alternative to mercury cadmium telluride [35]. It was suggested by [36] that it has potential for use as a solar cell material primarily because its energy gap may be altered from -0.15 eV to 2.25 eV by changing the alloy composition and thus the alloy may be tuned to absorb at any energy throughout the solar spectrum. It is also likely that the material has a direct bandgap and it is highly absorbing and that its electrical properties in the range of alloy compositions of interest are good.

[[35] have made mobility measurements on alloys with x < 0.34 and found the values to be comparable or better than those obtained with mercury cadmium telluride for the same bandgap]. [35] have also confirmed that the material may be made n-type or p-type thus allowing the formation of a variety of heterojunction types and also allowing the formation of homojunctions.

It should be noted that unlike mercury cadmium telluride the variation of the bandgap with alloy composition is not linear and [37] attribute this deviation as an effect of the change of the covalent radii of the Zn, Hg and Te atoms.

An alloy composition of x = 0.85 (where x = 1 for ZnTe and x = 0 for HgTe) corresponds to a bandgap of 1.71 eV which is a suitable value for the top absorber layer in a tandem cell and an alloy composition of x = 0.62 corresponds to a bandgap of 1 eV which makes the material suitable as a lower absorber in a tandem cell. Alternatively stacked solar cells using repeated homojunctions of different bandgaps or repeated heterojunctions with different bandgap absorbers could be used to absorb over the solar spectrum.

It should be noted that unlike mercury cadmium telluride where the lattice parameter does not change with composition the lattice parameter of mercury zinc telluride varies markedly with alloy composition ([29] and [37]). The electron affinity will also change with composition and thus it should be possible to investigate the influence of lattice matching and electron affinity matching to suitable window materials.

Given such potential we are currently producing mercury zinc telluride by thermal evaporation of ZnTe followed by the thermal evaporation of HgTe which is followed by an anneal in vacuum to homogenise the alloy.

Control of the alloy composition (and hence bandgap) is achieved by controlling the relative thicknesses of the ZnTe and HgTe deposited. We are currently correlating the way the bandgap, the lattice parameter, the carrier concentration, the mobility and the carrier type vary with alloy composition.

References

1. J.C.C. Fan and B.J. Palm, Solar Cells, 12 (1984), 401-420.

2. L.L. Kazmerski and Wagner, "Cu Ternary Chalcopyrite Solar Cells" in "Current Topics in Photovoltaics", (1985), 41-109, (Ed. T.J. Coutts and J.D. Meakin, Academic Press).

3. Solar Cells, 1986, Vol. 16 (Ed. T.J. Coutts, L.L. Kamerski and S. Wagner).

4. A.N.Y. Samaan, R. Vaidhyanathan and R. Noufi, Solar Cells, 16 (1986), 181-198.

5. R.A. Mickelson and W.S. Chen, Conf. Rec. I.E.E.E. Photovoltaic Spec. Conf. 15, 800-804 (1981).

6. E.R. Don, S.R. Baber and R. Hill, Proc. 3rd EC Photovoltaics Solar Energy Conference, Cannes, 1980, 897.

7. E.R. Don, PhD Thesis, Newcastle upon Tyne Polytechnic, 1984.

8. E.R. Don, R.R. Cooper and R. Hill, Proc. 6th EC Photovoltaics Solar Energy Conference, London, 1985, 768.

9. E.R. Don, G.J. Russell and R. Hill, Solar Cells, 16 (1986), 131-142.

10. R.A. Mickelson, W.S. Chen, Y.R. Hsiao and V.E. Lowe, I.E.E.E. Trans. Electron. Dev., 31 (1984), 542.

11. R. Hill and J.D. Meakin, "Cadmium Sulphide – Copper Sulphide Solar Cells", in "Current Topics in Photovoltaics", 1985, 223-271, [Ed. T.J. Coutts and J.D. Meakin, Academic].

12. R.E. Hollingsworth and J.R. Sites, Solar Cells, 16 (1986), 457-477.

13. R. Noufi, R.J. Matson, R.C. Powell and C. Herrington, Solar Cells, 16 (1986), 479-493.

14. R.J. Matson, R. Noufi, R.K. Ahrenkiel, R.C. Powell and D Cahen, Solar Cells, 16 (1986), 495-519.

15. S. Damaskinos, J.D. Meakin and J.E. Philips, 19th I.E.E.E. Photovoltaics Specialists Conference, May 1987, New Orleans, USA.

16. L.L. Kazmerski, N.A. Burnham, A.B. Swartlander, A.J. Nelson and S.E. Asher, 19th I.E.E.E. Photovoltaics Specialists Conference, May 1987, New Orleans, USA.

17. J.B. Mooney and R.H. Lamoreaux, Solar Cells, 16 (1986), 211-220.

18. J. Bougnot, S. Duchemin and M. Savelli, Solar Cells, 16 (1986), 221-236.

19. A.N.Y. Samaan, R. Vaidhyanathan, R. Noufi and R.D. Tomlinson, Solar Cells, 16 (1986), 181-198.

20. J.J. Loferski, J.S. Jain and S. Radhakrishna, Proc. Int. Workshop on the Physics of Semiconductor Devices, New Delhi, Wiley Eastern, New Delhi, 1981, 408-417.

21. N. Romeo, V. Canevari, G. Sberveglieri and A. Bosio, Solar Cells, 16 (1986), 155-164.

22. J.A. Thornton and T.C. Lommasson, Solar Cells, 16 (1986), 165-180.

23. V.K. Kapur, B.M. Basol and E.S. Tseng, 18th I.E.E.E. Photovoltaics Specialists Conference, (1985), 1429.

24. R.N. Bhattacharya and K. Rajeshwar, Solar Cells, 16 (1986), 237-243.

25. G. Hodes and D. Cahen, Solar Cells, 16 (1986), 245-254.

26. I. Shih and C.X. Qiu, 19th I.E.E.E. Photovoltaics Specialists Conference, New Orleans, USA, May 1987.

27. L.D. Laude, M.C. Joliet and C. Antoniadis, Solar Cells, 16 (1986), 199-209.

28. M.J. Carter, I. I'Anson, A. Knowles, H. Oumous and R. Hill, 19th I.E.E.E. Photovoltaics Specialists Conference, New Orleans, USA, May 1987.

29. J.C. Woolley and B. Ray, J. Phys. Chem. Solids, 13, 151 (1960).

30. D. Niculescu, J. Phys. C (Proc. Phys. Soc.) 1, (2), 804 (1968).

31. E. Cruceanu, D. Niculescu, I. Stamatescu, N. Nistor and S. Ionescu – Bujor Fiz. tverd. Tela 7, 2039 (1965).

32. A. Sher, D. Egger and A. Zemmel, Appl. Phys. Lett. 46, 59 (1985).

33. T. Tung, M.H. Kalisher, S. Sen, B.F. Zuck, E.J. Smith and W.M. Konkel, presented at the 1985 US Workshop on the Physics and Chemistry of Mercury Cadmium Telluride, San Diego, CA, 1985.

34. A. Sher, D. Egger, H. Feldstein and A. Raizman, J. Vac. Sci. Technol. A4, 2024 (1986).

35. J.P. Faurie, J. Reno, S. Sivanathan, I.K. Sou, X. Chu, M. Boukerche and P.S. Wijewarnasuriya, J. Vac. Sci. Technol. A4 (4), Jul/Aug 1986, p 2067.

36. R.W. Miles (Invited Talk) "VI National Seminar for Semiconductors and Devices", December 1986, Indian Association for the Cultivation of Science, Calcutta.

37. E.Z. Dziuba, D. Niculescu and N. Niculescu, Phys. Stat. Sol. 29, 813 (1968).

Table 1

Highest Reported Efficiencies for CdS/CuInSe₂ Solar Cells

Organisation	Efficiency (%)	Cell Area (cm²)	Jsc (mA/cm²)	Voc (V)	F.F.
Boeing	11.9	1.0	39.4	0.44	0.689
Arco	11.2	4.0	32.9	0.487	0.70
IEC	10.7	0.08	37.7	0.431	0.657
SERI	10.3	1.03	35.4	0.445	0.653

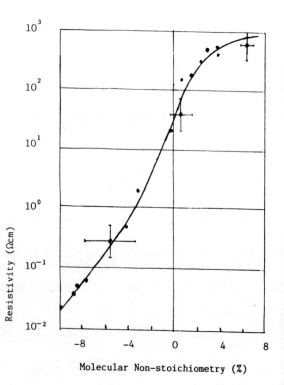

Figure 1

Resistivity of single crystal films of CuInSe₂ as a function molecular non-stoichiometry. Error-bars are shown on representative points.

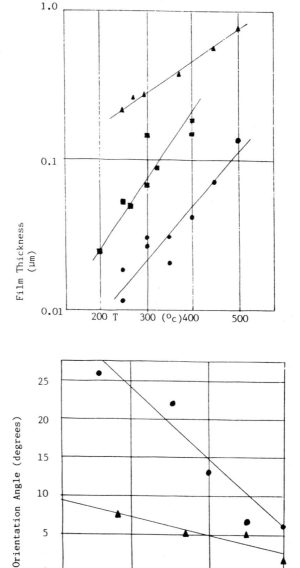

Figure 2

Grain size of films on glass substrates as a function of substrate temperature

▲ Cu-rich films
■ Slightly Cu-rich
● In-rich

Figure 3

Crystallite orientation angle of CuInSe₂ films on a glass substrate as a function of substrate temperature for:- ▲ Cu-rich films, ● In-rich films.

B R I G H T O N   P O L Y T E C H N I C
DEPARTMENT OF PHYSICAL SCIENCES
CRYSTALLOGRAPHY LABORATORY

Layer Compounds and their use as Secondary Battery Electrodes
by
A. A. Balchin
Moulsecoomb, Brighton, BN2 4GJ, England, U.K.

Layer compounds are a class of materials with highly anisotropic crystal structures. They usually crystallise in thin flakes or plate-like single crystals. The class includes a number of distinct crystallographic structure-types, which arise from different stackings of hexagonally close-packed planes of atoms. Three of these structures, cadmium iodide, gallium selenide, and molybdenum disulphide, are illustrated in Figure 1.

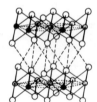 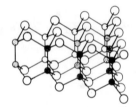 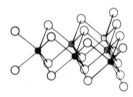

(a) CADMIUM IODIDE TYPE (CdI₂)   (b) GALLIUM SELENIDE TYPE (GaSe)   (c) MOLYBDENUM DISULPHIDE TYPE (MoS₂)

The di-sulphides, di-selenides and di-tellurides of tin, titanium, zirconium and hafnium are isomorphous members of the group of compounds which crystallise in space-group P3̄ml with the characteristic cadmium iodide layer structure. Their structures consist of stacked composite layers, each comprising two sheets of hexagonally close-packed chalcogenide atoms sandwiching a sheet of metal atoms in such a way that half of the octahedral interstices are filled. The atomic stacking sequence is :AcB:AcB:. Since primary valency is satisfied within the atomic sandwich :AcB:, adjacent layers are held together only by relatively weak van der Waal's forces. The resulting structure is markedly anisotropic, characterised by extended growth and pronounced cleavage perpendicular to the c-axis of the trigonal crystals.

Examination of these transition metal dichalcogenides has been stimulated by the work of Lee et al[1,2,3]. These workers report the observation of current controlled negative resistance and switching in single crystals of tin disulphide, zirconium disulphide, and hafnium disulphide (Figure 2) when an electric field is applied parallel to the three-fold axis.

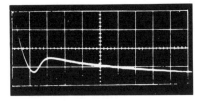

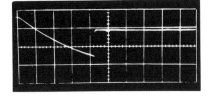

A single crystal of SnS_2 showing negative resistance.

A single crystal of ZrS_2 showing switching from a high (R_{II}) to a low (R_{III}) resistance state.

Figure 2.

Whittingham[4] has shown that the ability of titanium disulphide to take up mobile lithium ions into the small interstices in the van der Waal's gap between minimal layers under the action of a polarising voltage, and subsequently to release them during a process of electrical discharge, can provide the basis for a valuable secondary rechargeable battery of high power and high energy density. Foot and Nevett[5] have suggested a like role for nickel phosphorus trisulphide ($NiPS_3$), which has a structure similar to that of TiS_2.

Gallium sulphide and selenide comprise minimal sandwiches of four atomic sheets stacked :AcdB:. The co-ordination of the gallium is tetrahedral, and two tetrahedrons make up the thickness of each sandwich. Structural modifications of gallium selenide make up three polytypes :- a hexagonal β-structure of two anti-parallel layers : a hexagonal ϵ-structure of two parallel offset layers, and a rhombohedral γ-structure of three parallel offset layers. Gallium selenide is a most interesting material of device potential, exhibiting semiconducting, photoconducting, memory switching, negative resistance and low temperature photo- and electro-luminescent effects.

The Chemical Vapour Transport Technique of Crystal Growth

When grown by the chemical vapour transport method, layer compounds are usually obtained in the form of thin hexagonal plates, typically 2 cm^2 in superficial area and about 100 μ m thick — a form particularly suitable for making electrical measurements.

The halogen vapour transport technique is a cyclic reaction driven and maintained by a concentration or temperature gradient. In our growth experiments a sealed evacuated fused quartz ampoule some 20cm long x 1.6cm bore is loaded with a stoichiometric mix of the elements comprising the compound or solid solution required. Included in the contents of the ampoule is a small amount of a transporting material, in this case iodine, sufficient to give a concentration of about 5mg per cm^3 of ampoule volume. The ampoule is placed in a temperature gradient in a two-zone furnace, with the charged end of the ampoule towards the hotter zone. For a number of hours the reaction indicated by the route (a)-(b)-(c)-(d)-(e)-(f) of figure 3 is traced, during which time single crystals of dichalcogenide, MX_2, are deposited at the empty, cool end of the ampoule. The charge, originally present in the form of polycrystalline MX_2 at the reaction temperature T_1, is

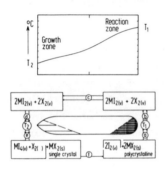

Figure 3.

transported via the intermediate gas phase MI_2 to the growth end of the ampoule. Here dissociation takes place at the growth temperature T_2. MX_2 is deposited in single crystal form, liberating MI_4 which cycles back to the reaction end of the ampoule. Chemically the process may be represented by the equations:-

$$2MX_2 + 2I_2 \rightleftharpoons 2MI_2 + 2X_2 \qquad \text{(Reaction)}$$
$$2MI_2 + 2X_2 \rightleftharpoons MI_4 + MX_2\downarrow + X_2 \qquad \text{(Growth)}$$

To achieve the steady thermal gradient required to obtain large single crystals of high quality furnace temperatures are stabilised to $\pm$ 1°C. Chromel-alumel thermocouples monitor independently the temperatures of the split heater windings for the hot reaction zone and the cooler

349

growth zone of the horizontal two-zone static furnaces used. The temperature gradient between reaction and growth zones is appreciably linear over the length of the ampoule.

A ten-fold increase in the sizes of crystals grown by the static process may result if the growth end of the ampoule is subjected to a periodic upward fluctuation of temperature during the growth time. This preferentially evaporates small nuclei and allows large crystals to grow at the expense of smaller ones. Due allowance must be made for thermal lag of the furnace by adjusting the duration of heating and cooling cycles. Best results are then achieved if the amplitude of the temperature fluctuation is of the same order as the height of the temperature gradient. By adopting this "Pendelofen" method of crystal growth, increases in crystal size of which Figure 4 is typical are obtained.

Comparison of single crystals of gallium selenide grown by the Pendelofen technique (left) and by the static iodine vapour transport method (right). Growth conditions are: $T_1 = 900°C$, $T_2 = 850°C$. Growth time for static vapour transport: 120h. Pendelofen crystals subjected to growth temperature fluctuation of $+40°C$ for 5 min in each hour for 240h.

Crystal Characterisation of Growth Products

Binary solid solutions have been successfully grown by the iodine vapour transport process of the following compounds:- $SnS_2/SnSe_2$, $TiS_2/TiSe_2$, $TiS_2/TiTe_2$, $TiSe_2/TiTe_2$, $ZrS_2/ZrSe_2$, $HfS_2/HfSe_2$, $GaS/GaSe$, SnS_2/ZrS_2, $SnSe_2/ZrSe_2$, $TaS_2/TaSe_2$. With the exception of the systems GaS/GaSe and $TaS_2/TaSe_2$, both of which are well-known for polytypism, all others form isomorphous trigonal solid solutions over the whole range of composition, isostructural with the end members. The lengths of the unit cell axes usually deviate only very slightly from Vegard's Law, varying almost linearly with the atomic concentration, x. The system $ZrS_xSe_{(2-x)}$ shown in Figure 5 is typical.

Lang x-ray topography of crystals grown by the static method shows that the crystals may have a high degree of perfection. Most are of sufficiently high quality for individual dislocations to be resolved. Large areas are completely free of defects. In crystals of tin disulphide line dislocations have a Burger's vector parallel to [1$\bar{2}$10] comparable with that found for undissociated dislocations in molybdenum disulphide by Amelincx and Delavignette[6]. These dislocations are curved, and from their radius of curvature a lower limit of yield stress of 10^4 N m^{-2} is obtained. Other crystals of SnS$_2$ show dislocations with Burger's vectors parallel to <11$\bar{2}$0>. Planar faults parallel to the basal plane arise from large area stacking faults with fault vector 1/3 $\bar{a}$ <1$\bar{1}$00> and bounded by partials similar to those observed by Amelincx and Delavignette. The faults arise from the glide of sulphur layers over each other, and due to the weak binding between them the faults have very low energy.

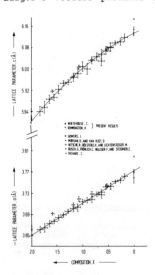

Values of the lengths of unit cell axes a and c plotted as functions of composition variable, x, for mixed compounds ZrS_xSe_{2-x}.

Figure 5.

[1$\bar{1}$00]

Figure 6. Large area stacking faults, revealed in {1$\bar{1}$00} type reflections.

1 m.m.

Intercalation complexes

Intercalated materials have many of the crystallographic and electrical properties which make them promising systems in which to investigate mechanisms of super-conductivity which rely on two-dimensional constraints. The effect of intercalation is to introduce into the van der Waal's gap metal ions or organic molecules which may modify the structural or electrical properties of the host material. Almost all layer compounds have this property of intercalating alkalis or organic complexes, usually with the result that the minimal sandwiches of the lattice are pushed further apart. The electron occupation of the relatively narrow d-bands of the transition metal compounds can be finely controlled by intercalation with alkali metals, organic electron donors (pyridine, hydrazine, ammonia, etc.) and 3d transition metals. The electronic properties of the host lattice may be greatly changed, and optical properties, conductivity and super-conductivity, metal-semiconductor-insulator transitions, magnetic properties and ionic conduction are all affected.

The use of lithium-intercalated titanium disulphide as a battery cathode gives a high voltage against a lithium anode. The energy density and the power density are much higher than a lead/acid cell. Electrochemically the reaction is totally reversible, but on intercalation of lithium the basal plane stacking distance of titanium disulphide alters, giving crystal fatigue and limited cycling life.

Nickel phosphorus trisulphide is a layer compound with a crystal structure related to those of TiS_2 and CdI_2. The layers are made up of double sheets of sulphur atoms with nickel atoms and pairs of phosphorus atoms located in the octahedral holes of the structure to form (P_2S_6) building units, largely covalently bonded, held together by the ionic attraction of the Ni^{2+} ions. The layers are then held together by van der Waal's bonds. The stacking arrangement is:-

$$\text{(gap) S} - Ni_{2/3}(P_2)_{1/3} - S \text{ (gap) S} - Ni_{2/3}(P_2)_{1/3} - S \text{ (gap)}$$

On intercalation with lithium, nickel phosphorus trisulphide gives a cell action with a better energy density than TiS_2. Three lithium ions per formula unit can be intercalated, two of them reversibly. The resulting cell voltage is higher than that of Li_xTiS_2, and on intercalation with lithium $NiPS_3$ retains a constant basal plane

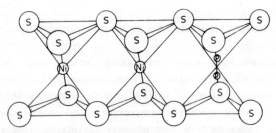

Figure 7. A minimal sandwich in the layer structure of NiPS$_3$.

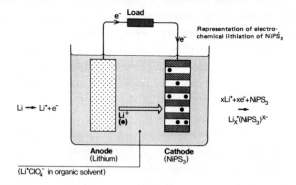

Figure 8. Electrochemical cell using lithium-intercalated NiPS$_3$.

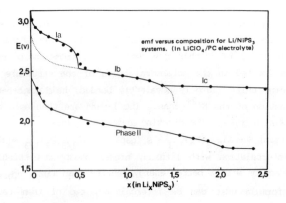

Figure 9. Emf vs. x during discharge of Li$_x$NiPS$_3$ cell.

separation, with improved cycling life. The octahedral interstices in the unoccupied regions of the van der Waal's gap are large enough to hold lithium ions without distortion, and they can migrate in and out of the crystal freely under the influence of a polarising voltage.

Secondary electric storage cells in which this migration mechanism is used to store power (Figure 8) may find application in load levelling systems, primary power systems for electric transport, or in medicine. The cell illustrated provides a voltage of the order of 2.5 V. (Figure 9).

Curve I is obtained at small current densities, and Curve II at higher ones. The discontinuity between curves I(a) and I(b) results when the lithium ions completely fill the 2d octahedral sites of the lattice (space-group C2/m) and then commence filling the 4h sites. Curve I(c) results from the formation of a two-phase system at high values of x. Similarly Curve II obtains when two intercalated phases $Li_{0.5}NiPS_3$ and $Li_{1.5}NiPS_3$ form in the cathode during lithium intercalation.

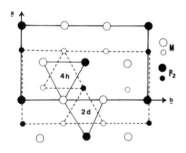

Figure 10. Cationic environment of lithium sites.

Conduction mechanisms in lithium substituted $NiPS_3$ have been studied. I-V characteristics for pure $NiPS_3$ substituted with lithium, indium, and gallium yield the trap concentrations, using the den Boer method[7], plotted in Figure 11. The trap energies are at the centre of the energy gap for the pure material and are probably associated with stacking faults or compositional defects. Pure $NiPS_3$ and Li_xNiPS_3 are p-type, and conduction arises from holes generated at nickel atoms and hopping between nickel sites. A small polaron mechanism may be proposed for Li_xNiPS_3.

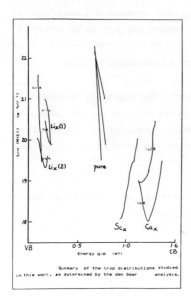

Summary of the trap distributions studied in this work, as determined by the den Boer analysis.

Figure 11.

NiPS$_3$ substituted with Sc, In, or Ga is n-type. The electron charge carriers probably originate from the phosphorus atoms. Charge transport is then due to large polarons in a narrow band environment.

At high lithium substitutions a compound Li$_2$NiP$_2$S$_6$ is formed. In an electrochemical cell this compound provides an open-circuit e.m.f. of over 2.9 V. Its structure is monoclinic (space-group C2) with:

a = 0.5926 nm, b = 1.097 nm, c = 0.6718 nm, β = 104° Atomic co-ordinates are listed in Figure 12, and the b-axis projection is shown in Figure 13.

Atomic positions for Li$_{0.5}$(Ni$_{0.5}$Li$_{0.5}$)PS$_3$ giving best fit between I_{obs} and I_{calc}.

Atom	x (± 0.0009)	y (± 0.0008)	z (± 0.0009)
Ni(1)	0.0000	0.3328	0.0000
Ni(2)	0.5000	0.8328	0.0000
Li(1)	0.5000	0.1672	0.0000
Li(2)	0.0000	0.6672	0.0000
P(1)	0.9429	0.0000	0.8292
P(2)	0.0571	0.0000	0.1708
P(3)	0.5571	0.5000	0.1708
P(4)	0.4429	0.5000	0.8292
S(1)	0.2481	0.8341	0.2479
S(2)	0.7481	0.3341	0.2479
S(3)	0.7519	0.1659	0.7521
S(4)	0.7508	0.0000	0.2475
S(5)	0.7481	0.6659	0.2479
S(6)	0.2492	0.0000	0.7526
S(7)	0.7492	0.5000	0.7526
S(8)	0.7519	0.8341	0.7521
S(9)	0.2519	0.3341	0.7521
S(10)	0.2519	0.6659	0.7521
S(11)	0.2481	0.1659	0.2479
S(12)	0.2508	0.5000	0.2475
Li(3)	0.5000	0.0000	0.5000
Li(4)	0.0000	0.5000	0.5000

Figure 12.

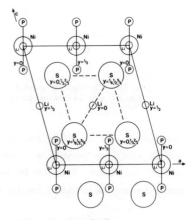

Li$_2$NiP$_2$S$_6$ Projection on (010)

Figure 13.

The structural bonding is intermediate between that of the empty van der Waal's gap in $NiPS_3$ and the fully lithium intercalated $Li_4P_2S_6$ (Mercier[8]). The partial lithium intercalation compensates the valency left unsatisfied by the substitution of univalent Li^+ for the divalent Ni^{2+}. The presence of intercalated lithium increases the bonding across the van der Waal's gap, but the covalent P-P bonds of $Li_4P_2S_6$ are absent in $Li_2NiP_2S_6$. This allows the lithium to remain mobile, and to move in and out of the van der Waal's region during electrical intercalation.

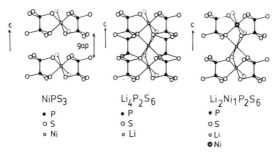

NiPS₃
• P
o S
o Ni

Li₄P₂S₆
• P
o S
o Li

Li₂Ni₁P₂S₆
• P
o S
o Li
o Ni

Comparison of $NiPS_3$, $Li_4P_2S_6$ and $Li_2Ni_1P_2S_6$ structures.

Figure 14.

References

(1)   P. A. Lee, G. Said and R. Davies ; Solid State Communications, 7 (1969) 1359.

(2)   S. Ahmed and P. A. Lee ; Journ. Phys. D (Applied Physics), 6 (1973) 593.

(3)   G. Said and P. A. Lee ; Phys. Stat. Solidi (a), 15 (1977) 99.

(4)   M. S. Whittingham ; Science, 192 (1976) 1126 ; Journ. Electrochem. Soc. 123 (1976) 315.

(5)   P. J. S. Foot and B. A. Nevett ; British Patent Application No. GB 2132181 (1984).

(6)   S. Amelincx and P. Delavignette in "Direct Observations of Imperfections in Crystals" Eds. Newkirk and Wernick (Interscience, New York, 1962) p. 295.

(7)   W. den Boer ; J. Physique, 42 (1981) C10-451.

(8)   R. Mercier, J. P. Molugani, B. Fahyo, J. Douglade and G. Robert; Journ. Solid State Chem., 43 (1982) 151.

OXIDE MATERIALS IN SEMICONDUCTOR SCIENCE AND TECHNOLOGY

C. A. HOGARTH

Department of Physics, Brunel University, Uxbridge, Middlesex, U.K.

Summary

Oxide materials are of great significance in the physics and technology of semiconductors. Some are semiconductors in their own right and others are used as insulators or as electrodes. Some have properties that may be used in discrete devices such as sensors. The field is reviewed with practical examples of recent activities.

1. INTRODUCTION

Many semiconductor devices are based on silicon in which the bonding is covalent, on gallium arsenide in which the bonding is partly covalent and partly ionic, and on other III-V and II-VI materials in which the bonding is mixed, the II-VI chalcogenide materials being in general more ionic than the III-V semiconductors[11]. Oxide semiconductors may be typified by cuprous oxide Cu_2O which was for many years the active component of the so-called copper oxide rectifier. Other oxides have been used as gas sensors, using the change of electrical conductivity with gas pressure as the sensing property. One major application of oxides in silicon technology relates to silicon dioxide SiO_2 which is used for passivating silicon surfaces and as the insulating material in the particular form of metal- insulator-silicon transistor known as the MOSFET or metal oxide silicon, field-effect transistor. Silicon dioxide has a large bandgap, about 10 eV, and this means that the free electron concentration at the normal operating temperatures of semiconductor devices will be very low and the resistance extremely high. Other oxides of significance in the semiconductor field include indium oxide, indium-tin oxide, cadmium oxide, tin oxide and di-cadmium stannate. When quoting the relevant properties of oxide semiconductors it is of the greatest importance to know the precise composition since even the smallest deviations from stoichiometry lead to enormous changes in electrical conductivity. For example, zinc oxide as taken from a reagent bottle will contain an excess of zinc which is present in the form of zinc ions (positively charged) and electrons and the material will be n-type. The greater the stoichiometric defect the greater will be the free-electron concentration. In the case of silicon oxide, fused quartz SiO_2 has a band gap of some 10 eV[2] whereas thin films of "pure" silicon monoxide have a bandgap, measured optically, of 2.3 eV[3]. The state of aggregation is also important. The properties of semiconductors are best characterised with the samples in single crystal form either as massive samples or thin monocrystalline layers, but the use of amorphous semiconductors, in particular hydrogenated amorphous silicon and glassy selenium, has widened the scope of semiconductors substantially and other potential applications of semiconducting glasses (chalcogenide and oxide) in electrical switches, printing systems, and formed dielectrics has brought many of these materials within the scope of semiconductor device physics. Indeed the boundary between semiconductor and insulator is not too tightly drawn and it is often convenient to regard some oxide materials as wide bandgap semiconductors rather than as lossy dielectrics.

2. CONDUCTION IN OXIDES

This is simply illustrated by taking zinc oxide ZnO as an example. At any real temperature there will be some dissociation of the oxide as follows.

$$2\,ZnO = 2\,Zn^{2+} + 2\,O^{2-}$$

If the oxygen ambient pressure in the system is reduced then some of the oxygen ions will dissociate into oxygen gas and electrons.

$$2\,O^{2-} \rightarrow O_2 + 4e$$

The total reaction may be described by the equation

$$2\,ZnO = 2\,Zn^{2+} + O_2 + 4e$$

and the application of the Law of Mass Action to such a process leads to the result that the free electron concentration n_f is proportional to a power of the oxygen pressure at a given temperature,

$$n_f \propto P_{O_2}^{-1/6}$$

The form of the variation has been tested on granular oxide material and the index found to have the value 1/4.3 based on measurements of electrical conductivity σ where $\sigma = n_f\,|\,e\,|\,\mu$, μ being the electron mobility. Since it is by no means certain how the mobility varies with oxygen pressure in such circumstances, the agreement between 1/4.3 (experimental) and 1/6 (theoretical) is not too bad. However the theory of the Seebeck effect α may be applied to these materials and the variation of n_f with oxygen pressure estimated from the results since to a first approximation α is proportional to $1/\ln n_f$ and from these results the index may be estimated as 1/6.1[4]. The closer agreement arises because the value of Seebeck coefficient is much less sensitive to the state of aggregation of the material than is the electrical conductivity.

For other oxide systems, e.g. Cu_2O, the usual state of the material is with excess oxygen and an analogous analysis to the above leads to the result that the hole concentration

$$p_f \propto PO_2^{-1/8}$$

Materials such as cadmium oxide may contain up to $1/2$% of excess cadmium (strongly degenerate n-type) and nickel oxide up to $1/2$% excess oxygen (black in colour and strongly degenerate p-type).

The variations with oxygen pressure of a variety of electrical properties depending on the free carrier concentration are less marked for bulk crystalline or glassy versions of semiconducting oxides than for thin film and granular assemblies. The more surface area that is available to interact with the surrounding gas, the greater will be the dependence of properties on the ambient. Thus care needs to be taken, particularly with thin granular film assemblies, about the ambient conditions (a) when the film is grown or deposited and (b) when it is used.

One other oxide material should be mentioned here, i.e. magnetic oxide of iron, Fe_3O_4 whose structure may be regarded as interpenetrating lattices of FeO and Fe_2O_3. The hopping of electrons between divalent and trivalent Fe sites means that the material is an electronic semiconductor of which a low-mobility hopping process is characteristic. Subsequently this process was found to be characteristic of conduction in phosphate glasses containing transition metal ions[5] and such materials provided the first of a series of high-resistance semiconducting oxide glasses which have shown interesting behaviour including electronic switching.

3. MEMORY SWITCHING

Controllable instabilities in semiconductors are of great importance in the development of switches and high frequency oscillators. Ovshinsky and his colleagues[6] demonstrated memory switching (Figure 1) in complex chalcogenide glasses and initiated a great deal of development work. Switches with quite complex properties could be constructed from these materials but the number of switching operations that could be carried out was limited to some 10^8 so that if used in the central processor unit of a large computer, they would last typically for a day or so. Drake and his colleagues[7] showed how to use copper-calcium-phosphate glasses for such memory switches and Moridi and I[8] were able to study the switching behaviour from a fundamental standpoint. By use of electron diffraction we were able to show that the effect of the energy dissipated during

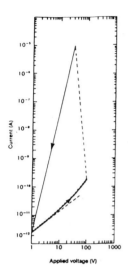

Fig.1 Typical voltage-current
characteristic for a thin oxide
glass showing memory switching

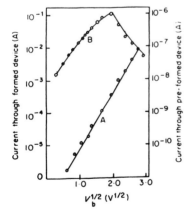

Fig.2 Circulating current as
a function of bias potential (V_b)
for an Al-SiO$_x$-Au device (A) pre-
formed and (B) after electroforming

the switching process was to crystallise the small region of oxide glass between the electrodes. A lower threshold voltage was recorded when the sample was at a higher temperature. A high-current pulse, used to return the switch to the high-resistance OFF-state, was found to re-vitrify the crystallised region. The large difference in carrier mobility could be a major factor in the memory switching. In these oxide glasses there was no obvious limitation to the number of switching operations, unlike the situation in the silicon-germanium-arsenic-tellurium glass in which the switching-on process was associated with tellurium coming out of solid solution and forming conducting bridges between the electrodes. After many such processes the device failed.

4.   ELECTROFORMING

One major use of oxides in semiconductor device technology is as insulators or as capacitor materials. As such any instabilities in these materials such as high-voltage breakdown at electric fields below the expected value can lead to device or circuit failures. Such premature breakdowns are often associated with voids or thinned regions in the oxide layers or with metallic inclusions. Care with the preparation of such layers and the use of chemical controls may prevent these premature destructive breakdowns but a less catastrophic form of high-voltage breakdown may be observed in those insulating oxides used in association with semiconductors, in SiO, Al_2O_3, Ta_2O_5, for example. Some years ago Wright and I[9] decided that we could probably improve thin film dielectrics for capacitors by using dielectric layers in the form of mixed films of for example SiO with B_2O_3. The object was to produce dielectric materials in which some of the dangling bonds in the SiO were satisfied by the boron atoms and certainly a mixed 70 mol.% SiO/30 mol.% B_2O_3 dielectric had a breakdown strength much greater than simple SiO films. Such complex materials are most readily formed when the components would form a glass if heated together at the glass-forming temperature. Although the temperatures reached at the substrate on which the oxides are deposited are only of the order of 100-200°C, the layers formed are amorphous, possessing a less open structure than amorphous SiO, and in many respects resemble borosilicate glass which normally forms at some 1700°C.

In spite of the high stability and normal breakdown strength, such films still show the electroforming effects originally described by Kreynina, Selivanov and Shamskaid[10] and which have been recently reviewed in depth by Ray and myself[11]. The phenomenon is well illustrated by reference to Figure 2 which shows the circulating current I_c as a function of bias potential V_b

for an Al-SiO-Al thin film sandwich with insulator thickness 40 nm in (A) the pre-formed and (B) the formed condition. After applying a forming voltage of some 5V the value of I_c increases dramatically by some three orders of magnitude and the device characteristics settle into a new mode when the sample is said to be electroformed. In this mode it is usual to observe a differential negative resistance after a peak current is reached and the value of V_b is continuously increased.

Similar phenomena have been observed in many MIM (metal-insulator-metal) structures and apart from the early work on SiO and Al_2O_3 such materials as GeO_2, Nb_2O_5, ZrO_2, polymers, halides and others have been studied as have our own complex oxide films made by vacuum evaporation from two controlled sources. A significant and regular variation between forming voltage and composition has been observed for SiO/Nb_2O_5 films[12].

Associated with the differential negative resistance are a series of memory states and these can be modified by varying the sample temperature and also the ambient atmosphere. Normally the samples are electroformed and studied while they are in a vacuum environment. It is also found that electrons can be emitted into vacuum or a neighbouring material from the electroformed sandwich samples from the positively biased electrode and the rate of emission rises sharply at about the position of the peak current of the V_b-I_c characteristic.

The electroforming clearly affects the structure of the MIM device and early experiments and recent work using X-ray photoelectron spectroscopy showed that metal from the negatively biased electrode was introduced into the oxide layer during the forming process. The various theories concerning how the incorporated metal affected the properties of the material have been reviewed in detail by Ray and myself and they may involve tunnelling between metallic centres, solid-state electrolysis, local avalanche effects, ionic motion and combinations of these processes. Knowing that breakdown and partial breakdown processes in high resistance materials are usually associated with localised conduction paths where the current density is high, Dearnaley, Morgan and Stoneham[13] proposed a theory whereby the metal introduced during the forming process formed a series of metallic chains interspersed to some extent by vacancies. Such chains could be regarded as filamentary or polyfilamentary and various experiments involving microscopy and other less direct techniques have demonstrated beyond reasonable doubt the presence and the importance of these filamentary paths in formed metal-oxide-metal structures. A study of the early technical literature on the early MOS devices suggests that this type of filamentary generation played some part in the instabilities and failures which were reported.

One simple experiment which I carried out with Nadeem[14] was to study the capacitance and the a.c. conductance of MIM structures before and after electroforming. The conduction properties were entirely consistent with conduction by metallic paths after the electroforming process. As the bias voltage across such devices increases, a situation is reached in which the current through filaments leads to such a high rate of localised Joule heating that the filaments start to rupture and although the applied voltage is increased the current is reduced. The dynamics of generation, rupture, re-joining and regeneration of filaments has been considered by many experimenters, and Ray and I[15] in a series of papers have studied the structure and distributions of filaments in oxides and have been able to obtain numerical solutions to our equations which lead to an almost perfect modelling of the characteristics of practical devices. Our calculations are based on a normal distribution of filamentary resistances.

The emission of electrons from formed MIM devices into a vacuum is associated with high-conductivity electrodes. The effect of electrode thickness was demonstrated by Gould and myself[16] who showed that an anode geometry with regular areas missing gave more efficient emission than a continuous electrode covering the same area. Certainly during the electroforming the anode is damaged and this is consistent with the predictions from a process involving some solid-state electrolysis along filamentary paths, leading to gaseous evolution of oxygen at the anode termination and to a metallic crowding-up effect at the cathode end, as I demonstrated with Rakhshani some years ago[17]. The highest reported emission current density is some 3×10^{-3} A cm^{-2} using SiO/B_2O_3 as the oxide, and copper or silver for the electrodes[18]. Such emitters excited by no more than 15 V operate at only 30°C above ambient but the emitted electrons can operate a cold-cathode vacuum ion gauge or can initiate chemical reactions in the gas phase.

5. TRANSPARENT ELECTRODES

In order that non-metallic materials may transmit light in the visible the bandgap needs to be of order 2 eV or greater. Many oxides meet this criterion and may in principle be used either as transparent electrodes for such devices as semiconductor solar cells and switching cells in which the switching is initiated by a beam of light, or as heating elements on the glass of indicating instruments used in cold or damp régimes to keep the glass clean and the instrument visible. The better conducting oxides including cadmium-tin and indium-tin oxides are used for this purpose.

6. ELECTROCHROMIC EFFECTS

Many papers have appeared to describe electrochromic effects in which a thin film of an oxide material may be coloured by the application of an electric field. One such material is di-cadmium stannate 2 CdO.SnO$_2$, (Cd$_2$SnO$_4$) which is known to have two stable species[19]. Normal

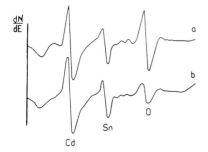

$\frac{dN}{dE}$

a

b

O

Sn

Cd

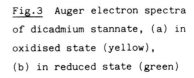

Fig.3 Auger electron spectra
of dicadmium stannate, (a) in
oxidised state (yellow),
(b) in reduced state (green)

300   400   500   600

KineticEnergy (eV)

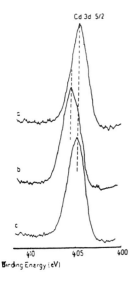

Cd 3d 5/2

a

b

c

410   405   400

Binding Energy (eV)

Fig.4 X-ray photoelectron spectra
of dicadmium stannate in the Cd $3d_{5/2}$
region: (a) in oxidised form,
(b) after electrochemical reduction,
(c) after heating in air at 250°C for
40 min.

preparation methods yield the high-stability yellow form. By heat treatment under a controlled atmosphere the green modification may be obtained but this can also be obtained by an electrochemical technique[20]. The green material has the higher electrical conductivity and is of great technical interest, for example as an electrode for alkaline secondary batteries, where its resistance to corrosion is a valuable property. The colours and indeed the composition of the material are of paramount importance in developing a technical process and the small differences in lattice parameter between 2 CdO.SnO$_2$ and CdO.SnO$_2$ barely show up by conventional X-ray diffraction. Hashemi, Golestani-Fard and the present author[21] have studied samples of both green (reduced) and yellow (oxidised) di-cadmium stannate by Auger and X-ray photo electron spectroscopy and have been able to show small differences in local bonding and valency changes. Figure 3 shows the Auger electron spectra of di-cadmium stannate in two forms where the relative oxygen content change may be seen. Figure 4 shows the XPS spectra of di-cadmium stannate. Significant changes have been established following heat and electric field treatments.

Recently it has been shown[22] that a variety of cadmium stannates may be produced in thin film form by the co-evaporation technique using two sources of Cd and SnO$_2$. Thin films of appropriate colour may be deposited by varying the relative rates of evaporation and hence the composition. The layers produced are characteristic of cadmium stannates when examined by X-ray diffraction and we believe that this technique may prove of wider use when preparing thin complex oxide films for use in electronic devices and systems.

REFERENCES

1. BAR-LEV, A. "Semiconductors and Electronic Devices" (Prentice Hall International, 1984).
2. MOTT, N.F. and DAVIS, E.A. "Electronic Processes in Non-crystalline Materials" (Clarendon Press, Oxford 1979).
3. AL-ANI, S.K.J., ARSHAK, K.I. and HOGARTH, C.A.
    J. Mater. Sci. 19 (1984) 1737.
4. HOGARTH, C.A.
    Nature (London) 161 (1948) 60.
5. DENTON, E.P., RAWSON, H. and STANWORTH, J.E.
    Nature (London) 173 (1954) 1030.
6. OVSHINSKY, S.R.
    Phys. Rev. Lett. 21 (1968) 1450.

7. DRAKE, C.F., SCANLAN, I.F. and ENGEL, A.
   Phys. Stat. Sol.(a) 32 (1969) 193.

8. MORIDI, G.R. and HOGARTH, C.A.
   Int. J. Electronics, 44 (1978) 297.

9. HOGARTH, C.A. and WRIGHT, L.A.
   Proc. IX Int. Conf. Phys. Semicond. (Moscow, 1968) Vol.2, Izv. Akad. Nauk SSSR (1968) p.1274.

10. KREYNINA, G.S., SELIVANOV, L.N. and SHAMSKAID, T.I.
    Radio Eng. Electron. Phys. (USSR) 8 (1960) 219.

11. RAY, A.K. and HOGARTH, C.A.
    Int. J. Electronics 57 (1984) 1.

12. AL-ISMAIL, S.A.Y., ARSHAK, K. and HOGARTH, C.A.
    Phys. Stat. Sol.(a) 89 (1985) 363.

13. DEARNALEY, G., MORGAN, D.V. and STONEHAM, A.M.
    J. Non-Cryst Solids 4 (1970) 593.

14. HOGARTH, C.A. and NADEEM, M.Y.
    Phys. Stat. Sol. (a) 56 (1979) K37.

15. RAY, A.K. and HOGARTH, C.A.
    Thin Solid Films 141 (1986) 201. [Contains references to earlier publications].

16. GOULD, R.D. and HOGARTH, C.A.
    Thin Solid Films 30 (1975) 131.

17. RAKHSHANI, A.E. and HOGARTH, C.A.
    J. Non-cryst. Solids 21 (1976) 147.

18. TAHERI, E.H.Z., HOGARTH, C.A. and GOULD, R.D.
    Phys. Stat. Sol.(a) 12 (1972) 563.

19. GOLESTANI-FARD, F., HASHEMI, T., MACKENZIE, K.J.D. and HOGARTH, C.A.
    J. Mater.Sci. 18 (1983) 3679.

20. HASHEMI, T., GOLESTANI-FARD, F. and MACKENZIE, K.J.D., High-Tech. Ceramics,
    Ed. P. VICENZINI, (Elsevier, Holland, 1987) p.2203.

21. HASHEMI, T., HOGARTH, C.A. and GOLESTANI-FARD, F.
    J., Mater. Sci. (in press).

22. HASHEMI, T., AL-DHHAN, Z., and HOGARTH, C.A.
    (Private communication).

TOPOLOGICAL REASONING OF THERMODYNAMIC AND KINETIC OF CVD PROCESSES

V.A.Voronin, V.A.Prochorov, M.I.Dronyuk, V.A.Goliousov

Lvov Polytechnical Institute, Lvov 290646, USSR

Investigations of multicomponent systems used in semiconductor materials CVD processes is suggested to unify the analysis of these systems. Physical chemical modelling serves as a basis for theoretical and experimental research of multicomponent vapor transport systems with the purpose of establishing thermodynamic parameters of vapor deposition. Chemical equilibrium research is concentrated on identification of all vapor phase components and their thermodynamic properties, as well as on possible phase and chemical transformations involving these components. A large number of atomic and molecular forms in both condensed and vapor phases, as well as existence of several types of phase and chemical transformations, necessiated the use of what we call "the structural topological model (STM) of chemical equilibrium". STM is supposed to prove a reliable method in chemical equilibrium research. The systems are described as twodimentional graph-models concluded all possible atomic and molecular forms for the given model as well as different physical and chemical processes within the system.

Physical chemical modelling of multicomponent vapor transport systems such as A-B-X, A-B-H-X (A - IIIA subgroup elements - Al, Ga, In; B - VA subgroup elements - P, As, Sb; H - Hydrogen; X - Halogen - Cl, Br, I) and structural patterns of these models are illustrated by figure. On this figure STM of chemical equilibrium for A-B-X system is given in the form of a function $G = (C_I...C_N, T_I...T_r)$, where C is a certain number of vertices, T is the number ribs connecting the vertices, N is the number of columns, and r is the number of rows of the stoichiometric matrix.

To solve the equations, which describe equilibrium in A-X systems, we have suggested the solution algorithm on the basis of modified Nelder-Mead method (I). The equations involving tensimetric measurement data are to be solved for the realization of inderect equilibrium problem. The solution of the direct equilibrium problem is given in (2). The same procedure was made with A-H-X systems. To determine thermodynamic characteristics of components, we have used

$$
\begin{array}{ccccc}
& (I) & & (2) & \\
4AX_{3(s)} + 4B_{(s)} & \rightleftharpoons & 4AX_{3(g)} + 4B_{(s)} & \rightleftharpoons & 2A_2X_{6(g)} + 4B_{(s)} \\
\Vert (3) & (6) & \Vert (4) & (7) & \Vert (5) \\
4AX_{2(s)} + 4B_{(s)} + 2X_{2(g)} & \rightleftharpoons & 4AX_{2(g)} + 4B_{(s)} + 2X_{2(g)} & \rightleftharpoons & 2A_2X_{4(g)} + 4B_{(s)} + 2X_{2(g)} \\
\Vert (8) & (II) & \Vert (9) & (I2) & \Vert (I0) \\
4AX_{(s)} + 4B_{(s)} + 4X_{2(g)} & \rightleftharpoons & 4AX_{(g)} + 4B_{(s)} + 4X_{2(g)} & \rightleftharpoons & 2A_2X_{2(g)} + 4B_{(s)} + 4X_{2(g)} \\
\Vert (I3) & (I6) & \Vert (I4) & (I7) & \Vert (I5) \\
4A_{(s)} + 4B_{(s)} + 6X_{2(s)} & \rightleftharpoons & 4A_{(g)} + 4B_{(s)} + 6X_{2(g)} & \rightleftharpoons & 4AB_{(g)} + 6X_{2(g)} \\
\Vert (I8) & (2I) & \Vert (I9) & (22) & \Vert (20) \\
4AB_{(s)} + I2X_{(g)} & \rightleftharpoons & 4AB_{(s)} + 6X_{2(g)} & \rightleftharpoons & B_{4(g)} + 4A_{(s)} + 6X_{2(g)} \\
\Vert (23) & (26) & \Vert (24) & (27) & \Vert (25) \\
4B_{(s)} + 4A_{(s)} + 6X_{2(g)} & \rightleftharpoons & 4B_{(g)} + 4A_{(s)} + 6X_{2(g)} & \rightleftharpoons & 2B_{2(g)} + 4A_{(s)} + 6X_{2(g)} \\
\Vert (28) & (3I) & \Vert (29) & (32) & \Vert (30) \\
4BX_{(s)} + 4A_{(s)} + 4X_{2(g)} & \rightleftharpoons & 4BX_{(g)} + 4A_{(s)} + 4X_{2(g)} & \rightleftharpoons & 2B_2X_{2(g)} + 4A_{(s)} + 4X_{2(g)} \\
\Vert (33) & (36) & \Vert (34) & (37) & \Vert (35) \\
4BX_{2(s)} + 4A_{(s)} + 2X_{2(g)} & \rightleftharpoons & 4BX_{2(g)} + 4A_{(s)} + 2X_{2(g)} & \rightleftharpoons & 2B_2X_{4(g)} + 4A_{(s)} + 2X_{2(g)} \\
\Vert (38) & (4I) & \Vert (39) & (42) & \Vert (40) \\
4BX_{3(s)} + 4A_{(s)} & \rightleftharpoons & 4BX_{3(g)} + 4A_{(s)} & \rightleftharpoons & 2B_2X_{6(g)} + 4A_{(s)}
\end{array}
$$

the results of A-H-X systems tensimetric research (3,4).

To describe transport processes in CVD systems graph-models are also suggested. In this case the tops of the graph-model correspond to the vapor system elements which transform the masses of substances included, the inner and outer sources and the outputs of substances in the system. The graph-model ribs are equivalent to the integrated material flows of substances which correspond to different substances transport mechanisms in the vapor phase.

REFERENCES

(I) J.A.Nelder, R.Mead, Computer Journ. $\underline{7}$, 308-313, 1965.

(2) V.A.Prochorov, V.A.Voronin, Journ. Phys. Chimii, $\underline{53}$, 500-501, 1979.

(3) V.A.Voronin, V.A.Prochorov, M.Chub, V.A.Goliousov, A.V. Klymkiv, Neorganicheskie materialy, $\underline{22}$, 1450-1452, 1986.

(4) V.A.Voronin, V.A.Prochorov, M.Chub, V.A.Goliousov, L.N. Luchka, Neorganicheskie materialy, $\underline{22}$, 1453-1456, 1986.

DLTS INVESTIGATIONS OF THE METASTABLITY OF THE DX CENTRE IN MBE GROWN ALGAAS

Y. N. Mohapatra*, V. Kumar*, S. Subramanian+, and B. M. Arora+

* Department of Physics, Indian Institue of Science, Bangalore 560012

+ Tata Institue of Fundamental Research, Homi Bhabha Road, Bombay 400005

ABSTRACT: Trapping characteristics of the two peaks associated with the well known DX centre in MBE grown, silicon doped $Al_xGa_{1-x}As$ (x = 0.36) are studied by DLTS as function of filling pulse width. Several features are noted including interdependencies of the two peaks, logarithmic growth as function of filling pulse width, shift in the peak temperature of one of the peaks the other remaining constant and reduction in the peak width. These features are explained by a comprehensive model having the following features. The DX center is the dominant donor having two states A and B both of which can emit to the conduction band. Only the metastable state B can capture electrons directly with a distributed energy barrier. After capturing an electron the metastable state B converts into the ground state A. The emmision from B is field dependent. Model calculations are presented which show all the essential features of the experimental data.

1. INTRODUCTION:

It is well known that most of the commonly used donor

impurities such as Si, Sn, Ge, Se, Te [1] give rise to a large concentration of a deep level defect referred to as the DX center [2] which controls the electrical and optical properties of $Al_xGa_{1-x}As$ material for $x > 0.2$. This centre gives rise to a variety of interesting physical phenomena such as persistent photoconductivity. The earlier proposed model [2] of the DX centre as a donor-vacancy complex is now being critically reviewed. A number of new models are being proposed as new experimental results are being discovered [3-7]. The dominant electrical characteristics of this defect appear to be independent of the growth technique as well as the dopant [1] although differences in details have been observed. Sample-to-sample variation of the nature and shape of DLTS spectra have been reported though not correlated with any fabrication or experimental parameter [2]. Several authors have reported two well resolved DLTS peaks or a main peak with a shoulder [1,2,8,9,10]. The energy levels for the DX centre obtained from DLTS and Hall measurements do not match. Clearly the DX centre is a complex centre about which much is yet to be learnt.

Although several authors have studied the emission characteristics of the DX centre, little attention has been paid to its capture kinetics [8,11]. It has been well established [2] that the electron capture is thermally activated with a large barrier which results in persistent photoconductivity at low temperatures. Thus all models for the DX centre invoke large lattice relaxation [2]. Zhou et al [8] studied the capture from the two DX peaks treating them as independent traps and assuming them to be the dominant donors. Caswell et al [11] noted the slow capture of electrons by the dominant DX donor from MODFET channel over several decades in time with roughly ln(t) dependence. This was explained by introducing a capture barrier with gaussian distribution in energy.

In this work, we report a study of the trapping kinetics of the DX centre by DLTS. The filling pulse width is varied from 1 microsecond to 100 miliseconds. Two peaks are observed. A model for the DX centre is proposed to explain the observed features.

2. Experimental

The samples used in this study consist of about 1 micron thick GaAs buffer layer, about 1 micron thick AlGaAs layer and about 0.5 micron thick GaAs layer sequentially grown by MBE on Si Doped n+ GaAs substrate. The epilayers are also silicon doped. The doping in the AlGaAS layer is about 4 x 10^{17} cm^{-3}. The Al fraction is estimated to be 0.36 from the band edge photoluminescence at 77 K. All the measurements reported in this study have been done by etching off the top GaAs layer and depositing 1000 A Au dots to form Schottky barriers. Since the phosphoric acid etching solution is not selective, the actual thickness of the AlGaAs layers varies slightly from sample to sample.

The capacitance measurements are performed using a Boonton 72B capacitance meter. The DLTS system uses double box car window scheme [12]. The capacitance meter is disconnected from the sample using fast relays while applying short pulses for capture cross section measurements [13].

A typical set of DLTS spectra is shown in Fig. 1. The quiescent reverse bias is 4 volts. The pulse width is varied over nearly five decades. The rate window is fixed at 28.2 msec. Similar curves are obtained with time constants upto 2820 msec. These spectra showing two peaks labeled A and B, have the following features not reported earlier.

a. The increase in the peak heights is seen to be slow and apparently logarithmic with increasing pulse width.

b. The peak temperature of A remains essentially constant as the filling time is increased.

c. The peak B apparently shifts to higher temperatures with increasing time.

d. The peak width also decreases with increasing time as seen on the low temperature side of the spectrum.

3. **Discussion:**

Realising that the capacitance transients are nonexponential, a complete calculation of the emission transient is carried out on the lines suggested by Subramanian et al [10]. The time and temperature dependence of the width of the space charge layer is obtained from the solution of the Poisson equation. The ionized trap concentration in the neutral region and the edge region is calculated from the charge neurtrality condition since the DX centre itself is the dominant donor i.e. $N_{DX} \gg N_D$. The DLTS spectra calculated from this model alone do not show any of the observed features. however, a forced fit of our experimental spectra with this model shows that the concentration of peak A increases linearly with ln(t) while that of peak B initially increases and then decreases. This has made us believe in the interdependence of the two levels associated with the DX.

It has been suggested by several authors [14,15] that the energy level associated with a trap in an alloy semiconductor such as under consideration here, is distributed due to alloy scattering. Thus the emission time constant must be distributed. However, our attempts to fit the DLTS spectra showed that the half width of the energy level must be less than 10 meV to obtain the kind of peak width observed. This is consistent with the observation of Mooney et al [16] that the activation energy of the DX centre is independent of Al fraction x.

Calculations of DLTS spectra with different concentrations such that $z = N_{DX}/N_D$ varied. from 0.01 to 10,000 showed that the peak will always shift to the lower temperature with increasing z. Thus feature (c) could not be explained as being due to filling of the centre.

The capture of electrons by the DX centre is well described by the model proposed recently by Caswell et al [11]. They have assumed that the DX centre captures electrons from the L minimum only. The capture barrier is about 0.2 eV and has a gaussian distribution with a width of 0.045 eV. Thus the electrons being captured are above the conduction band edge and distributed in energy. With these features this model predicts a slow capture continuing over more than four decades in time. They did not consider any emission from the DX centre during the capture. We have included the emission which plays significant role at long times. With the electron capture modelled along these lines, the observed slow growth of the DLTS peaks could be explained. However, an attmpt to fit the observed DLTS spectra assuming two independent traps still failed to explain features (c) and (d) mentioned in Section 2.

Hasegawa and Ohno have recently proposed [6] a model for the DX centre assuming a deep shallow instability. We have adapted this model as illustrated in Fig. 2. The level B captures electrons from the conduction band according to the model described above. It is a shallow state which can emit directly to the conduction band. However after capturing an electron the metastable level B converts into the ground state A which cannot capture directly. The ground state A emits electrons directly to the conduction band leaving behind an ionized donor in state B. The major variation of this model from that of the Hasegawa and Ohno model is that the state B is not assumed to be hydrogenic here. The interconversion rate from the metastable state B to the ground state A is assumed to be thermally activated with

$$R = R_0 \exp\left(-E_r/kT\right).$$

Further, we assume that the emission energy of the level B is field dependent due to the Poole-Frenkel effect. This is consistent with the results of Hwang and Choe [17]. As the concentration of the filled traps increases with capture time, the electric field in the depletion layer during subsequent emission gets reduced leading to features (c) and (d).

The numerical calculations are quite involved. This model introduces two new parameters namely R_0 and E_r. We have studied the affect of these parameters on the calculated DLTS spectra. Since several other values such as the enrgy levels, capture cross sections etc. are also not well established, it is not possible to obtain exact values for R_0 and E_r and nor can we obtain one to one fit with the experimental curves. However, the model gives reasonable values and a qualitative fit.

4.   **Conclusions**

A comprehensive model for the DX centre has been presented here. It includes the essential features of the models proposed by Lang and Logan [2], Hasegawa and Ohno [6] and Caswell et al [11]. The interconversion rate between the meatstable state and the ground state has been quantified. This model explains all the features observed experimentally and attributed to the DX centre.

Acknowledgements   We thank Professor J. R. Arthur of Oregan State University, U.S.A. for providing the samples. Part of this work has been funded by the Department of Science and Technology.

References

1. O. Kumagai, H. Kawai, Y. Mori, and K. Kaneko, Appl. Phys. Lett. 45, 1322, 1984.
2. D. V. Lang, R.A. Logan, and M. Jaros, Phys. Rev. B19, 1015,1979.
3. M. Mizuta, M. Tachikawa, H.Kukimoto, and S. Minumura, Jpn. J. Appl. Phys. 24, L143, 1985.
4. N. Iwata, Y. Matsumoto, T.Baba, and M. Ogawa,Jpn. J. Appl. Phys., 25, L349, 1986.
5. Y. Ashizawa and M.O. Watanabe, Jpn. J. Appl. Phys. 24, L883, 1985.
6. H. Hasegawa and H. Ohno, Jpn. J. Appl. Phys.,25, L319, 1986.
7. R.Legros, P. M. Mooney, and S. L. Wright, Phys. Rev. B35, 7505, 1987.
8. B.L. Zhou, K. Ploog, E. Gmelin, Q. Zheng and M. Schulz, Appl. Phys., A28, 223, 1982.
9. A.J. Valois and G.Y. Robinson, J. Vac. Sci. Technol., B3, 649, 1985.
10. S. Subramanian, U.Shculler, and J.R.Arthur, J. Appl. Phys., 58, 845, 1985.
11. N.S.Caswell, P.M.Mooney, S.L.Wright, and P.M.Solomon, Appl. Phys. Lett., 48, 1093,1986
12. L. Jansson, V. Kumar, L-A Ledebo, and K. Nideborn, J. Phys. E: Sci. Instrum., 14, 357, 1981.
13. M. Mohan Chandra and V. Kumar, J. Phys. E: Sci. Instrum., 17, 949, 1984.
14. P. Omling, L. Samuelon, and H. G. Grimmeiss, J. Appl. Phys., 54, 5117, 1983.
15. B. M. Arora, Solid St. Comm., 61, 105, 1987.
16. P. M. Mooney, E. Calleja, S.L. Wright, and M. Heilblum, Intl. Conf. on Defects in Semiconductors, Paris, 1986.
17. I. D. Hwang and B. Choe, Jpn. J. Appl. Phys., 25, L891, 1986.

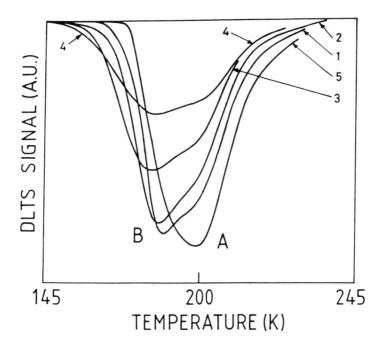

Fig. 1.   DLTS spectra of the DX centre in GaAlAs for
different filling pulse widths: 1) 1.5 ms, 2)0.15
ms, 3) 15 µs, 4) 2 µs, 5) 60 ms. The rate window
is 28.2 ms and the reverse bias 4V.

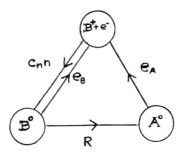

Fig. 2.   Schematic representatoion of the proposed model of
the DX center showing different transition rates.

DRY OXIDATION OF SILICON:
GROWTH KINETICS FOR THIN OXIDES

J.Vasi, S.S.Moharir and A.N.Chandorkar

Department of Electrical Engineering
Indian Institute of Technology, Bombay
Bombay 400076 India

1. INTRODUCTION

Oxidation of silicon is a vital processing step in the fabrication of metal-oxide-semiconductor (MOS) integrated circuits. A study of the kinetics of growth of silicon dioxide is useful not only because of its technological importance, but also because a fuller understanding of the silicon/silicon-dioxide system can well lead to a better appreciation of oxidation processes in general.

Present-day VLSI circuits use gate oxides of the order of 300 Å; further scaling of MOS devices is likely to result in even thinner gate oxides. Although the kinetics of the growth of silicon dioxide on silicon in a dry oxygen ambient has been studied for a long time, there is still considerable controversy over the models applicable for oxidation, especially for thin oxides. Indeed, no well-accepted model exists which adequately explains the rapid growth rate in the initial regime of oxidation. The aim of this paper is firstly, to review the existing models which attempt to explain the 'anomalously' rapid initial oxidation rate, and secondly, to offer two new models for the growth kinetics of thin oxides.

In Section 2, the classical Deal-Grove model for the oxidation of silicon, which was developed over two decades ago and describes the experimental data well for oxide thicknesses more than about 500 - 800 Å, is briefly presented. Section 3 describes some of the different models which have been suggested to explain the rapid initial oxidation rate, which is not explained by the Deal-Grove model. Section 4 presents two new models based on spatially varying diffusivities of oxygen in silicon dioxide to explain the data for thin oxides.

2. THE CLASSICAL PICTURE: THE DEAL-GROVE MODEL

Oxidation of silicon in a dry oxygen ambient (as opposed to a wet ambient which contains water) has been extensively studied since 1960. The Deal-Grove (D-G) model [1] presented in 1965 described remarkably well the observed data for dry oxidation for relatively thick oxides. Consequently, the D-G model has enjoyed wide acceptance for many years. The D-G model, as is well known, is a first order model based on three equal steady-state fluxes: the flux F_1 of oxidant from the gas to the oxide outer surface controlled by the gas phase transport coefficient h , the flux F_2 of the oxidant diffusing through the oxide proportional to the diffusivity D of the oxidant, and the flux F_3 corresponding to oxidation reaction at the oxide-silicon interface determined by the reaction rate constant k. This is shown in Fig. 1.

The resulting differential equation relating the oxide thickness X and time t is

$$\frac{dX}{dt} = \frac{B}{2X + A} , \qquad \qquad .....(1)$$

where B is the parabolic rate constant related to the diffusivity D , and B/A is the linear rate constant related (if h is large) to the reaction rate constant k . The solution of this is the familiar Deal-Grove equation

$$X + AX = B (t + \tau) . \qquad \qquad .....(2)$$

Here τ is an arbitrary constant introduced by Deal and Grove to circumvent the problem of the rapid initial oxidation. It essentially represents the time before t = 0 when the oxidation should have started in order to fit the data, as shown in Fig. 2. Several attempts have been made, as described in the next two sections, to explain the origin of the rapid initial oxidation phase for dry oxidation. As found by Deal and Grove and others, this initial phase is absent for oxidation in a wet ambient.

A study of the dependence of the constants B and B/A on the oxidant partial pressure p has proved useful as a way to identify possible models of oxidation. The D-G model would predict that both B and B/A , which depend on the solubility of oxidant in the oxide C* , vary linearly with p , due to Henry's law. However, it is found experimentally that although B is indeed proportional to p [1-4], the linear rate constant B/A is found to vary as p^n where $0.5 < n < 1.0$ [2,3,5] with n tending to 0.5 at lower temperatures and pressures, and to 1.0 at higher temperatures and pressures.

3. MODELS FOR THE GROWTH OF THIN OXIDES

Several models have been proposed to explain the rapid initial oxidation of silicon in dry oxygen. Deal and Grove [1] themselves invoked the space-charge model of Cabrera and Mott [6] since their model did not fit the data for small thicknesses. Subsequently, many other models have been suggested with widely differing physical origins, but no single model so far enjoys universal acceptability. This section reviews the major existing models.

3.1 Diffusion of Charged Species

Cabrera and Mott [6], describing oxidation processes in metals, postulated the possibility of the build-up of a space charge layer within the oxide by electron tunneling or thermionic emission. The electric field thus created would aid in the transport of the oxidant species which are charged. They derived an inverse logarithmic dependence of X on t , that is $1/X = C_1 - C_2 \log t$.

Deal and Grove used this model, postulating the motion of a charged species during oxidation. Based on the linear pressure dependence of B , they suggested that the oxidant species is a singly charged oxygen molecule O_2^- . A calculation of the Debye length of the space charge region gave about 150 Å for dry oxidation and about 5 Å for wet oxidation, which agreed reasonably well with their observation that for dry oxides an initial rapid oxidation phase exists for about 300 - 500 Å, whereas for wet oxides no such initial phase is observed. However, no quantitative comparison of the X - t data with the Cabrera-Mott equation was attempted.

The existence of charged species to explain the rapid initial oxidation was also invoked by Kamigaki and Itoh [3] who obtain a fit of their data of oxidation at low partial pressures (0.01 - 0.001 atm) to the Cabrera-Mott inverse logarithmic law. Hu [7] has postulated that the negatively charged species diffusing through the oxide is O^- . Tiller [8,9] also considers it not unlikely that a charged oxygen species is involved. No quantitative fit to data (except under restricted conditions by Kamigaki and Itoh) has been obtained using the charged species models.

3.2 Different Reacting Species

The D-G model assumes that the oxidant species diffusing and reacting at the silicon-oxide interface is molecular oxygen. Consequently, it predicts that both B and B/A , and hence dX/dt vary linearly with pressure.

Experiments by van der Meulen [2] established that B/A , in general, varies as p^n , with n tending to 1.0 at higher temperatures and pressures and to 0.5 at lower temperatures and pressures. Ghez and van der Meulen [10] explained this data by assuming that the reacting species at the interface include both molecular and atomic oxygen. Their derived expression for dX/dt is shown to qualitatively account for the p^n variation. At lower temperatures, reaction with atomic oxygen dominates giving rise to n = 0.5 , and at higher temperatures, reaction with molecular oxygen dominates giving n = 1.0 . The expression for dX/dt , which now includes additional terms to describe the possibility of atomic oxygen reaction, only qualitatively explains the increase of oxidation rate for small thicknesses.

Blanc [11], in a modification of the above, assumes that only atomic oxygen reacts at the interface. However, Blanc does perform a quantitative comparison with available data for thin oxides, and the fit is good; the value of the linear rate constant is, of course, different from that obtained from the D-G model. Unfortunately, Blanc's model predicts that the linear rate constant varies as the square root of the partial pressure for all temperatures, which is contrary to the data of van der Meulen as well as others.

3.3 Stress in the Oxide Film

Fargeix, Ghibaudo and Kamarinos [12] pointed out that since the inverse oxidation rate dt/dX can be written as

$$\frac{dt}{dX} = \frac{A}{B} + \frac{2X}{B} \qquad \ldots\ldots(3)$$

and the plot of dt/dX versus X shows larger slopes for small X , the parabolic rate constant B must be smaller for thin oxides. This implies that there is a decrease of diffusivity for small oxide thicknesses, rather than an increase as postulated by many models (such as charged species diffusion and microchannels) to explain the increased oxidation rate and as seems intuitively more reasonable.

The reduction in diffusivity for small thicknesses is attributed by them to stress near the silicon-oxide interface caused by the conversion of silicon to silicon dioxide, with the stress decreasing away from the interface. This relaxation of stress is suggested as being due to viscous flow of the oxide as time proceeds [13,14]. The effective diffusivity D' for an oxide thickness X is defined as

$$D'(X) = \left[\frac{1}{X} \int_0^X \frac{dx}{D(x)} \right]^{-1} \qquad \ldots\ldots(4)$$

where D(x) is the diffusivity which varies with distance x in the oxide. If the relationship between D and oxide stress s is assumed to be D = D_0 exp (- as/kT), Fargeix and Ghibaudo [13] show that the stress-diffusivity model fits the experimental data.

A possible problem with the stress model, however, is that it predicts that the thickness L for which the rapid oxidation phase is active should decrease continuously with increasing temperature. This seems at variance with the data of Massoud et al. [15] which shows almost the same L for temperatures between 800° C and 1000° C. Furthermore, one would expect that since viscous flow of silicon dioxide is high at temperatures above about 950° C, no stress and consequent rapid oxidation should occur above 950° C , which also does not agree with experiment. It may be noted, however, that the observation that diffusivity is smaller for small thicknesses is still valid.

3.4 Special Interface Regions

Murali and Murarka [16] have proposed that oxidation takes place not just at the silicon-oxide interface but in a finite region of the silicon into which also the oxygen diffuses. For thin oxides, since the interfacial concentration C_i of oxygen is high, not all the oxygen reacts at the interface, and there is a large diffused zone in the silicon in which oxidation also occurs. As the thickness grows, C_i decreases and the size of the diffused zone diminishes, until ultimately all the reaction does take place at the interface, and the classical D-G model obtains. This model has been used to fit the thin oxide data of Irene [17] and others.

Stoneham et al. [18] postulate the existence of a 'reactive' layer near the interface which is, unlike the Murali-Murarka diffused zone, a silicon-rich oxide layer. Oxidation takes place at the outer (silica) interface of this layer, with silicon diffusing through it. The layer grows in the initial phase and then reaches a constant value of about 20 Å. This layer serves to reduce the level of stress at the silicon-oxide interface during oxidation. No explicit comparison with growth rate data has been made for this model.

A model has been proposed by Schafer and Lyon [19] which invokes the layer of fixed positive charge that builds up at the interface during oxidation. They postulate that the build-up of charge as oxidation proceeds reduces the concentration of holes at the silicon-oxide interface which in turn reduces the number of broken Si - Si bonds there. Since the reaction rate is determined by this number, the growth rate decreases with increasing thickness. The model fits their data fairly well but only if a specific (albeit plausible) variation of charge within the layer is assumed. None of the models coming under this category would be able to explain partial pressure effects easily.

3.5 Microchannels

Revesz and Evans [20] suggested the possibility of the occurrence of microchannels in thin oxides, along which the oxidant species can diffuse rapidly and speed up the growth rate. Irene [17] has examined this possibility more carefully, correlating initial growth rate, breakdown strength and TEM studies for both dry and wet oxides. He concludes that the evidence suggests that microchannels, about 50 Å in diameter, are present in dry oxides whereas wet oxides form a protective layer which enhances the breakdown strength and also prevents the formation of microchannels through which the oxidant can preferentially move.

4. NEW MODELS BASED ON DIFFUSIVITY VARIATION

In spite of the large number of models to explain the initial rapid oxidation phase, no single model has been uniformly successful in explaining all aspects of the initial regime. There is still room for new ideas, and we present here two interlinked models.

4.1 Power Law X - t Dependence

Recently, extensive data on the growth of silicon dioxide in a dry oxygen ambient for different temperatures and partial pressures and different silicon orientations and resistivities has been reported by Massoud [21] and Massoud, Plummer and Irene [15]. The measurements were taken using an in situ ellipsometer connected to a computer, with the result that the oxide thickness could be (almost) continuously monitored, and growth rates dX/dt accurately obtained by numerical differentiation. This wealth of data, besides providing a test for any model, may also help to suggest a new model.

Figure 3 shows the data of X versus t on a log-log plot, obtained from Massoud [21]. This data, for (100) and (111) silicon for temperatures between 800° C and 1000° C , is almost closely enough spaced to be represented on the plot by continuous lines. Only data greater than 50 A has been plotted since data for lower thicknesses may be unreliable, partly due to the fact that the starting initial oxide thickness varied from 7 - 22 Å [21].

As can be seen, the plots are remarkably good straight lines for all temperatures and both orientations for thicknesses ranging from 50 Å to about 800 Å. This has also been mentioned by Massoud [21]. Reisman and Nicollian [22] have also shown that the existing data in the literature can be fitted over a wide range of thicknesses by a power law. This indicates that the X - t relationship should be

$$X^n = K t . \qquad \dots (5)$$

For the case of the (100) and (111) oriented wafers, the value of n is almost exactly 3/2 . For (110) oriented wafers (whose data has not been shown here), the value of n is somewhat more, about 1.65 . This power law dependence suggests that the growth rate dX/dt should be given (for (100) and (111) silicon) by

$$\frac{dX}{dt} = \frac{K'}{\sqrt{X}} , \qquad \dots (6)$$

where $K' = 2K/3$. This equation is to be compared with Eq. (1) from the Deal-Grove model.

Based upon values of K obtained from Fig. 3, we have calculated dX/dt for various X from Eq. (6). These are plotted in Fig. 4 for three temperatures together with Massoud's data. Again, the fit is good, lending credence to the validity of Eq. (6). This relation is much simpler than that used by Massoud et al. [15] to fit the data:

$$\frac{dX}{dt} = \frac{B}{2X + A} + C_1 \exp(-X/L_1) + C_2 \exp(-X/L_2)$$

which, however, has the advantage of reducing to the D-G model for large X.

One possible reason for the $1/X^{1/2}$ dependence of Eq. (6) is a variation of the diffusivity $D(x)$ with distance x in the oxide. If $D(x)$ varies as $x^{1/2}$, which means that diffusivity is small near the silicon-oxide interface and increases into the bulk of the oxide, then using Eqs. (3) and (4) and assuming that the linear rate constant is large, we get Eq. (6). A more reasonable variation for $D(x)$ is $D_0 x^{1/2} / (x^{1/2} + a)$ so that $D(x)$ tends to D_0 at large x ; the resulting equation then is

$$\frac{dX}{dt} = \frac{B}{2 X + 4a X^{1/2}} . \qquad \dots(7)$$

This equation tends to Eq. (6) for small X and would therefore fit Massoud's data, but also asymptotically approaches the D-G model for large X.

The reason for the variation of diffusivity with x is uncertain; at lower temperatures, it could be stress as suggested by Fargeix and Ghibaudo. Another possibility is put forward in the next section.

4.2 Diffusion by Adsorbed Oxygen

We present here a silicon oxidation model based on the diffusion through the oxide of adsorbed (rather than free) oxygen species. We assume that oxygen molecules get adsorbed at the outer surface of the oxide layer, and the adsorption isotherm applicable is the Langmuir isotherm [23]. The adsorbed molecules do not desorb, but move through the oxide along 'micropores' (not to be confused with the microchannels of Section 3.5, which are much larger) in the adsorbed state. Diffusion does not now follow the ideal Fickian behaviour. We assume that the flux is proportional to the gradient of the chemical potential. This gives rise to diffusivity variation given by Darken's law [24].

To derive an equation for the growth rate based on these considerations, consider Fig. 5 which shows the variation of concentarations. C_b is the concentration of the oxygen in the gas phase adjacent to the outer oxide surface, and q_o and q_i are the adsorbed concentrations at the outer surface and the interface, respectively. Assuming the Langmuir isotherm to be valid, we get

$$q_o = \frac{K_1\, C_1}{1\, +\, K_2\, C_1} \qquad \ldots\ldots(8)$$

where K_1 and K_2 are constants and K_1/K_2 is the saturated concentration q_s of oxygen adsorbed at the outer oxide surface. Darken's law for diffusivity is

$$D = D_o \{\, d \ln C_1\, /\, d \ln q_o\, \}\,, \qquad \ldots\ldots(9)$$

which gives, for the Langmuir isotherm,

$$D = D_o\, /\, \{\, 1\, -\, (\, q/q_s\,)\, \}\,. \qquad \ldots\ldots(10)$$

This indicates a diffusivity variation with D small near $x = 0$. The reacting flux is given by $k.q_i$, where k is the reaction rate constant. The reaction rate is proportional to the adsorbed concentration at the interface, which is similar to Hu's [25] assumption. The final equation obtained for q is

$$(q_i - q_s)\, \exp\left(-\,\frac{k\, q_i\, X}{D_o\, q_s}\right) = \frac{K_1\, C_b}{1 + K_2 C_b}\, -\, q_s \qquad \ldots\ldots(11)$$

This is solved numerically for q_i. The resultant plot of dX/dt is shown in Fig. 6 together with Massoud's data for $1000\,^\circ$ C. The parameters K_1, K_2, k and D_o have been optimized to give best fit. The results are quite encouraging, and further consequenses of the model are being worked out.

5. CONCLUSIONS

The recent interest in thin MOS oxides has led to renewed efforts to explain the 'anomalous' rapid initial oxidation phase. This paper has reviewed the several models which have been suggested, and presented two new interpretations of the data. These two depend on the variation of the diffusivity with distance in the oxide. We have suggested that a reason for the variation could be the diffusion of adsorbed rather than free oxygen species.

386

Acknowledgements.   The authors wish to thank Dr. A.S. Moharir,
Dr. R. Lal and Dr. D. Sharma for useful discussions.

REFERENCES

1. B.E. Deal and A.S. Grove, J. Appl. Phys. 36, 3770 (1965).
2. Y.J. van der Meulen, J. Electrochem. Soc. 119, 530 (1972).
3. Y. Kamigaki and Y. Itoh, J. Appl. Phys. 48, 2891 (1977).
4. M.A. Hopper, R.A. Clarke and L. Young, J. Electrochem. Soc. 122, 1216 (1975).
5. L.N. Lie, R.R. Razouk and B.E. Deal, J. Electrochem. Soc. 129, 2828 (1982).
6. N. Cabrera and N.F.Mott, Rep. Prog. Phys. 12, 163 (1948).
7. S.M. Hu, Appl. Phys. Lett. 42, 872 (1983).
8. W.A. Tiller, J. Electrochem. Soc. 127, 619 (1980).
9. W.A. Tiller, J. Electrochem. soc. 127, 625 (1980).
10. R. Ghez and Y.J. van der Meulen, J. Electrochem. Soc. 119, 1100 (1972).
11. J. Blanc, Appl. Phys. Lett. 33, 424 (1978).
12. A. Fargeix, G. Ghibaudo and G. Kamarinos, J. Appl. Phys. 54, 2878 (1983).
13. A. Fargeix and G. Ghibaudo, J. Appl. Phys. 54, 7153 (1983).
14. G. Ghibaudo, Phil. Mag. B 55, 147 (1987).
15. H.Z. Massoud, J.D. Plummer and E.A. Irene, J. Electrochem. Soc. 132, 2685 (1985).
16. V. Murali and S.P. Murarka, J. Appl. Phys. 60, 2106 (1986).
17. E.A. Irene, J. Electrochem. Soc. 125, 1708 (1978).
18. A.M. Stoneham, C.R.M. Grovenor and A. Cerezo, Phil. Mag. B 55, 201 (1987).
19. S.A. Schafer and S.A. Lyon, Appl. Phys. Lett. 47, 154 (1985).
20. A.G. Revesz and R.J. Evans, J. Phys. Chem. Solids 30, 551 (1969).
21. H.Z. Massoud, Ph.D. Thesis, Stanford University (1983).
22. A. Reisman and E.H. Nicollian, J. Electronic Mat. 16, 45 (1987).
23. I. Langmuir, J. Chem. Soc. 40, 1361 (1918).
24. L.S. Darken, Trans. AIME 175, 184 (1948).
25. S.M. Hu, J. Appl. Phys. 55,4095 (1984).

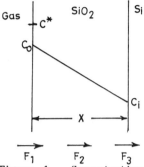

Figure 1. Concentrations and fluxes in the silicon-oxide system. X is the oxide thickness.

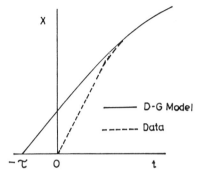

Figure 2. Plot comparing the observed initial oxidation phase with the Deal-Grove (D-G) model.

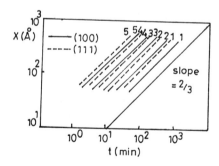

Figure 3. Experimental data on X versus t for different temperatures. (1) 800°C (2) 850°C (3) 900°C (4) 900°C (5) 1000°C. Data from Massoud [21].

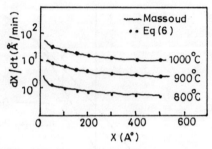

Figure 4. dX/dt versus X for (100) Si. Solid line is data from Massoud [21]; dots are from Eq. (6).

Figure 5. Definition of concentrations for the adsorbed oxygen model.

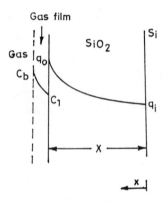

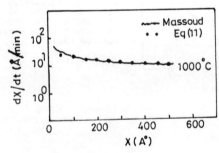

Figure 6. dX/dt versus X for (100) Si. Solid line is data from Massoud [21]; dots are from Eq. (11).

The Changing Nature of Reliability Physics (*)

J.W. Lathrop

Center for Semiconductor Device Reliability Research
Clemson University, Clemson, SC 29634-0915

ABSTRACT

Increased semiconductor device complexity has emphasized the need for greater understanding of the physical nature of device failure. This paper reviews the basic tenets of semiconductor device reliability and how technological change is causing failure mechanism research to shift from extrinsic to intrinsic phenomena. A review of the research status in three areas - electromigration, charge injection, and electrostatic discharge - is presented, with emphasis on describing the physical phenomena under investigation.

CLASSIFICATION OF FAILURE MECHANISMS (See Figure 1)

Wearout -- Wearout refers to degradation of the main population of non-defective devices and ideally will only occur a considerable time after manufacture. Wearout mechanisms can be classified as operational if they require the external application of voltage to the chip, and as environmental if they do not. Examples of environmental wearout are corrosion, surface ionic inversion, and diffusion, while examples of operational wearout are electromigration and charge injection. Wearout of the main population is often termed a device's "intrinsic" reliability.

Defect Accelerated Wearout -- Defective devices, i.e. those involving freak distributions of one or more physical parameters, will also be affected by environmental and operational wearout mechanisms. However, because of the presence of flaws, either introduced during manufacture or present in the starting material, defective devices will degrade faster than those of the main population. Wearout of a defective freak distribution is often termed the device's "extrinsic" reliability. Defect accelerated wearout leads to the "infant mortalities" of Figure 1.

Overstress -- The third category of failure mechanism is that of over-stress. The term can include mechanical or thermal overstress, but most frequently is used to designate electrical overstress in the form of electrostatic discharge (ESD) or electrical overstress transients. Exposure to ESD may occur at any time during life, but is more likely soon after manufacture when components are handled, tested, inserted in boards, etc. ESD is shown schematically in the bathtub curve of Figure 1 as a series of discrete spikes.

IMPACT OF TECHNOLOGICAL CHANGE ON SEMICONDUCTOR DEVICE RELIABILITY

The reliability of semiconductor devices has always been of concern, but the continued evolution of technology, particularly over the past decade, has greatly changed the nature of this concern. (Although this paper deals in general with the reliability of all types of devices, technology has had its greatest impact on the integrated circuit (IC), and since IC's also represent

---

(*) Supported in part by the Semiconductor Research Corporation

by far the greatest quantity of commercial devices, the specific failure mechanisms to be discussed will relate to IC's.) Traditionally, IC reliability has been concerned with infant mortality. The main approach to reliability assurance has been to manufacture devices as well as possible by strict adherence to specifications, tight quality control, clean room fabrication, etc. and then to utilize accelerated burn-in screening procedures to weed out any weak devices that might have slipped past final test. Wearout, in general, has not been a problem because, by extrapolating accelerated test results to use conditions, it almost always can be shown that devices would almost certainly become obsolete long before the main distribution would be affected.

Figure 2 illustrates the exponential increase in complexity that has occurred as a result of technological developments in the integrated circuit industry. Single-chip IC devices, which currently contain as many as 10-million transistors in 100+ lead packages, by the mid to late 1990's can be expected to contain as many as 1-billion transistors in 400+ pin packages. As can be seen from this figure, overall chip reliability has shown only modest improvement over the decades. Although this translates into spectacular improvements in per-transistor reliability, it is the per-chip reliability which must ultimately be of concern from a systems standpoint.

Development of submicrometer technology has introduced a number of new factors into the reliability equation: there is less need for concern regarding defect-associated extrinsic failure, but greater need regarding intrinsic wear-out failure. Improvement in fabrication methods and equipment, which has led to feature sizes less than 1 μm and oxide thicknesses less than 10 nm, has also resulted in fewer defects. Put simply, processing demands for these high complexity devices are so great that they cannot be made at all unless they are made correctly. This is confirmed by the excellent yields currently being achieved in the IC industry -- yields that only a few years ago would be unthinkable are now commonplace. Since yield at final test and reliability enjoy a symbiotic relationship, this is evidence that the combination of smaller devices and fewer processing defects means a lower probability of extrinsic failure.

On the other hand, as device dimensions shrink, high current density in the fine-line metal interconnects may cause opening or shorting to occur, and current flowing across the thin oxide layers may cause them to lose their insulating qualities. In other words, the device wears out. The specter of IC's wearing out in the same fashion as vacuum tubes is perhaps a little extreme, but even a small amount of wearout of the main device population could have disastrous effects on overall IC reliability.

THE CHANGING NATURE OF RELIABILITY RESEARCH

Because extrinsic failure is dependent on a wide variety of quality defects such as pinholes, foreign particles, and various other process irreg-ularities, its study is highly dependent on the particular conditions under which the test devices were processed. Data reproducibility is a serious problem and large quantities of devices are necessary for statistical analysis. Elevated temperature is generally used as the accelerating stress for most defect accelerated failure mechanisms and extrapolation to use conditions involves use of the familiar Arrhenius relationship.

As concern shifts from extrinsic to intrinsic failure, emphasis is no longer on testing and statistically analyzing large numbers of devices, but on understanding the basic nature of the failure mechanisms. As a result, fewer samples are required because only the main population is being studied. This is fortunate because, with the exception of memory chips, as devices become more complex fewer are manufactured. If it were necessary to acquire statistically significant quantities of freak populations of advanced application specific integrated circuit (ASIC) chips it is conceivable that in many cases more devices would be required for reliability testing than for the particular application itself! Increased device complexity shifts emphasis from failure statistics to failure physics.

When extrinsic reliability problems are discovered, corrective action for the most part involves the fabrication process, e.g. institution of better quality control procedures or modification of the process sequence. On the other hand, when an intrinsic mechanism becomes limiting more fundamental action is required -- it is necessary to alter the basic design of the device rather than to correct flaws. In theory this could involve changing the basic fabrication process, although this is difficult because of built in inflexibility associated with the equipment. A more viable possibility is to alter the device's electrical specifications or its topological layout.

Intrinsic reliability research involves the following sequence of events: 1) characterizing the effect, 2) mathematically modeling the effect, 3) incorporating the model in appropriate computer aided design and manufacturing tools, and 4) using these tools to optimize design. Understanding the physical principles of various failure mechanisms will be the key to successful implementation of this sequence. It is important to note that implementation will involve a profound change in the way reliability is viewed by the IC designer. Making the most reliable device possible will no longer necessarily be the goal, but rather achieving an optimum tradeoff of reliability with regard to cost and performance. Three failure mechanisms of particular concern to very large scale integrated (VLSI) devices, electromigration, charge injection, and electrostatic discharge (ESD), will now be discussed with particular emphasis on the physics of each.

ELECTROMIGRATION

Electromigration is a clear example of a wearout mechanism. It originates as a result of momentum transfer between current carrying electrons and metal atoms of the interconnect metallization. This momentum transfer results in an atomic flux which is proportional to the dot product of mobility and the driving force [1]. Mass transport can take place through either point defects, such as vacancies or interstitials, or gross defects, such as grain boundaries or surfaces. At normal device operating temperatures, which are less than half the melting temperature of the metal interconnects, mass transport will be primarily along the grain boundaries. For an ideal grain structure, the atomic flux, J_b, can be expressed as [2]

$$J_b \propto \frac{1}{kT} \frac{N_b \delta}{d} D_{bo} \, e^{-E_a/kT} \cdot j\rho ez^* \quad ,$$

where D_{bo} and E_a are respectively, the diffusivity constant and activation energy for grain boundary diffusion, j the current density, ρ the electrical resistivity and ez^* the effective charge. The quantity N_b is the atomic density, δ is the effective grain boundary width (≈ 10 Å) for mass transport and d is the average grain size.

For a given current density, it can be seen that the grain boundary diffusional flux can be reduced by reducing the grain boundary diffusion, D_{bo}, or the grain boundary effective charge number, Z^*, or by increasing the grain size, d. It has been found that addition of a few atomic percent of Cu into the Al metallization can reduce the EM flux. To be effective, however, the Cu must go into the grain boundaries, where it forms the intermetallic compound Al_2Cu, rather than into the grains themselves, since Cu within the grains merely increases resistance. "Plugging" the grain boundaries with Al_2Cu is achieved by first depositing Cu on the metal film and then, through a series of heat treatments, drive the Cu to the grain boundaries where it is formed into the Al_2Cu alloy. The effect of Cu is to reduce D_{bo} by increasing the activation energy of Al diffusion at the grain boundaries.

Divergence, originating from changes in grain size, when combined with grain boundary diffusional flux, can result in either accumulation of metal (hillocks and whiskers) or depletion of metal (voids). Thus EM can result in both open and short electrical circuit failure modes. The use of single crystal interconnects (very large d) is not practical although a large grain bamboo-type microstructure has been found effective [3]. At the other extreme, Ghate [4] has shown that a film of very fine grains, relative to the line width, can be used to minimize changes in grain size, with consequent reduction in the flux divergence and localized voiding. Another approach to the reduction of voiding is to avoid the propagation of EM cracks completely through the conduction path by using multilayer films. If one of the layers has a higher resistance, so that its current density is less, and/or if it has a higher atomic number so that momentum transfer is less efficient, then current could still flow even if the other layer were to open up. While such a structure should be very effective against voids, shorts would still be expected.

EM has been studied extensively under dc conditions and Black has presented an empirical equation which describes the conductor lifetime due to voids (median time to failure) [5]

$$t_{50} = A(W) \; J^{-n} \exp (E_a/kT),$$

where J = current density
E_a = activation energy ($0.6 < E_a < 0.7$ for Al-Cu alloy)
n = exponent (normally $n \approx 2$, but $1 < n < 15$ possible)
$A(W)$ = material constant = f(conductor width, W)

Current flowing in actual digital IC's is pulsed rather than dc. Consequently it is important when assessing the impact of EM to know how pulsing affects the phenomenon. The simplest approach is to assume that the dc equation of Black is still valid, but with the dc current density replaced by the average current density, $J_{av} = J_{peak} \cdot r$, where r = duty factor. If this assumption holds true then the pulse magnitude, J_{peak}, could be allowed to increase as the pulse width is reduced as long as the product is constant, with no resulting change in t_{50}. This reasoning would seem to imply that the peak current density could approach ∞ as W_p approached 0. Experimental work is currently underway to determine how far this relationship will hold.

Modeling from first principles of pulsed EM effects is proceeding simultaneously with the experimental work. One approach being developed by Harrison at Clemson [6] approximates the metallurgical state of the line using Monte-Carlo statistical methods. A log-normal grain size distribution is assumed which is characterized by a median value and a standard deviation. The modeled line is partitioned into sections whose length is equal to the average grain size. Boundaries between segments are called nodes and the mass flux divergence of each node is calculated. By combining the effect of momentum transfer induced mass transport along the line with back diffusion effects [7], it is possible to calculate the loss of mass at each node. The increase in local temperature and degradation of heat transfer due to current crowding as mass is lost are also included in the model. Finally, the increase in resistance of the line due to the reduction in cross-sectional area is calculated to allow a direct comparison with experiment. Calculated median time to failure vs duty factor for a typical set of model parameters is shown in Figure 3. It can be seen that the deviation from the J_{av} - constant assumption becomes significant at low duty factors.

Once a model has been developed which can account for EM under pulsed conditions and in leads of different geometry (steps, vias, corners, etc.), it can be included in available computer aided design tools to calculate reliability due to EM. Preliminary work by Frost and Poole [8] has indicated the feasibility of reliability simulation and at the same time point out the need for better models. As more information is received regarding the experimental behavior of EM test vehicles, the first order tools can be refined and limiting constraints eliminated. Analysis capable of locating critical areas [9] can then be followed by more thorough design optimization.

CHARGE INJECTION

As IC feature size is reduced, a corresponding reduction in thickness of the oxide dielectric layers must also take place. At thicknesses greater than 10 nm current flow through oxide is primarily by hot electrons, but below about 10 nm, oxide films are so thin that appreciable numbers of thermal electrons (or holes) are able to transit by means of Fowler-Nordheim tunneling. Passage of electrons through an oxide can change its insulating and dielectric properties and consequently the characteristics of devices which utilize these properties. As shown in Figure 4, taken from Ning, Osburn, and Yu [10], hot electrons may be injected into the oxide from the channel, the avalanche plasma near the drain region and, as a result of the existence of a high transverse field, from thermally generated carriers in the bulk. As shown in the figure, these hot electrons will charge the oxide increasing the subthreshold leakage and changing the threshold voltage. However, the contribution of hot electrons to device degradation is expected to lessen as devices shrink for two reasons: 1) geometrical changes such as buried channels and lightly doped drains (LDD) have been incorporated in devices to minimize the injection of hot electrons, and 2) the threshold voltage shift is inversely proportional to capacitance per unit area, C_0, which increases as the oxide layer gets thinner. For these reasons, concern relative to oxide wearout mechanisms is shifting from the injection of hot electrons into the oxide conduction band to the tunneling of thermal electrons.

Tunneling current in non-defective oxides can lead ultimately to breakdown of the oxide, a phenomenon referred to as time dependent dielectric breakdown (TDDB). TDDB can be studied either by the application of a constant voltage stress and observing I vs t or by the application of a constant current stress and observing V vs t. Another technique often referred to (inappropriately) as time-zero dielectric breakdown (TZDB) involves the use of a voltage ramp. Figure 5 illustrates the type of current vs voltage characteristic obtained with a voltage ramp [11]. Three types of current can be clearly identified in the figure: displacement, Fowler-Nordheim tunneling, and impact ionization. Breakdown is defined as the point at which the current rapidly increases due to impact ionization.

By taking data at different ramp rates it is possible to determine either the total fluence density (electrons/cm^2) or the total charge density (coulombs /cm^2) to breakdown. It can be seen from Figure 6 that these values are not constant, but increase as the ramping rate is decreased [11]. If charge-to-breakdown were a constant development of a reliability CAD tool would merely require integrating the current density over life. However, it is obvious from Figure 6 that the situation is more complicated.

Several physical mechanisms relating to charge generation and trapping in oxides have been reported in the literature: impact ionization [12], charge injection and trapping [13], trap generation [14], atomic hydrogen diffusion and bond breaking [15], and resonance tunneling due to local quantum states in the vicinity of the injecting surface [16]. The most likely explanation of the observed effects is that an initial buildup stage occurs, during which local-ized high field regions are formed, followed by a rapid runaway stage that occurs after a critical field is reached.

Localized high field regions come about due to such micro-irregularities as interface roughness, nonuniform trap distribution, or dopant impurity seg-regation at grain boundaries. If free holes are created in the oxide, either by direct Fowler-Nordheim tunneling or by the generation of hole-electron pairs from impact ionization by hot electrons, they will migrate to the vicinity of the cathode, primarily in the localized high field regions, where they aggre-gate, further enhancing the localized field at the cathode. This positive feedback will result in breakdown when a critical field value is reached. It is not known at present whether impact ionization or direct tunneling is primarily responsible for supplying the necessary free holes. Also involved are such effects as trap generation and trapping and detrapping of electrons.

Whereas the model that has been described is able to predict breakdown under accelerated conditions of voltage, such as those shown in the TZDB curve of Figure 5, it is not at all clear to the reliability engineer how results can be extrapolated to use conditions of 5 volts or less. Figure 6 would seem to imply a threshold, since at ramp rates above about 0.1 volts/sec essentially all the charge through the oxide leads to breakdown, but considerably smaller amounts at lower ramp rates. (At a ramp rate of 0.01 V/sec only about 10% of the current flowing through the oxide appears to contribute to breakdown.) This would seem to support the impact ionization theory since only a fraction of the injected electrons (tail portion of the energy distribution) might have sufficient energy (> 9 eV) for electron-hole pair generation. If this is so, then it will be necessary to derive an appropriate derating factor, which will be a function of voltage and thickness.

ELECTROSTATIC DISCHARGE

Protecting semiconductor devices from ESD damage is an area involving much reliability effort. One aspect involves avoiding exposure to ESD pulses, activity which encompasses a vast area of physics relating to triboelectricity. In this paper, however, discussion will be restricted to the interaction of ESD pulses with semiconductor devices rather than generation and avoidance of electrical overstress. This "after the fact" approach to ESD involves the use of an on-chip protection circuit at each input and output pin which is capable of dissipating excess energy without damage. Exposure to ESD, which is of concern today, will be of even more concern in the future for two reasons: 1) greater chip complexity, which may necessitate as many as 400 pins per chip, will require that large amounts of chip area be devoted to protection circuits, and 2) reduced feature size/oxide thickness will make devices more susceptible to voltage overstress.

Historically designers of on-chip protection circuits have subjected test structures, incorporating different geometries and utilizing different process technologies, to injection pulses and then subsequently analyzed them for ESD damage. Empirical interpolation of this data has then been used to meet particular design objectives. Because a certain amount of silicon area (volume) is required in which to dissipate the pulse energy, protective circuit design involves a tradeoff between area usage and the degree of protection desired. However, in order to effectively quantify this tradeoff two things are needed: 1) a precisely defined, repeatable, injection impulse testing system which can be related to ESD exposure in the field, and 2) a theoretical basis for ESD design methodology which will allow optimization. Neither of these presently exist and both will require a greater degree of understanding of the physical mechanisms involved.

Metal-oxide-silicon capacitors subjected to ESD pulses exhibit a number of phenomena depending on the pulse shape and amplitude: hot electron trapping, oxide breakdown, filamentation, contact damage, and thermal runaway (second breakdown). Complete breakdown, which occurs when sufficient voltage is applied, consists of two phases: electrical breakdown of the oxide, which occurs within a few nanoseconds, followed by thermal runaway, which requires roughly 10X as long. Thermal runaway causes melting of the contact and substrate and subsequent ejection of molten material [17]. Figure 7 is a micrograph showing damage to the corner of a square field plate caused by the injection of a single 75 volt impulse [18]. The curved pattern of the deposited material, a mixture of silicon and aluminum, indicates that the molten material was ejected under the influence of the crossed electric and magnetic fields associated with the discharge current flow. This means that the time scale for the ballistic sequence must be of the same order as the electrical events. Furthermore it can be calculated that the average trajectory speed is greater than 500 m/sec.

When pulses insufficient to cause complete breakdown are applied, filamentation can occur. Sub-breakdown pulses applied to commercial protection circuits can result in filamentation, which degrades rather than destroys the circuit. Progressive degradation of the forward characteristic of a commercial protection circuit caused by the successive application of rectangular pulses is shown in Figure 8. (Little change was observed in the reverse characteristic.) Subcatastrophic filamentation is a potential source of latent failure. It has also been determined that protection circuit response depends critically on rise time. Information gained in the study of MOS capacitors can be used to develop circuit models which will allow protection circuits to be tailored to specific applications [19].

396

CONCLUSIONS

Three examples have been presented indicating increased emphasis on the intrinsic reliability of complex IC's. In each case - electromigration, charge injection, and ESD - knowledge of the physical mechanisms involved will permit the development of mathematical models which can be incorporated in computer aided design tools. Existence of a new class of CAD tools, which have been modified to include reliability factors, should allow designers to achieve an optimized tradeoff between reliability, cost, and performance.

REFERENCES

1. H.B. Huntington, "Diffusion in Solids - Recent Developments," p. 303, Academic Press, New York (1975)
2. T. Kwok, Proc. 4th Int'l VLSI Multilevel Intercon. Conf., p.456 (1987)
3. E. Kinsborn, "A Model for the Width Dependence of Electromigration Lifetimes in Aluminum Thin-Film Stripes," Phys. Lett. vol. 36, p. 968 (1968)
4. P.B. Ghate, "Aluminum Alloy Metallization for Integrated Circuits," Thin Solid Films, vol. 53, p. 117 (1981)
5. J.R. Black, "Electromigration Failure Modes in Aluminum Metallization for Semiconductor Devices," Proc. IEEE, v. 57, p. 1587 (1969)
6. J.W. Lathrop, et al, "Design Rules Hold Key to Future VLSI Reliability," Proc. 7th UGIM Symp., pp. 91-94, Rochester, NY (1987)
7. J.D. Venables and R.G. Lye, "A Statistical Model for Electromigration Induced Failure in Thin Film Conductors," Proc. 10th IEEE Int. Rel. Phys. Symp., p. 159, New York (1972)
8. D.F. Frost and K.F. Poole, " A Method for Predicting VLSI Device Reliability Using Series Models for Failure Mechanisms," IEEE Trans. on Rel., vol. R-36, pp. 234-242 (June 1987)
9. C.H. Stapper, "LSI Yield Modeling and Process Monitoring," IBM J. Res. Dev. p. 228 (May 1976)
10. T.H. Ning, C.M. Osburn, and H.N. Yu, "Effect of Electron Trapping on IGFET Characteristics," J. Electron. Mater., vol. 6, p.65 (1977)
11. C.-F. Chen, et al, "The Dielectric Reliability of Intrinsic Thin SiO_2 Films Thermally Grown on a Heavily Doped Si Substrate - Characterization and Modeling," IEEE Trans. Elect. Dev., vol. ED-34, pp. 1540-1552 (July 1987)
12. N. Klein and P. Solomon, J. Appl. Phys., vol. 47, p. 4364 (1976)
13. I.C. Chen et al, "Electrical Breakdown of Thin Gate and Tunneling Oxides," IEEE Trans. Elec. Dev., vol. ED-32, pp. 413-422(Feb. 1985)
14. M.S. Liang and C. Hu, "Electron Trapping in Very Thin Thermal Silicon Dioxides," IEDM Tech. Dig., pp. 396-399 (Dec. 1981)
15. F.J. Feigl, et al, J. Appl. Phys., vol. 52, pp. 5665-5682 (Sept. 1981)
16. B. Ricco et al, Phys. Rev. Letts., vol. 51, p. 1795 (1983)
17. D.G. Pierce, "Electro-thermomigration as an Electrical Overstress Failure Mechanism," Proc. 7th EOS/ESD Symp., pp. 67-76, Minneapolis, MN (Sept.1985)
18. M.A. Bridgwood and R.H. Kelley, "Modeling the Effects of Narrow Impulsive Overstress on Capacitive Test Structures," Proc. 7th EOS/ESD Symp., pp. 84-91, Minneapolis, MN (Sept. 1985)
19. M.A. Bridgwood, "Breakdown Mechanisms in MOS Capacitors Following Electrical Overstress," Proc. 8th EOS/ESD Symp., pp. 200-207, Minneapolis, MN (Sept. 1986)

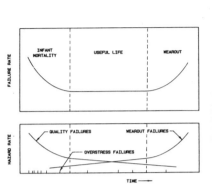

Fig. 1   Classical failure rate vs
time curve for devices.

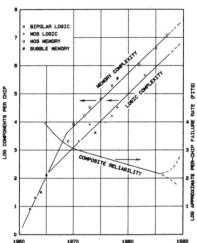

Fig.2   Maximum IC complexity and
reliability vs time.

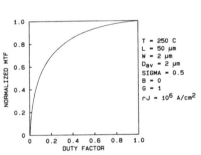

Fig. 3   Example of calculated electromigration MTF vs duty factor
for constant r·J.

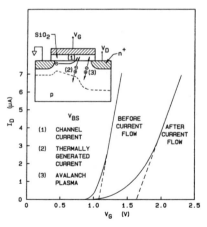

Fig. 4   Oxide charging effect
(after Ning, Osburn, and Yu)

398

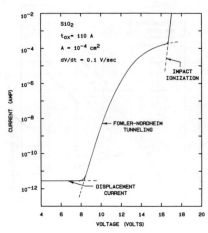

Fig. 5 Ramp-voltage stressed IV
characteristic. (after Chen)

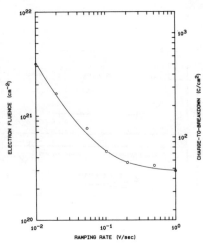

Fig. 6 Charge-to-breakdown vs
ramping rate. (after Chen)

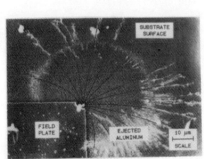

Fig. 7 Electronmicrograph of ESD
damage to MOS capacitor.

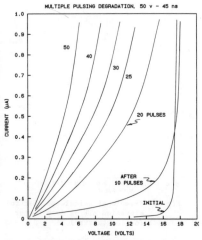

Fig. 8 Forward IV characteristic of
commercial protection circuit
after sequential pulsing.

PHASE DIAGRAM STUDIES IN THE Te-RICH HgCdTe SYSTEM

S.C. Gupta, F.R. Chavda and A.K. Garg
Solid State Physics Laboratory
Lucknow Road, Delhi-7, India

ABSTRACT

In this paper we review and present the work on the phase diagram in the Te-rich corner of the Hg-Cd-Te ternary system. The liquidus temperatures have been determined for $(Hg_{1-z}Cd_z)_{1-y}Te_y$ compositions with $0.65 \leqslant y \leqslant 1$ and $0.03 \leqslant Z \leqslant 0.18$ by direct visual method and differential scanning calorimeter experiments. Liquidus isotherms have been drawn. Supercooling below the liquidus temperature has also been studied. For fixed amount of Hg in the ternary system, the solubility of Cd increases with the increase in the liquidus temperature. Also at fixed temperature, the solubility of Cd has a maximum with respect to the changes in the Hg content. Supercooling has been measured to be more than $25^{\circ}C$ in small ($< 2g$) sample melts and observed to be much smaller ($2-10^{\circ}C$) in larger melts (~ 100 gms). It is found that to grow $Hg_{1-x}Cd_xTe$ epilayers with compositional uniformity $\Delta x = 0.001$ by LPE from Te-rich solution, temperature control better than $0.4^{\circ}C$ is required.

INTRODUCTION

The pseudo-binary alloy $Hg_{1-x}Cd_xTe$ (MCT) is one of the most important material used for the fabrication of infrared photo-conductive and photovoltaic detectors in the 3-5 μm and 8-18 μm wavelength regions [1]. It is an alloy semiconductor whose energy gap can be tailored from 0. to 1.6 eV at $4.2^{\circ}K$ by varying the content of CdTe and temperature of operation. It has high electron mobility, high mobility ratio, long minority carrier lifetimes and relatively low dielectric constant [2].

The technological importance of MCT has stimulated an extensive research on the growth of both bulk and thin epitaxial

layer and a number of growth techniques have been developed
[3]. The bulk material grown at melt temperature suffers from
technical problems such as Te inclusion, lateral and vertical
non-uniformity stress induced defects etc. It requires complex
long annealing to overcome some of these problems. Among
the epitaxial techniques, liquid phase epitaxy (LPE), Vapour
Phase Epitaxy (VPE), Molecular Beam Epitaxy (MBE) and Metallo-
organic Chemical Vapour Deposition (MOCVD) are the most important. In VPE technique, compositional grading is the major
problem, MOCVD and MBE are not yet well developed for MCT.
At present LPE is the most successful technique for the growth
of device quality MCT layers [4]. There are two ways by which
LPE can be achieved, (i) from Te-rich solution [5-7] and
(ii) from Hg rich solution [8]. The LPE growth from Hg-rich
solution has not been persued much. This is because of the
low solubility of Te and Cd in Hg and high vapour pressure
of Hg at the growth temperature [9].

In order to grow high quality epilayers, it is essential
to know the phase diagram of the Hg Cd Te system, the liquidus
temperature, the slope of the liquidus surfaces and the relationship between the source compositions and the deposited
solid.

In this paper, we present some work carried on the phase
diagram studies in the Te rich corner of HgCdTe system. Both
differential scanning calorimeter (DSC) experiments and direct
visual methods have been used for determining the liquidus
temperature.

ALLOY PREPARATION

Each alloy of $(Hg_{1-z} Cd_z)_{1-y}Te_y$ composition with $0.65 \leqslant y \leqslant$
1 and $0.03 \leqslant Z \leqslant 0.18$ is prepared from 7N purity Hg, QZR6N
purity Te and DZR6N Cd purchased from Cominco Electronic
materials Canada. The required amounts of Hg, Cd and Te total-
ling about 100 gms are taken in a clean quartz tube. The
quartz tube is cleaned, etched in aqueous HF and degassed
at $1000^{\circ}C$ in vacuum to remove residual contaminants before
loading. The loaded quartz ampoule is evacuated to 10^{-6} mm

of Hg and sealed. The ampoule is then placed in a furnace and the temperature is raised 50°C above the anticipated melting point. The furnace is periodically rocked to homogenise the melt and the ampoule is kept in the furnace for about 24 hrs. Next the ampoule is rapidly cooled to room temperature.

LIQUIDUS TEMPERATURE AND SUPERCOOLING STUDIES

(a)   Small Melts

A Dupont 1090 differential thermal analyser is used to study the thermal events in small ($<$2 gm) Hg-Cd-Te alloys. The specimen is encapsulated in a capsule. Differential Scanning Calorimeter measurements are taken at different heating and cooling rates. A typical plot for $y = 0.822$ and $Z = 0.013$ at heating and cooling rated 10°C/min is shown in Fig.1. It gives the liquidus temperature as 455°C and eutectic temperature as 412°C. The supercooling is 31°C at the liquidus temperature and 42°C at the solidification temperature as indicated by the separation between the heating and cooling curves. Similar measurements are taken for a few compositions. In every case supercooling obtained is more than 25°C. Supercooling is much less than that expected for homogenous nucleation (0.2 x melting temperature) of pure metals [10].Nucleation may be heterogenous in our case.

(b)   Large Melts

Fig.2 shows the schematic of the apparatus used for supercooling studies and measurement of liquidus temperature in large melts ($\sim$100 gm) by direct visual method. Pt - Pt Rh (13%) thermocouple with Eurotherm temperature controller and indicator are used to control and measure the temperature respectively. A sealed ampoule (in which alloy is prepared) is placed in a transparent furnace. The temperature of the furnace is raised to 20°C below the anticipated liquidus temperature. Next the temperature of the furnace is increased very slowly (say 1°C in 4-5 hrs) and the ampoule is rotated every now and then with the rod attached to it. As the liquidus temperature is approached, the temperature is increased by $1^{o}C$ and kept at that temperature for about a day. The ampoule is rotated frequently and seen that no solid is left below the liquid surface. This temperature where the entire solid melts is then taken as the liquidus temperature T_1. After determining the liquidus temperature, the furnace is cooled at the rate of $1^{o}C$ in 5 hrs. The ampoule is rotated and the surface of the liquid is continuously watched. The temperature T_c at which the precipitate appears, before the spontaneous nucleation starts, is noted. The supercooling is then $\Delta T = T_1 - T_c$. The liquidus temperature and super-cooling are thus determined for $(Hg_{1-z}Cd_z)_{1-y}Te_y$ compositions with $0.65 \leqslant y \leqslant 1$ and $0.03 \leqslant Z \leqslant 0.18$ and also for $y = 0.822$, $Z = 0.013$, and are shown in Fig.3. The T_1 values interpolated from Fig.3 are $7-8^{o}C$ higher than the values obtained by Bowers et.al [11] from LPE growth experiments, and $6-7^{o}C$ lower than the values obtained by Harman [12] for two ternary liquid compositions $Hg_{0.18}Cd_{0.016}Te_{0.804}$ and $Hg_{0.157}Cd_{0.018}Te_{0.825}$. For other compositions also our values differ from Harman's.

The liquidus temperature and supercooling studies are also made for the alloys for which DSC measurements are taken. The liquidus temperature (in the two sets of experiments, viz., for $y = 0.822$, $z = 0.013$) agrees within the experimental error limits; $T_1 = 453^{o}C$ DVM and $455^{o}C$ by DSC measurements. Supercooling in large melts is in the range $2-10^{o}C$. This can be contrasted with $25^{o}C$ in Small (< 2 gm) and with values of upto $14^{o}C$ reported by Harman [13] and 5 to $15^{o}C$ by Mroczkowski and Vydyanath [6].

During the determination of liquidus temperature by DVM, it is visually seen that in all alloys, a portion melts at $410-412^{o}C$. DSC measurement also substantiates that the first melt appear at $412^{o}C$ (Fig.1). Since this tempeature is independent of the amount of Cd in the alloy, we attribute this to HgTe-Te eutectic. We have established the composition of this eutectic. It is $Hg_{1-y}Te_y$ with $y = 0.85$. It agrees well with the reported value [14]. However this temperature is $6-7^{o}C$ lower than the value $418^{o}C$ reported by Harman [12].

Fig.3 also shows that (there is a minimum temperature at which HgCdTe system melts), liquidus temperature is a function of z and y of the $(Hg_{1-z}Cd_z)_{1-y}Te_y$ system; and corresponds to a eutectic tempeature of this system. It increases with the increase in the Cd content in the alloy.

LIQUIDUS ISOTHERMS AND TIE LINES

Fig.4 shows the liquidus isotherms for the Te rich corner of the HgCdTe system at temperatures $460^{\circ}C$, $500^{\circ}C$, $560^{\circ}C$ and $600^{\circ}C$. These isotherms have been obtained from the curves of Fig.3. These isotherms represent the composition of $(Hg_{1-z}Cd_z)_{1-y}Te_y$ liquids that are in equilibrium with $Hg_{1-x}Cd_xTe$. Each isotherm shows the variation of atomic portion of Hg with the atomic fraction of Cd. For fixed amount of Hg in the ternary system the solubility of Cd increases with the increase in the liquidus temperature. Also at fixed temperature, the solubility of Cd has a maximum with respect to the changes in the Hg content. Fig.5 shows the liquidus isotherms in the Te - HgTe - CdTe region of the phase diagram. The shape of the isothermal sections $460^{\circ}C$-$560^{\circ}CC$ is interesting since they are concave with respect to the CdTe - HgTe section, whereas at higher temperatures the shape of the Hg-Te and Cd-Te liquidus lines would indicate a convex isothermal section in the intermediate field. The transition from convex to concave occurs in the region at $600 - 650^{\circ}C$. The concavity of the surface at lower tempeatures appears to be related to the presence of the Te rich side of the HgTe-Te eutectic. Fig.6 shows Schmit's data [15] on the Tie lines and segregation coefficient in the Te-rich side of the Hg-Cd-Te system. This data shows that 1 at. % change of Hg in the solution produces 2 at. % change in Cd_x in solid $Hg_{1-x}Cd_x$Te. Therefore to achieve control on Cd_x to the extent 0.1 at. % requires melt composition be controlled to within 0.05 at. % Hg. Using the data from the Fig.3 we find $(\Delta z/\Delta T)=0.12$ at. % $/^{\circ}C$. This corresponds to a required temperature stability of better than $0.4^{\circ}C$.

SUMMARY

The liquidus temperatures have been determined for$(Hg_{1-z}Cd_z)_{1-y}Te_y$ compositions with $0.65 \leqslant Y \leqslant 1$ and $0.03 \leqslant z \leqslant 0.18$ by direct visual method and DSC experiments. Supercooling has been observed $> 25^{\circ}C$ in small (< 2 gm) melts and observed much smaller ($2-10^{\circ}C$) in large melts (~ 100 gm). Melting temperature is a function of z and y of $(Hg_{1-z}Cd_z)_{1-y}Te_y$ system and it increases with the increase in Cd content in the alloy. It is concluded that to grow MCT epilayers with compositional uniformity $\Delta x = 0.001$ by LPE from Te-rich solution, temperature control better than $0.4^{\circ}C$ is required.

ACKNOWLEDGEMENTS

The authors are thankful to Dr A.K. Sreedhar, Director SPL for his constant encouragement, guidance and helpful discussions. It is a pleasure to thank the Director for his kind permission to present this paper. Authors also wish to thank Dr S.S. Singh, Defence Science Centre for the DSC measurements.

REFERENCES

1. R.K. Willardson and A.C. Beer, Semiconductors and Semimetals (Academic Press, New York) Vol.18,1981.

2. R.Dornhaus and G. Nimtz in Springer Tracts in Modern Physics (Springer-Verlag Berlin) Vol.98,1983.

3. J.L. Schmit, J. Crystal Growth 65, 249,1983.

4. E.R. Gertner, Ann. Rev. Mater. Sci. 15, 303, 1985.

5. C.C. Wang, S.H. Shin, M.Chu, M.Lanir and A.H.B. Vanderwyck, J. Electrochem. Soc. 127, 175, 1980.

6. J.A. Mroczkowski and H.R. Vydyanath, J. Electrochem. Soc. 128, 655, 1981.

7. R.A. Wood and R.J. Hager, J.Vac. Sci. Technol.A1,1608,1983.

8. M.H. Kalisher, J. Crystal Growth 70.365, 1984.

9. P.E. Herning, J. Electron.Mater. 13, 1, 1984.

10. D. Turnbull, J. Metals 188, 1144, 1950.

11.   J.E. Bowers, J.L. Schmit, C.J. Speerschneider and
      R.B. Maciolek, IEEE Trans. Electron. Dev. ED27, 24, 1980.

12.   T.C. Harman, J. Electron. Mater.9, 945, 1980.

13.   T.C. Harman, J. Electron. Mater. 10, 1069, 1981.

14.   T. Tung, L. Golonka and R.F. Brebrick, J. Electrochem.
      Soc. 128, 1601, 1981.

15.   J.L. Schmit (Private Communication).

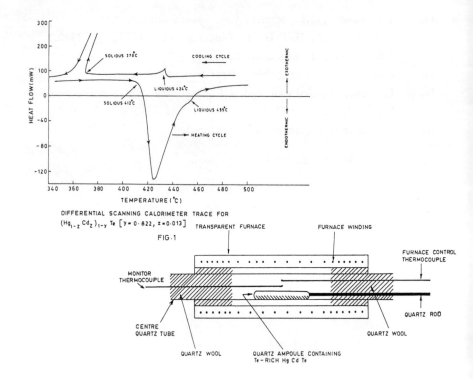

DIFFERENTIAL SCANNING CALORIMETER TRACE FOR
$(Hg_{1-z}Cd_z)_{1-y}Te\ [y=0.822,\ z=0.013]$

FIG.1

SCHEMATIC OF THE DIRECT VISUAL METHOD APPARATUS USED TO MEASURE THE LIQUIDUS
TEMPERATURE OF Te-RICH Hg Cd Te SYSTEM.

FIG.2

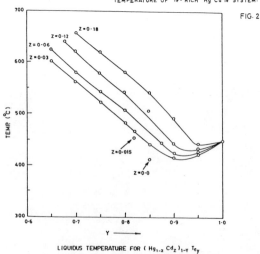

LIQUIDUS TEMPERATURE FOR $(Hg_{1-z}Cd_z)_{1-y}Te_y$
BY DIRECT VISUAL METHOD.

FIG.3

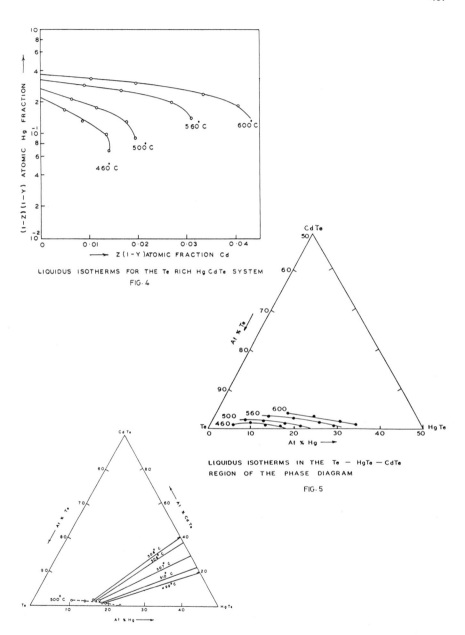

LIQUIDUS ISOTHERMS FOR THE Te RICH Hg Cd Te SYSTEM

FIG. 4

LIQUIDUS ISOTHERMS IN THE Te — HgTe — CdTe
REGION OF THE PHASE DIAGRAM

FIG. 5

TIE-LINES ON THE Te-RICH SIDE OF THE Cd-Hg-Te
PHASE DIAGRAM

FIG. 6

GROWTH, CHARACTERISATION AND HEAVY-DOPING
EFFECTS IN INDIUM PHOSPHIDE

D. N. Bose   and   B. Seishu[+]
Materials Science Centre
Indian Institute of Technology
Kharagpur 721302

INTRODUCTION

Indium Phosphide, a direct band-gap semiconductor with
E_g = 1.35 eV is becoming of increasing importance in optoelectronics as a lattice-matched substrate for fibre-optic
sources and detectors, and also high speed electronic devices.
In this paper LPE growth of InP with Se doping varying from
10^{18} - $10^{20}/cm^3$ is described. The variation of the photoluminescence edge with doping has thus been determined both theoretically and experimentally. Hall mobility has been studied as
a function of temperature (77-300K) and doping to establish
dominent scattering mechanisms. LEDs have been fabricated using
p-n junctions. Finally the effect of heavy doping in reducing
the semiconductor-metal transition pressure has been determined
for the first time.

GROWTH

Liquid phase epitaxial (LPE) growth was accomplished
using a conventional sliding-bin graphite boat in a transparent
horizontal furnace under hydrogen ambient. Undoped, Se-doped n
type and Zn-doped p type layers were grown using the supercooling technique from solutions supercooled by about 3-5°C in the
temperature range 650°C-615°C. For heavy-doping in the range
$N_D-N_A \geq 1 \times 10^{20}$ cm^{-3} it was found necessary to reduce the
growth temperature to 525°C-475°C to take advantage of the retrograde solid solubility. A 10-15 sec. pre-etch in Indium prior to
growth was found to be essential for removal of surface damage.
To assess the crystalline quality n$^+$p junctions were fabricated
on p-type substrates whereas for transport measurements Fe-doped
(100) semi-insulating substrates were used. For long growth runs
_____
+ Present address : Semiconductor Complex Ltd., Chandigarh.

and thick epi layers (> 5 μm) the measured layer thickness fell short of the values theoretically predicted by Hsieh /1/. On the other hand shorter growth periods and higher supercooling gave layers with accurately predicted thicknesses. This is partly due to the fact that for large amounts of Se in the melt $[X_{Se}(liq) > 1$ at %] it was found that the incorporation of Se does not follow the linear relation $X_{Se}(liq)/X_{Se}(sol) = k_{Se}$ as is true when $X_{Se}(liq)$ 5 x 10^{-2} at %. It was found that under these conditions the free electron concentration was less than that of selenium atoms in the epi layer.

These results can be explained considering that the substrate-melt interface at the growth temperature acts as a Schottky barrier. On the basis of the theory of Vogel and Zschauer /2/ relating the diffusion coefficient of the impurity in the solid and the Debye length, the diffusion coefficient of Se in InP at 650°C, $D_{Se} = 10^{-14}$ cm^2/s. The active donor solubility limit for Se was found to be $\simeq$ 2 x 10^{20} cm^{-3}.

CHARACTERIZATION

Surface morphological studies were carried out on all the epitaxial layers. Etch pit densities (EPD) were determined using the Huber etch. Apart from dislocation pits, saucer pits, protrusions and growth striations usually revealed by the etchant, precipitates of Se and dislocation loops were also observed. It was concluded that

i) the EPD in epilayer is generally equal to or greater than that of the substrate

ii) in layers heavily doped with Se EPD was less than that of the substrate while Te doping to $N_D - N_A \simeq$ 1 x 10^{19} cm^{-3} showed only a small reduction in the EPD of the epilayers.

The Se concentration in the epilayers were determined by electron microprobe analysis. For heavy doping this was not directly proportional to the carrier concentration as discussed later.

ELECTRICAL AND OPTICAL PROPERTIES

The resistivity, carrier concentration and Hall mobility of the layers were determined between 77-300K using the Van der Pauw technique on clover-leaf shaped samples. The results are presented in Table I.

TABLE I - Electrical properties of Se-doped InP LPE Layers

Carrier Conc. (cm^{-3})	Resistivity (Ω.cm)		Mobility (cm^2)/V-s	
	300K	77K	300K	77K
1.2×10^{18}	3.01×10^{-3}	3.10×10^{-3}	1728	1676
6.1×10^{18}	7.21×10^{-4}	6.63×10^{-4}	1421	1545
1.0×10^{19}	6.49×10^{-4}	5.87×10^{-4}	963	1064
5.5×10^{19}	1.89×10^{-4}	1.7×10^{-4}	600	669
1.0×10^{20}	1.38×10^{-4}	1.27×10^{-4}	451	478
1.83×10^{20}	8.54×10^{-5}	7.16×10^{-5}	400	477

Mobility value of layers with $N_D - N_A \leq 1 \times 10^{19}$ cm^{-3} have been compared with theoretical values calculated on the basis of ionised impurity scattering at low temperature and optical phonon scattering at high temperature. In this doping range the compensation ratio $C = N_A^- / N_D^+$ has been found to be in the range 0.2 - 0.3. The relatively high value of μ at $n = 1.83 \times 10^{20}$ cm^{-3} compared with a value of 150 cm^2/V-s obtained by Hawrylo is attributed the strain in the epi layers as discussed later.

For free carrier concentrations $N_D - N_A \geq 1 \times 10^{19}$ cm^{-3}, theoretical value of mobility have been calculated with the compensation ratio as a variable parameter. The relaxation time approximation has been used to calculate the effective mobility due to polar optical phonon scattering and ionised impurity scattering. Reasonable agreement with theory has been found at 77K and 300K for $C = 0.2 - 0.3$, the mobility varying with doping concentration as $n^{-1/3}$.

Photoluminescence spectra of the samples were studied at 77K and 300K to compare with the results of Hawrylo /3/ who showed emission occurring in the red with the heaviest doping.

Theoretical determination of the band-edge shift was carried out as reported earlier /4/ taking into account (a) Moss-Burstein shift (b) band-gap shrinkage due to electron-donor and electron-electron interactions, (c) non-parabolicity of the Γ band and (d) presence of higher conduction bands.

The latter two factors were found to be essential for good agreement between theory and experiment. It was also found that whereas for light doping k-conservation rule was obeyed, for heavy doping due to severe disorder this was violated. Comparison between theoretical calculation and experimental results are given in Table II.

TABLE II   Comparison between theoretical and experimentally determined photoluminescence edge at 300K

THEORY		EXPERIMENT	
$n(cm^{-3})$	$E_{g(opt)}$ eV	$n(cm^{-3})$	$E_{g(opt)}$ eV
1×10^{18}	1.319	1.2×10^{18}	1.318
6×10^{18}	1.374	6.1×10^{18}	1.362
1×10^{19}	1.399	1.0×10^{19}	1.398
4×10^{19}	1.538	3.46×10^{19}	1.562
6×10^{19}	1.686	5.5×10^{19}	1.626
1×10^{20}	1.842	1.73×10^{20}	1.91

In addition to the band-edge luminescence at 77K a strong emission peak at 1.21 eV was found as also reported by other workers /5/. This deep trap is considered to be due to a $V_{In}Se_3$ complex. LEDs were fabricated by growing $5\,\mu$m thick layers with doping 1.05 - 2.5 x 10^{17} on p^+ substrates /6/. The diodes of 1.0 mm^2 were found to exhibit emission peaks at 1.341 eV and 1.335 eV respectively at 50 mA and 75 mA forward bias. The V_{bi} and ideality factor n were found to be 1.00 eV and 1.96 respectively, the reverse saturation current being 85 μA.

Finally the resistivity of samples with doping concentrations varying from 6.7 x 10^{18} / cm^3 to 1.8 x 10^{20} / cm^3 were studying as a function of hydrostatic pressure /7/. It

is known /8/ that bulk InP undergoes a semiconductor –
metal transition pressure at 10.8 GPa . In the present case
the transition pressure was found to decrease with increasing
doping according to the relation

$$P_c = P_o \left[1 - k(\frac{n}{n_m}) \right]^{\alpha}$$

where P_o = 7.41 GPa , k = 3.74, n_m = 1.98 x 10^{28} m^{-3} and
α = 0.63.

The study of the heavy doping effects in InP revealed
several interesting phenomena not previously observed in GaAs,
probably due to the   fact that the heaviest reporting in
GaAs has been less than 10^{20} cm^{-3}. In the present case shift
of the photoluminescence edge by almost 50% from 1.34 eV to
1.91 eV has been found in excellent agreement with theory.
Further due to the presence of large strain in the epi layers
enhancement of electron mobility and decrease in the semiconductor–metal transition pressure have also been observed. This
may be attributed to the difference in atomic sizes between P
and Se doped which results in strain increasing with doping
concentration.

REFERENCES

1) J.I.Hsieh, Semiconductor Handbook Vol.3 p.415-497.North
   Holland, New York (1980).

2) K.H. Zschauer and A. Vogel. Proc. Gallium Arsenide. Symp.
   p.100, Ints. of Phys. London (1971).

3) F.Z. Hawrylo, Appl. Phys. Lett. 31(11), 1038 (1980).

4) B. Seishu and D.N. Bose, Appl. Phys. Lett. 42(3), 287(1983).

5) S.H. Groves and M.C. Planko, J. of Cryst. Growth 54,81 (1981).

6) B. Seishu and D.N. Bose, Proc. All Indian Symp. on Commn.
   Bangalore 97, 81 (1984).

7) D.N. Bose, B. Seishu, G. Parthasarathy and E.S.R. Gopal,
   Proc. Roy. Soc. Lond. A, 405, 345 (1986).

8) C.S. Menoni and I.L. Spain, Phys. Rev. B 35, 14, 7520 (1987).

Liquid Phase Epitaxial Growth of $In_{1-x}Ga_xAs_yP_{1-y}$ $(y \approx 2.2x)$

Lattice Matched to InP

S.S.Chandvankar and B.M.Arora
Tata Institute of Fundamental Research
Homi Bhabha Road, Bombay 400005

Abstract

Epitaxial growth of $In_{1-x}Ga_xAs_yP_{1-y}$ alloy materials by liquid phase epitaxy is described. Problems related to the epi-growth are discussed. Doping with group IV element Ge enhances n type conductivity. A new photoluminescence band is produced by Ge and Sn.

414

Introduction.

Quaternary alloys like $In_{1-x}Ga_xAs_yP_{1-y}$ have emerged as materials of great interest for optical communication [1] in the wavelength range 1.3 μm - 1.6 μm where the optical fibres have low loss. Suitability of these materials is due to the fact that (1) they can be grown to a very high degree of perfection on InP substrates to which they are lattice matched over compositions for which

$$y \lesssim 2.2x \qquad (1)$$

and (2) over these compositions ranging from InP (x=0, y=0) to $In_{0.53}Ga_{.47}As$ (x=0.47, y=1), the band gap of the material can be continuously varied from 1.35 ev (λ_{Eg} 0.92 μm) to 0.75 ev ($\lambda_{Eg} \approx 1.65$ μm) according to the relation [2]

$$Eg_{300K}(y) \approx 1.35 - 0.775\ y + 0.149\ y^2 \quad ev \qquad (2)$$

while keeping the lattice constant fixed by maintaining y≈2.2x.

This property of band gap tailoring allows fabrication of heterostructure lasers and detectors for efficient room temperature operation. In particular, the materials are well suited for sources and detectors at about 1.55 μm wavelength where the currently available optical fibers have very low loss (<1db/km) and nearly zero dispersion.

Liquid Phase Epitaxy (LPE) is an established technique [3,4] for growing these materials in multiple layered structures for device purposes. In our laboratory, we have used this technique for growing single epitaxial layers of $In_{1-x}Ga_xAs_yP_{1-y}$ lattice matched to InP substrates, over the entire composition range from

InP (x=0, y=0) to $In_{0.53}Ga_{0.47}As$ (x=0.47, y=1.0). A two phase technique has been used for most of the growth experiments. Various growth aspects and some characterisations of the epitaxial layers are reported here.

Experimental

Figure 1 shows a schematic of the growth apparatus. Details of the growth reactor and the growth procedures are reported elsewhere [5]. Salient experimental aspects of the liquid phase epitaxial growth of $In_{1-x}Ga_xAs_yP_{1-y}$ lattice matched to InP are as follows:

1) Growth of $In_{1-x}Ga_xAs_yP_{1-y}$ on InP substrates is a hetero-epitaxy. In LPE, a lattice mismatch between the substrate and the epi-layer greater than 1-2% commonly results in rough and uneven growth. In InGaAsP on InP, mismatch greater than 1% prohibits epitaxial nucleation to occur. Therefore, it is desirable to match the lattice constants of the substrate and the epilayer closely. By using Vegard's law it has been shown [7] that lattice matching to InP restricts the compositions of $In_{1-x}Ga_x$ As_yP_{1-y} epilayers to those for which x and y values satisfy equation (1).

2) A desired solid composition, $In_{1-x}Ga_xAs_yP_{1-y}$, can be grown from a saturated solution containing In, Ga, As and P in a particular proportion at a specific temperature. This information can be generally obtained from phase diagrams constructed by detailed measurements of the liquidus and solidus. Detailed tables of liquidus and solidus of this system at different

temperatures have been reported by Nakajima et al. [8] and
others [9,10] R.Sankaran et al [11] have measured the distribution coefficients of Ga, As and P in the temperature range
600 - 750°C. They have shown that the distribution coefficient
of Ga remains practically constant with temperature while those
of As and P change rather drastically by about an order of
magnitude. Because of this, great care is necessary in weighing
the amounts of solutes.

Generally, Indium is used as a solvent. Gallium is added
in the form of GaAs lump. Arsenic is obtained from GaAs as well
as from InAs lumps and Phosphorous is added in the form of InP
lump. It may be noted that the distribution coefficients depend
significantly on the substrate orientation [10] and hence the
liquid composition to grow a particular epilayer composition
changes with the substrate orientation. Our growth experiments
are performed on <100> substrates. Typical quantities which we
have used to grow InP, $In_{0.53}Ga_{0.47}As$ and $In_{1-x}Ga_xAs_yP_{1-y}$ epilayers with $\lambda_{Eg} \approx 1.1$ μm, 1.3 μm and 1.55 μm are listed in Table I.
Out of these, we have extensively used the compositions for InP
and quaternary layers with composition corresponding to $\lambda_{Eg} \approx 1.1$μm
and 1.3 μm. About seven layers have been grown successfully from
a single solution with small change in the layer composition in
successive growths. Success rate of growing good quality 1.55 μm
quaternary layers and $In_{0.53}Ga_{0.47}As$ ternary layers is relatively
low and will be treated in future publication.

3) In principle, thin layer epitaxial growth can be
carried out by heating together the solvent and the solute
components in ultrahigh purity hydrogen flow to obtain a homogeneous saturated solution, bringing this solution in contact with
the InP substrate and allowing it to cool by a few degrees and
then separating the solution from the substrate. Several different
variants are generally used for growth experiments. These are
(a) Growth from a supersaturated solution ($\sim$ 5-10°C supercooled
below the saturation temperature) without further cooling i.e.
step cooling (b) Growth from a saturated solution with ramp
cooling (c) Combination of (a) and (b) (d) A two phase technique
in which an extra lump of InP is kept floating on top of the
solution. This last scheme of growth has been used by many
workers [12,13] and is reported to be less sensitive to the
exactness of the proportion of the phosphorous content in the
solution. We have verified this result in our growth experiments.
Since we use very small quantities of materials (e.g. total
weight of the solution used is $\approx$ 200-250 mg) all our growth
experiments of the quaternary layers employ two phase technique.

4) Another important aspect for uniform and smooth epitaxial
growth is the preparation and protection of the InP substrate
prior to growth.

We have used InP n^+ substrates with $N_d \approx$ 2-4x10^{18} cm^{-3} and
semi-insulating substrates with $\rho \sim 10^7 \Omega$-cm in our growth experiments. In the beginning, the substrates are cut from a 2" diameter
wafer to the required size of 4 mm x 4.5 mm and then lapped to

about 200-280 μm to suit the recess in our boat. Substrate to be used in the growth run is first cleaned in organic solvents like trichloroethylene, acetone and methanol; then etched for one minute in $20H_2SO_4:1H_2O_2:1H_2O$, rinsed in water and then etched in 1% Br_2-CH_3OH mixture for 3 minutes, followed by rinse in methanol. The substrate is then dried in a soxhlet apparatus having isopropyl alcohol. Surface prepared with this procedure is smooth, clean with no marks on it. The substrate is then quickly loaded into the graphite boat and covered with a graphite disc. Uncovered InP substrate surface degrades on heating to about 600°C due to phosphorous loss from the surface. Our growth temperatures are higher ($\sim$ 650°C) and hence the substrate surface needs to be protected. While some workers use InP cover piece over the substrate, we find a smooth surface intimately touching the graphite cover to be adequate. To remove any residual surface damage, it is common to back-etch the substrate before the growth is initiated. We use pure indium for back etch since its etch rate is not very high at the growth temperature ($\sim$ 635°C)

Growth Procedure:

In a typical growth run InP substrate is put in the recess labelled growth substrate (see Fig.1). A graphite disc is put on this substrate through a hole in the slider as seen in the inset of Fig.1. The middle well contains pure indium as an etchant. The front well contains solution consisting of indium and InAs, GaAs lumps and platelets of InP floating on the solution, weighed so as to prepare a saturated solution at ∿635°C. The weight of InP platelets is however much more than required for saturation. The solutions are initially homogenised at about 675°C for 30 minutes in high puirty H_2 ambient. Then the temperature of the boat is lowered to about 635°C and slow cooling of the solution is initiated by using an Electronic Programmable Cooler. The cooling rate employed is typically 0.4°C/minute. Just when the programmed cooling is commenced at ∿635°C, the etching solution is moved over the growth substrate for 15-20 seconds and then quickly moved away. At this stage, the slider is so positioned that the growth solution comes over the growth substrate. After cooling by a desired temperature drop the growth is terminated by sliding the solution away from the substrate. Figure 1(a) shows a plot of thickness of epi-layers versus growth time. With this scheme, we have grown a number of layers of these quaternary materials without any intentional dopant. Also we have grown many layers with group IV dopants such as Ge and Sn.

Difficulties Encountered in the Growth:

It has been reported in the literature [14] that growing epitaxial layers of this system is relatively easier on <111> P surface of InP rather than on <100> surface of InP. Our growth experiments are mainly on < 100 > surfaces. The most common problem which we have faced is that at the end of the growth run, all the solution does not wipe off and a metallic layer remains sticking on top of the growth. This problem is related to

morphology of the epi-growth which in this case is nearly
always non-uniform. Figure 2(a) and (b) show the pictures of
the interfacial region between the epi and the substrate. Fig.2a
shows that the interface is not smooth but kinky with triangular
pyramidal growth islands and metal filled in between as against
a smooth, uniform layer as seen in Fig.2(b).

Some crystal growers have suggested [15] that lowering
the cooling rate of the solution may overcome this problem.
However, we have not been able to observe this improvement by
reducing the cooling rate upto $\sim$0.4°C/min. We have a much
higher rate of successful uniform growths having clean wipe-
offs when we have used an excess piece of InP floating on top
of the solution (two phase technique) than when just accurately
weighed InP piece is used (single phase technique). One likely
explanation of this improvement by the two phase technique over
the single phase technique is that the solution in contact with
the InP platelets on top of the solution comes to an equilibrium
composition which initiates very nearly lattice matched growth
on another InP substrate under the solution. The solution is
normally heated much above the growth temperature by about 40°C
during homogenisation. At this high temperature, an excess amount
of phosphorous is dissolved in the solution. When the temperature
is lowered for growth, this excess phosphorous will be removed
by precipitation of quaternary material on the remaining InP
platelets and the solution achieves a quasi-equilibrium with

this solid. If no InP platelets are used on top of the solution, the composition of the solution may be extremely critical to the accuracy of the actual weighings of the solute components.

Still we are far from a thorough understanding of the growth at the very initial times when heterogeneous nucleation occurs. For example, it is well known that growing InP on top of InGaAsP which is As rich, is much more difficult than growth of the InGaAsP on top of InP substrate [16]. Thus the surface kinetics and nucleation are very important to the understanding of the growth phenomena in the heteroepitaxy system.

Results and Discussion.

Well lattice matched layers of InGaAsP on InP substrates are very smooth, shiny with a mirror like finish. Under microscope, features like facets, terraces, ripples, S-shaped marks, solution pull marks, trapped indium globules are seen which are characteristic features of the LPE growth. Fig.3 shows a photograph of a typical surface morphology of these materials.

To confirm the lattice matching of the quaternary alloy, back reflection x-ray diffraction measurements are done. Most of the layers are lattice matched with $|\frac{\Delta a}{a}| \leqq 0.2\%$; and the diffraction curve gives single peak. In case where mismatch is more, the diffraction curve is either broadened or additional peaks are observed.

Band gap of the epi-layer material is determined by transmission measurements using Cary 17D double beam spectrophotometer. Figure 4 shows 300K transmission plots in the wavelength range from 900 nm to 1.7 μm for several $In_{1-x}Ga_xAs_yP_{1-y}$ layers grown in our laboratory from x=0, y=0 for InP to x=0.47, y=1 for $In_{0.53}Ga_{0.47}As$. Transmission of four quaternary layers one ternary and one binary layer is shown in Fig.4.

We have performed a number of photoluminescence and Hall measurements on quaternary samples with nominal composition of band gap $\sim 1.3\mu$m. The as grown layers without any intentional doping are n type with background concentration of $\sim 10^{17} cm^{-3}$. In accordance with the experience of other growers we find that baking the solution for about 18 hours at $\sim 700°C$ reduces this background to about $2 \times 10^{16} cm^{-3}$. Using such a solution, we have carried out doping experiments with Ge. We find that upto n $\sim 3 \times 10^{18} cm^{-3}$, the electron concentration increases linearly with the Ge content in the solution, giving a distribution coefficient of about 5×10^{-3} (Fig.5). Another interesting result is that doping with Ge introduces a new band of luminescence which is shifted from the band edge towards longer wavelength by about 0.2 μm as shown in Fig.6. We have observed a similar band introduced by Sn as well. Some details of the above results are given in an accompanying paper [17].

Summary

We have presented a brief report on the growth of epitaxial layers of the semiconductor alloy $In_{1-x}Ga_xAs_yP_{1-y}$ lattice matched to InP substrates by LPE. These layers are tuned for band gaps at wavelengths $\lambda_{Eg} \approx 0.92$, 1.1, 1.3, 1.55 and 1.65 μm. Doping these materials with group IV elements is undertaken. A new band

of luminescence has been seen in the germanium doped material.
Further efforts are being made for improving the reproducibility
of the grown materials.

Acknowledgements

We are thankful to several members of our group,
Dr. S. Subramanian, Dr. A.K. Srivastava, S. Chakravarty and
R. Rajalakshmi, who have participated in discussions and the
work of characterisation of the materials. The detailed works
related to various properties of the materials will appear in
the future publications separately.

References

1. H.C.Casey Jr. and M.B.Panish, Heterostructure Lasers, Part B, Materials and Operating Characteristics, (Academic Press, New York, 1978).

2. T.P. Pearsall in GaInAsP Alloy Semiconductors, Ed.T.P.Pearsall (Wiley, New York, 1978).

3. H.Kressel and H.Nelson in Physics of Thin Films, Ed. M.H.Francombe, (Academic Press, New York, 1973).

4. L.R.Dawson in Progress in Solid State Chemistry, Vol.7. Co-Ed. H.Reiss and J.O.McCaldin, (Pergamon Press, Oxford,1972).

5. S.Subramanian and B.M.Arora, Indian J. of Pure and Appl.Phys. 17, 348 (1979).

6. J.J. Hsieh in Handbook on Semiconductors Vol.3, Vol. Ed. S.P.Keller (North Holland, Amsterdam 1980).

7. R.E.Nahory, M.A.Pollack, W.D.Johnston, Jr. and R.L.Barns, Appl.Phys.Lett., 33, 659 (1978).

8. K.Nakajima, T.Kusunoki and K.Akita Fujitsu Scientific and Technical Journal, 16, 59 (1980).

9. M.A.Pollack, R.E.Nahory, J.C.DeWinter and A.A.Ballman, Appl.Phys.Lett. 33, 314 (1978).

10. K.Nakajima and T.Tanahashi, J.Cryst.Growth, 71, 463 (1985).

11. R.Sankaran, G.A.Antypass, R.L.Moon, J.S.Escher and L.W.Jame-, J.Vac.Sci. and Technol., 13, 932 (1976).

12. T.Yamamoto, K.Sakai and J.Akiba, Jap.J.Appl.Phys. 16, 1699 (1977).

13. M.Feng, L.W.Cook, M.M.Tashima, and G.E.Stillman, J.Electron Mater. 9, 241 (1980).

14. K.Nakajima, T.Kusunoki, K.Akita and T.Kotani,
    J.Elec.Chem.Soc. 125, 123 (1978).

15. R.Sankaran, R.L.Moon and G.A.Antypass,
    J.Cryst.Growth 33, 271 (1975).

16. K.Nakajima in Semiconductors and Semimetals, Vol.22, Part A
    Ed. W.T.Tsang (Academic Press, New York, 1985).

17. R.Rajalakshmi, S.S.Chandvankar, S.Chakravarty, A.K.Srivastava
    and B.M.Arora, IVth International Workshop on Physics of
    Semiconductors, Madras, 1987.

Table I

Typical solution compositions used for growing $In_{1-x}Ga_xAs_yP_{1-y}$ layers (y 2.2x) epitaxially on <100> InP substrates at about 635°C.

Composition attempted		λ_{Eg} Expected (μm)	λ_{Eg} [a] (μm)	Solution Composition				Remarks
x	y			Indium mg	InAs mg	GaAs mg	InP mg	
0	0	0.92	0.930	226.80	-	-	*	Good growth
0.17	0.38	1.10	1.100	150.40	7.00	1.0	6.6	Seven layers grown from same solution
0.28	0.60	1.30	1.230	183.40	12.10	2.30	3.8	Six layers grown from same solution
0.40	0.88	1.55	1.50-1.588	244.0	15.80	7.33	4.0	Single layer grown from this solution
0.47	1.00	1.65	1.60	281.32	19.70	9.84	-	Single layer grown from this solution

a) λ_{Eg} is the wavelength where the transmission falls rapidly (see Fig.4).

* Undoped InP piece used for solution saturation.

Table II

Electrical data of $In_{1-x}Ga_xAs_yP_{1-y}$ ($y \approx 2.2x$) layers grown on <100> InP at 635°C.

Sample number	Composition		Nominal band gap λ_{Eg} (μm)	Dopant	Carrier Concentration n cm^{-3}		Hall Mobility cm^{2}V^{-1}s^{-1}	
	x	y			300K	77K	μ_{300K}	μ_{77K}
855FSL5	0	0	0.92	Undoped	1.40×10^{17}	-	1650	-
86d$_1$SL2	0.17	0.38	1.10	Undoped	1.90×10^{17}	1.70×10^{17}	1970	2613
86d$_1$SL6	0.17	0.38	1.10	Undoped	4.00×10^{16}	3.38×10^{16}	1634	3368
grown from solution baked for 18 hours								
86ZSL1	0.28	0.60	1.30	Sn doped $x_{Sn}^{\ell} = .0045$	5.20×10^{17}	4.90×10^{17}	1839	2139
86b$_1$SL2	0.28	0.60	1.30	Undoped	2.20×10^{16}	1.19×10^{16}	1563	2787
grown from solution baked for 17 hours								
86b$_1$SL3	0.28	0.60	1.30	Germanium $x_{Ge}^{\ell} = .0023$	3.70×10^{17}	7.8×10^{16}	1477	1703
86b$_1$SL4	0.28	0.60	1.30	$x_{Ge}^{\ell} = .0049$	9.80×10^{17}	9.8×10^{17}	1330	1403
86b$_1$SL6	0.28	0.60	1.30	$x_{Ge}^{\ell} = .0134$	2.30×10^{18}	2.34×10^{18}	1233	1335
86b$_1$SL17	0.28	0.60	1.30	$x_{Ge}^{\ell} = .0183$	3.0×10^{18}	3.0×10^{18}	1044	1105
86CSL2	0.47	1.00	1.65	Undoped	1.46×10^{17}	-	3413	-

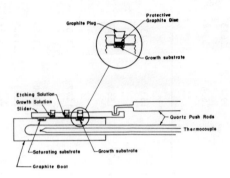

Figure 1   : Schematic of the Growth Apparatus.

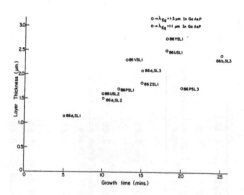

Figure 1(a)   : Plot of thickness of the grown layers versus
               growth time.

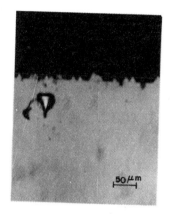

Figure 2(a)   : Epi-substrate interface with pyramidal
               non-uniform growth.

Figure 2(b)   : Smooth, uniform junction line of epi-substrate
               interface.

430

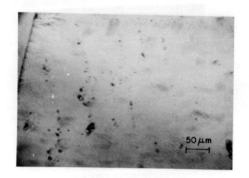

Figure 3   : Typical surface morphology of the grown layer.

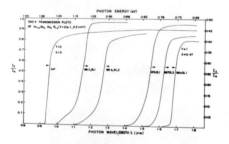

Figure 4   : 300K transmission plots of several epi-layers.

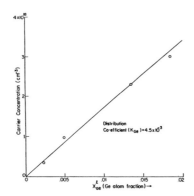

Figure 5 : Plot of carrier concentration versus atom fraction of Ge in the solution.

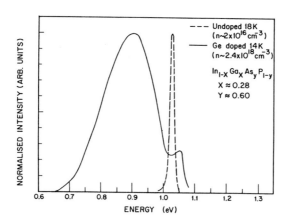

Figure 6 : Photoluminescence of undoped and Ge doped layers.

PHOTOVOLTAICS

TECHNOLOGY AND CHARACTERISTICS OF AMORPHOUS SILICON FOR SOLAR CELLS APPLICATIONS

by

F.P.Califano-University of Rome-La Sapienza-Dept. of Electronics
Via Eudossiana 18 Rome

INTRODUCTION

Although photovoltaic conversion of sunlight into electricity is an old technology,it was only in 1954, with the first silicon cell made at Bell Labs.,that its pratical use became possible . For many years,after 1954,solar cells use was limited to space applications,with high qualified small industries producing less than 50kWp/year.In 1974 ,National Science Foundation carried out a feasibility study which demonstrated the possibility of terrestrial applications of solar cells.After this study many industries became involved in solar cells manifacturing and,as a result,their price has decreased and the quality is increased.

However the prediction that solar cells could contribute to reduce the use of fossil fuels has failed and,at the present time,the use of solar cells is limited to special applications and the ipothesis of grid connected systems has now a limited support.

The fact is that at present price,a very limited number of applications are cost effective and ,at the same time, the learning curve for the crystalline technology shows no relevant cost reductions also in presence of a consistent market increase. On the other hand,new technologies are becaming more and more attractive for cost reduction:between these the amorphous silicon is,at present,the most promissing one and to this we shall reserve the present paper.

AMORPHOUS SOLAR CELLS

Amorphous solar cells are different from crystalline cells in

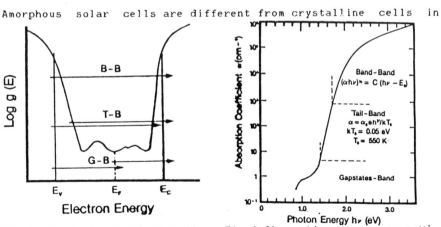

Fig.1 Band structure of a-Si   Fig.2 Absorption coeff. of a-Si

436

many regards.The main differences can be summarized as follows:

Cristalline cells
-PN junction
-carrier diffusion dominate

-Voc determined by carrier diffusion over the barrier
-Poisson's equation satisfied
by free charge and ionized dopants

Amorphous cells
-PIN structure
-field assisted collection dominate
-Voc determined by carrier recombination in tail states
-Poisson's equation satisfied
by large density of trapped
charge

A schematic of the band structure of amorphous silicon is reported in fig.1,where the possible carriers transitions are indicated.In fig.2 the influence of these transitions on the optical absorption coeffi-cient are reported. Finally,in fig.3,the optical absorp-tion coefficient of amorphous,micro-crystalline and crystalline silicon are compared.Notice the low value of the absorption coefficient of mono-crystalline silicon,compared to that of the other materials.

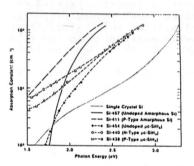

Fig.3 Absorption constants of semiconductors

ELECTRICAL PROPERTIES OF a-Si

Amorphous silicon can be doped and both n and p type can be prepared using the same dopants as for mono-crystalline silicon. Fig.4 reports the dark conductivities vs tempera-ture for intrinsic and doped a-Si. The main difference between a-Si and x-Si is in the carriers mobilities, which are very low for a-Si.In fig.5 a comparison is made between these values.

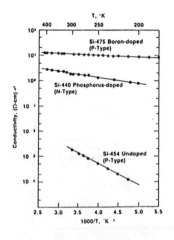

Fig.4 Dark conductivities

GROWING AMORPHOUS MATERIALS

The techniques used to grow amor-phous silicon can be summarized as follows:

1-G-D of silanes
-RF glow discharge
-DC glow discharge
-MW glow discharge

2-CVD of silanes
-hetero-CVD
-homo -CVD

3-Reactive sputtering
-RF diode sputtering
-magnetron sputtering

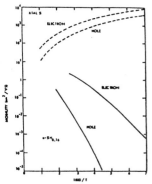

Fig.5 Mobilities vs T
for Si and a-Si

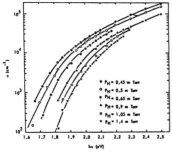

Fig.6 a-Si absorption co-
stant vs pressure

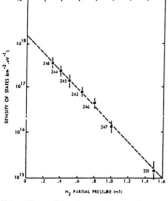

Fig.7 a-Si density of
states vs pressure

-ion beam sputtering
4-Evaporation
-reactive e-beam evaporation
-ionized-cluster beam deposition

The previously indicated methods can
be also summarized as follows:

Plasma methods

SiH4+electric field (DC/RF/MW)->SiHx
Si + Ar + H2 +electric field--->SiHx

Non-plasma methods

SiH4 + heat/photons(UV/IR)----->SiHx
Si + H2 +heat/photons(UV/IR)--->SiHx

The methods which have major success
in a-Si preparation are the RF glow
discharges and the UV photo CVD.A
comparison between the essential
properties of the films prepared
using these methods can be summarized as follows:
Advantages of photo CVD:
-low temperature process
-few species in film growth region
-possible selection of primary reactions
Disvantages of photo CVD
-low deposition rate
-if Hg is used,possible film contamination
-window absorption and coating
-high cost of lasers ,if used
-for direct CVD have to use higher
silanes.
The glow discharge is,at the present
time ,the technology most widely used
in research apparatus and the only
one used in production.The properties
of the materials prepared by glow discharge strongly depend from the operating pressure,as shown in fig.6- 7,
where the absorption coefficient and
the density of states are reported vs
the pressure.The effective role of
the pressure in glow discharge has
not yet been understood,however it is
clear now that it has to be moderately low if good quality material is
required.In fig.8 the schematic of a
glow discharge apparatus is reported:
it shows the growing of hidrogenated

438

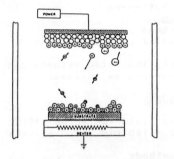

Fig.8 Schematic of a glow discharge apparatus

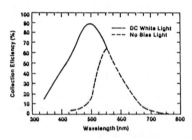

Fig.9 Spectral response of a-Si cell

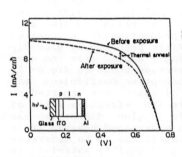

Fig.10 Characteristics degradation

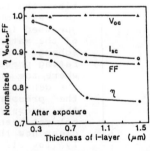

Fig.11 Parameters degradation

silicon on the substrate.

THE ROLE OF DEFECTS IN a-Si

Amorphous silicon,not only presents high concentration of traps,as shown in fig.9,where the spectral response is plotted in presence and without bias light,but it also presents a high instability,which is now understood as a bulk effect enhanced by recombination processes (1).The effect of this degradation is indicated in figs.10-11,in which the degradation of the external characteristics and of the single parameters of a typical a-Si solar cells are reported;in fig.11,in particular the effect of light exposure has been measured for different thicknesses of the i layer. Notice how as the cell becames thicker the effect of degradation is more evident.The effect of light soaking on a-Si solar cells is also repored, with more details in figs.12-13,where the change in the external characteristics and the variations of the photocurrent and of the dark current are reported.

The a-Si degradation is a important limit to the use of this material in photovoltaic applications:there have been many attempts to reduce this degradation,including the use of other materials,such as fluorine,in the preparation of the materials.Up to now,not too many progresses have been made and the only possibility to reduce the instability remains that of reducing the active layer thickness to about 3000 A. With such a strong reduction of the active layer the cell is no more able to absorp the main part of the solar radiation

and,due to this its efficiency becomes too low.To avoid it
stacked cells have been made using materials with different or
equal optical bandgap.

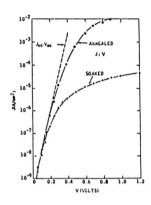

Fig.12 external characteristic
variation after light soaking

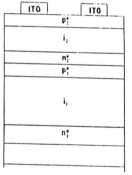

Fig.13 Currents variation in
soaked cells

TANDEM CELLS

The basic structure of a tandem cell is reported in fig.14:in
this structure tunnel effect is used
at the contact between the two cells.
If semiconductors having different
gaps are used for the two cells,high
efficiencies can be obtained,which
became even higher if three or more
cells are used.Figs.15-16,show,both
for the crystalline and the amorphous
case the maximum efficiencies which
can be obtained using variable gap
multijunction cells.The figures also
indicate the values of the optimal
bandgaps.It should be pointed out that
making multijunction solar cells is a
sophisticated technology which has not
yet given reasonable results both with
crystalline and amorphous materials.

Fig.14 Structure of a
multijunction cell

The main difficulties are connected,
in both cases,to the technology of
growing different semiconductors one on top of the other;however
some results have been obtained and they are shown in figs.17-18.
In these figs. are reported the external characteristic of a
triple cell made with variable gap materials and the contribution
to the total efficiency of the single cells.Notice that the
external characteristic shown is that of the a-Si cell which,up

to now,has shown the highest efficiency.A proper design of multi-

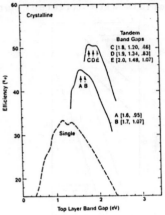

Fig.15 Monocrystalline tandem cells efficiency

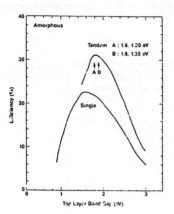

Fig.16 Amorphous materials tandem cells efficiency

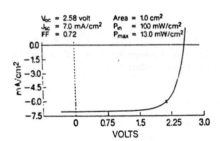

Fig.17 External characteristic of a triple cell

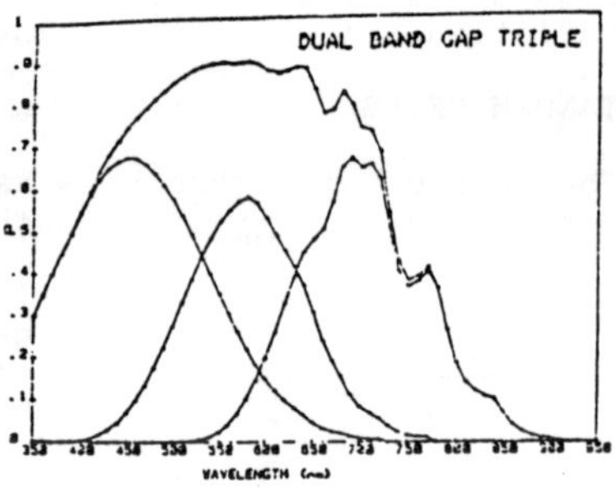

Fig.18 Contributions of the single cells

junction cells is necessary to avoid that as the spectral distribution of the source changes,the short circuit currents of the single cells vary differently;if it happens,the efficiency drastically drops and a triple cell shows an efficiency lower than that of a single cell.In fig.19 the short circuit currents of the single elements of the cell of fig.17 are reported,for different solar spectra.Notice how,in this case,the matching has been optimized.
Making amorphous material other than silicon is still a problem; however,some alloys between Si and Ge have been prepared,and in fig.20 the absorption coefficients of these materials are shown:

a increasing concentration of germanium lowers the gap,but,at the same time the quality of the material get worst and it forbids to use low gap materials.

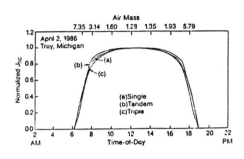

Fig.19 Matching between single elements

Fig.20 a-Si-Ge absorption coefficients.

ACTUAL AMORPHOUS SOLAR CELLS

The historical efficiency trend for a-solar cells is reported in fig.21:the diagram shows how very large improvements have been obtained in recent years and how there is a delay of about 4 years between the maximum efficiencies obtained with small area devices and that of large area cells.There is also a delay between the large area cells produced in laboratories and that made in production:this delay is of about 2 more years. In fact,actual production of amorphous solar cells has an average efficiency of about 5% ; in conclusion,there is a delay of about 6 years between the best results obtained to day and the cells coming out from produ-ction.As a consequence it will be only in 1991 that 13% solar cells will be produced. Due to this monocrystalline solar cells will still be,for a long time , the only ones which will be used in medium and large size sys-tems.

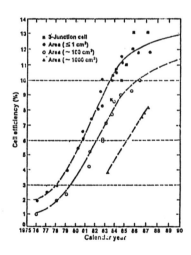

Fig.21 Historical efficiencies

THE PHOTOVOLTAIC MARKET

Amorphous silicon solar cells have been the first thin film devices to enter the market,with the last year's production equivalent to more than 10 MW.

Its market share has been continously growing and, in fact, it is passed from 3% in 1981 to more than 37% during last year:the rest of the market is accounted by crystalline silicon.In fig.22 is indicated the growing of the amorphous silicon share on the total photovoltaic market.

The growth of the photovoltaic market during these last 10 years, is shown in fig.23,were the shipments of the last 7 years are reported together with the sharing beetwen the different technologies.The market growing rate has been much lower of what it was expected and is showing now a saturation effect.

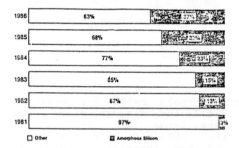

Fig.22 World PV market share

-developing countries have always been indicated as a large potential market:in fact the main part of the systems installed

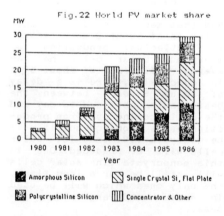

Fig.23 World PV shipments

The reasons of this slowly growing market are :
-the cost decrease of PV modules has been much lower of what was extimated in the past
-the limited engineering capabilities of the PV industry had as a consequence that systems of limited releability have been installed and to day it is hard , some time,to convince people that PV systems can be pratically maintenance free

in these countries have donations of international organizations and/or of countries involved in solar cells production
-the learning curves for solar cells modules and systems as shown in fig.24 apply to industrialized countries where installation, maintenance and shipment are much cheaper.However it is a common procedure to use these curves for every country and as a consequence the economical evaluations are generally wrong.
-developing countries have limited economical resources and more urgent problems than using photovoltaic, in non competitive

applications.

However,it should be kept in mind that photovoltaic can really contribute to improve the quality of life in many countries,but to reach this result the actual policy used to introduce it in developing countries has to be completely changed.

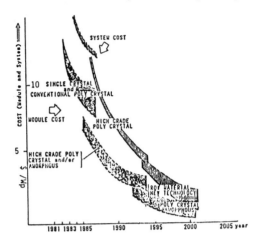

The new policy should include:
-the involvement of the developing countries has to be much higher than in the past
-design technique has to be completely transferred to the countries which are willing to use PV technology
-systems should be standardized so that a minimum training of the local people would be required
-photovoltaic should be properly used, avoiding to install it when other energy sources or technologies could be more appropriate than photovoltaic.

Fig.24 Learning curve for solar cells modules and systems.The impact of new technologies is shown.

CONCLUSIONS

The photovoltaic technology has made relevant progresses in these last years and the time is coming that it will be used for energetic applications with a cost of the kWh competitive,in many applications, to that of conventional power plants.
To reach this result it is now clear that the amorphous solar cell technology has to be still improved to meet the following goals:
-stability improved to 20 years,within 10% efficiency decay
-efficiency per square meter should reach at least 15%

At the same time a larger market has to be developed and it requires:
-cost reduction of the peak watt at less than 1$,what also means the concentration of modules production in a limited number of plants and the use of highly automated production techniques
-systems standardization to drastically reduce the engineering costs and improve their releability
-new policies to spread the PV technology in developing countries,avoiding that the benefits coming from the reduction of modules cost are limited to indusrialized countries.

444

The problem of the cost reduction is a very hard one to be met for very large scale applications, as shown in fig.25, where the cost of the kWh is quoted for various cells efficiencies and for different modules cost. Notice that at low efficiencies the cost of the peak watt becames extremely low.

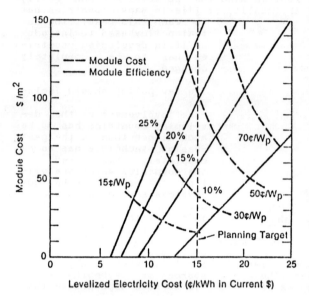

Fig.25 Levelized energy costs

While reaching the very low costes indicated in fig.25 may be too optimistic,it is secure that a drastic cost reduction can be obtained using the thin film technology:this will make reasonable other and more important applications which will contribute to spread around the PV technology and further reduce the costs for new applications.

BIBLIOGRAPHY
(1) D.L.Staebler.C.R.Wronski.Appl.Phys.Lett.,31,292,1977
Japanese a-Si Program-1987 Subcontractors Review Meeting-Palo Alto -Ca-USA

Five Year Research Plan- DOE USA -1987

Progress,World Competition,and Markets Drive :a-Si Technology-E.S.Sabisky - SERI report TP-211-3151

Amorphous Silicon Alloys for Solar Cells-K.W.Mitchell,K.J.Tourjan Ann.Rev.Energy ,1985,10.1-34

In Review -Vol.VIII,N.3

Solar cells,17,1986

Figs.9,17,18,19,20 are courtesy of S.Hudgens from ECD
Figs.3,4,11,15,16 are courtesy of C.R.Wronski

NEW TYPE OF HETEROSTRUCTURES FOR SOLAR CELLS

B.L. Sharma, G.C. Dubey and Vinod K. Jain
Solid State Physics Laboratory
Lucknow Road, Delhi-110007
INDIA

INTRODUCTION

Various approaches have been used to enhance the collection efficiency and open circuit voltage of Si solar cells [1,2]. Heterostructure
solar cells [1,3,4] have also been reported to improve the spectral response and open circuit voltage. As heavily doped emitters limit the efficiency of solar cells due to shorter lifetime of minority carriers,
surface recombination and band narrowing, it has been improved by using
shallower junction and surface passivation or by using a high-low junction
emitter. Further improvements have also been achieved by providing a
back surface field (i.e. a heavily doped region at the rear of the cell)
and by antireflection coating in the front of the cell.

In this paper a new type of heterostructure is proposed in which
the 'dead layer' problem (often encountered in solar cells) is overcome
by depositing a heterolayer of a suitable semiconductor on top of a solar
cell. In this case the two contacts are still taken from the emitter
and base of the solar cell. The effect of such a heterolayer on various
homojunction and heterojunction solar cells is theoretically analyzed
here with the help of energy band diagrams. As the heterolayer parameters
depend on the structure under consideration, an estimate of the absorption of incident solar radiation in different regions of the heterostructure are also included for all cases in the text. Preliminary experimental results of an $(i)\mu cSi:H-(n^+)Si-(p)Si-(p^+)$ solar cell are also
reported in this paper.

THEORETICAL MODEL

The use of wide bandgap semiconductors to avoid photogeneration and recombination in certain regions and as aids in collection has been discussed by Fonash and Rothwarf [5]. Wide bandgap semiconductor is employed in these cases either as a window in front of a homojunction to form a heteroface solar cell (refer Fig. 1) or as an emitter or base in a heterojunction solar cell. In the proposed heterostructure a layer of wide bandgap semiconductor is deposited on top of a homojunction or heterojunction solar cell to form a heterojunction. A heterostructure in which a p-type aSi:H layer is deposited on top of a $(n^+)Si-(p)Si-(p^+)Si$ back surface field (BSF) solar cell is considered here to demonstrate the advantage of such a heterojunction. The equilibrium energy band diagrams of a BSF Si solar cell and a heterolayer deposited BSF Si solar cells are shown in Fig. 2 and 3 respectively. It can be seen from Fig. 3 that an electric field is formed across the emitter due to positive space charges present at the two edges of the emitter. The necessary condition for deposited layer to be effective is that the positive space charge Q_1 at the front end of the emitter has to be greater than Q_2 at the other end under all conditions. The strength of this electric field under equilibrium condition can be written as

$$E = \frac{V_{D1} - V_{D2}}{d} \qquad (1)$$

where d is the thickness of the emitter and V_{D1} and V_{D2} are the built-in voltage of the heterojunction between deposited layer and emitter and of the solar cell junction respectively. The actual electric field strength under illumination will, however, depend on the photogeneration of carriers in different regions. Ideally this strength will be maximum when the carriers are primarily generated in the base region of the solar cell.

This electric field will assist the collection of photogenerated carriers and thus increase the short circuit current.

The photogenerated carriers in different regions can be determined by knowing the percentage absorption of incident solar radiation in each layer. If the thickness of the base region is such that the incident radiation is completely absorbed in it then one needs to know the percentage absorption in the top layer and the emitter. By assuming specular reflection, the percentage absorption of the incident solar radiation of $h\nu > E_g$ (aSi:H) in the top layer of thickness t can be estimated from [6]

$$[I(o) - I(t)]/I(o) = [1 - \exp(-2\,\alpha_1\, t)] \qquad (2)$$

where $I(o)$ and $I(t)$ are the intensities of the incident radiation and the radiation transmitted through the layer of thickness t respectively and α_1 is the average absorption coefficient for the radiation of $h\nu > E_g$(aSi:H) in aSi:H. The estimation of the percentage absorption of this incident radiation in the emitter can be made by using the expression

$$[I(t) - I(t+d)]/I(o) = \exp(-2\,\alpha_1\, t)\, [1-\exp(-\,\alpha_2\, d)] \qquad (3)$$

where α_2 is the average absorption coefficient for radiation of $h\nu > E_g$ (aSi:H) in Si. Since the top layer acts as a window for incident solar radiation of E_g(aSi:H) $> h\nu > E_g$(Si), the percentage absorption of this radiation in the emitter can be estimated from the expression (2) by replacing $2\alpha_1 t$ by $\alpha_2' d$ where α_2' is the average absorption coefficient for this radiation in Si. As mentioned earlier, in both these cases the radiation transmitted through the emitter is, however, absorbed in the base region.

By using the above expressions for the percentage absorption one can optimize the thickness of the top layer for which the electric field strength under illumination is maximum.

CALCULATIONS FOR PRACTICAL SOLAR CELL STRUCTURES

The V_{D1}, V_{D2}, Q_1, Q_2, E and percantage absorption in different regions for a typical (p)aSi:H-(n$^+$)Si-(p)Si-(p$^+$) Si solar cell are calcu-lated and tabulated in Tables 1 and 2. For the sake of comparison, these values are also calculated and tabulated for heterostruc tures formed by deposition of p-type GaAs or n-type degenerate ITO instead of p-type aSi:H. In these calculations the effective barrier height ϕ_B between n-type ITO and n-type Si (refer Fig. 4) is estimated on the basis of the model proposed by Jain et al [9]. It can be seen from the tables that from equilibrium electric field considerations the p-type aSi:H layer solar cell appears to be better than both p-type GaAs and n-type degenerate ITO deposited solar cells while from the point of view of absorption in the top layer the transparent ITO deposited solar cell is better.

Since (p)Al$_x$Ga$_{1-x}$As(x = 0.8) - (p)GaAs - (n)GaAs - (n$^+$)GaAs hetero-face solar cells have resulted in best single cell efficiencies reported so far [10], heterostructures formed by deposited of n-type Al$_x$Ga$_{1-x}$As on p-n-n$^+$ GaAs solar cell (refer Fig. 5) and p-type Al$_x$Ga$_{1-x}$As on n-p-p$^+$ GaAs solar cell (refer Fig. 6) are also considered here. Both x = 0.4 and x = 0.8 are taken for calculations and the calculated values for all cases are also included in the tables. In can be seen from these calculations that, inspite of very large equilibrium electric field strength in the case of (n)Al$_x$Ga$_{1-x}$As - (p)GaAs - (n)Ga As - (n$^+$)GaAs solar cell, due to absorption considerations (refer Table 2) it may not prove as good as the reported heteroface $(p)Al_xGa_{1-x}As-(p)GaAs-(n^+)GaAs$ solar cell.

A case of wide bandgap semiconductor layer deposited on a hetero-junction solar cell (refer Fig. 7) is also considered for the sake of completeness. Although the dark saturation current density I_0 in this case is larger than $(n^+)Si$ - $(p)Si$ - $(p^+)Si$ homojunction solar cell, it is included here as a efficient GaAs-Ge-Si solar cell is reported in literature [11].

EXPERIMENTAL RESULTS

The preliminary experimental results of a heterostructure formed by deposition of (i) μcSi:H layer on $(n^+)Si$ - $(p)Si$ - $(p^+)Si$ solar cell is discussed below. This n^+-p-p^+ Si solar cell was formed by diffusion of phosphorus at the front and aluminium at the back of a 2 ohm-cm p-type (100) oriented Si wafer. The top contact fingers (having covered area between 7-8%) were formed by deposition of a three layer Ti/Pd/Ag structure and photolithography. The large area back contact was made by depositing a four layer Al/Ti/Pd/Ag BSFR structure. This fabrication was carried out under highly controlled conditions used for fabricating space quality solar cells.

A (i)μcSi:H layer of thickness $\sim$ 400 $\overset{o}{A}$ was deposited on the top of the solar cell by rf glow discharge using 1 % SiH_4 + 99 % H_2 at higher power ratings [12].

A typical voltage and current output from an illuminated $(n^+)Si$ - $(p)Si$ - $(p^+)Si$ solar cell before and after deposition of (i) μcSi:H layer are shown in Fig. 8. It can be seen from this figure that the short circuit current changed from 112 mA to 148 mA while the open circuit voltage remained same after deposition. The relative spectral response normalized with respect to the standard solar cell (obtained from Spectro Lab, USA) before and after deposition are shown in Fig. 9. As expected from refractive index considerations, the reflectively measurements indicated that the spectral response change was not due to antireflection coating effect.

CONCLUSIONS

The preliminary experimental results clearly indicate that there is considerable improvement in the short circuit current due to the electric field created by the deposition of an intrinsic $\mu cSi:H$ on top of a n^+-p-p^+ Si solar cell. Theoretically it is expected to increase further by using a p-type amorphous or microcrystalline Si layer and using an appropriate antireflection coating. It is finally concluded that the 'dead layer' problem can, in general, be overcome to a great extent and collection efficiency increased by using a heterolayer of wide bandgap semiconductor on top of a solar cell.

ACKNOWLEDGEMENT

Authors are thankful to Dr.A.K. Sreedhar, Director, Solid State Physics Laboratory for suggesting this problem and his continuous guidance.

REFERENCES

1. H.J. Hovel, Semiconductors and Semimetals (Eds. R.K. Willardson and A.C. Beer), Vol. 11, Academic (1975).

2. A. Neugroschel, F.A. Lindholm, S.C. Pao and J.G. Fossum, Appl. Phys. Lett. 33, 168 (1978).

3. M.A. Green,Solar Cells : Operating Principles, Technology and System Applications (Ed. N. Holonyak, Jr.) Prentice-Hall (1982).

4. S.J. Fonash, Solar Cell Device Physics, Academic (1981).

5. S.J. Fonash and A. Rothwarf, Current Topics in Photovoltaics (Eds. T.J. Coutts and J.D. Meakin), pp. 1-39, Academic (1985).

6. G.D. Cody, Semiconductors and Semimetals (Ed. J.I. Pankove), Vol.21, p.16, Academic (1984).

7. F.Demichelis, E. Minetti-Mezzetti, A.Tagliaferro, E. Tresso, P. Rave and N.M. Ravindra, J. Appl. Phys. 59, 611 (1986).

8. D.E. Aspnes, S.M. Kelso, R.A. Logan and R. Bhat, J. Appl. Phys. 60, 754 (1986).

9. V.K. Jain, R.K. Purohit and B.L. Sharma, Solar Cells, 6, 335 (1982).

10. J.M. Woodall and H.J. Hovel, Appl. Phys. Lett. 30, 492 (1977).

11. B.Y. Tsaur, C.C. John, G.W. Turner, F.M. Davis and R.P. Gale, IEEE 16th Photovoltaic Specialists Conf. Proc., p.1143 (1982).

12. G.C. Dubey, R.A. Singh, S.N. Mukerjee, S. Pal and M.G. Rao, Bull. Mater. Sci. 8, 267 (1986).

TABLE 1

Cell structure	Carrier concentration (cm^{-3})			Solar cell junction I_o (A/cm^2)	Space charge (C/cm^2)		Built-in voltage (eV)		Electric field stregnth* E (V/cm)
	Top layer	Emitter	Base		Q_1	Q_2	V_{D1}	V_{D2}	
(p)aSi:H-(n$^+$)Si-(p)Si-(p$^+$)Si	10^{16}	10^{19}	10^{16}	6.9×10^{-13}	7×10^{-8}	5.6×10^{-8}	1.45	0.93	1.76×10^4
(p)GaAs-(n$^+$)Si-(p)Si-(p$^+$)Si	10^{16}	10^{19}	10^{16}	6.9×10^{-13}	6.7×10^{-8}	5.6×10^{-8}	1.21	0.93	9.3×10^3
(n)Al$_x$Ga$_{1-x}$As(x=0.4)-(p)GaAs-(n)GaAs-(n$^+$)GaAs	10^{18}	10^{17}	10^{16}	9.4×10^{-17}	2.5×10^{-7}	6.3×10^{-8}	1.88	1.21	2.2×10^4
(p)Al$_x$Ga$_{1-x}$As(x=0.4)-(n)GaAs-(p)GaAs-(p$^+$)GaAs	10^{18}	10^{17}	10^{16}	9.9×10^{-17}	2×10^{-7}	6.3×10^{-8}	1.24	1.21	1×10^3
(n)Al$_x$Ga$_{1-x}$As(x=0.8)-(p)GaAs-(n)GaAs-(n$^+$)GaAs	10^{18}	10^{17}	10^{16}	9.4×10^{-17}	2.9×10^{-7}	6.3×10^{-8}	2.48	1.21	4.4×10^4
(p)Al$_x$Ga$_{1-x}$As(x=0.8)-(n)GaAs-(p)GaAs-(p$^+$)GaAs	10^{18}	10^{17}	10^{16}	9.9×10^{-17}	2.2×10^{-7}	6.3×10^{-8}	1.35	1.21	4.6×10^3
(p)GaAs-(n$^+$)Ge-(p)Si-(p$^+$)Si	10^{16}	10^{19}	10^{16}	0.9×10^{-12}	6.6×10^{-8}	5.1×10^{-8}	1.19	0.81	1.3×10^4
(n)ITO**-SiO$_x$-(n$^+$)Si-(p)Si-(p$^+$)Si	Degenerate	10^{19}	10^{16}	6.9×10^{-13}	1.8×10^{-6}	5.6×10^{-8}	1.01	0.93	2.6×10^3
(n)ITO***-SiO$_x$-(p$^+$)Si-(n)-Si-(n$^+$)Si	Degenerate	10^{19}	10^{16}	1.1×10^{-12}	2×10^{-6}	5.5×10^{-8}	1.21	0.9	1×110^4

*   calculated by taking emitter thickness as 0.3 μm

**   ITO deposited by spraying

*** ITO deposited by sputtering

TABLE 2

Cell Structure	Layer thickness (Å)		Incident Photon energies (eV)	Average absorption coefficient (cm^{-1})		Percentage absorption of incident radiation %		
	Top layer	Emitter		α_1	α_2	Top	Emitter	Base
(p)aSi:H-(n$^+$)Si-(p)Si-(p$^+$)Si	300	3000	hυ>1.7	5x10^4	4x10^3	26	8	66
			1.7>hυ>1.1	-	10^3	-	3	97
(p)GaAs-(n$^+$)Si-(p)Si-(p$^+$)Si	500	3000	hυ>1.43	4x10^4	4x10^3	19	9	72
			1.43>hυ>1.1	-	8x10^2	-	2.4	97.6
(n)Al$_x$Ga$_{1-x}$As(x=0.4)-(p)GaAs-(n)GaAs-(n$^+$)GaAs	300	3000	hυ>1.92	6.7x10^4	1.1x10^5	19	77	4
			1.92>hυ>1.43	-	3.1x10^4	-	46	54
(n)Al$_x$Ga$_{1-x}$As(x=0.8)-(p)GaAs-(n)GaAs-(n$^+$)GaAs	300	3000	hυ>2.55	2.7x10^4	1.42x10^5	8	91	1
			2.55>hυ>1.43	-	4.2x10^4	-	72	28
(p)GaAs-(n$^+$)Ge-(p)Si-(p$^+$)Si	500	3000	hυ>1.43	4x10^4	8x10^3	19	17	64
			1.43>hυ>0.67	-	8x10^3	-	22	78
(n)ITO-SiO$_x$-(n$^+$)Si-(p)Si-(p$^+$)Si	700	3000	3.5>hυ>1.1	-	10^3	-	3	97

α for aSi:H taken from reference [7]

α's for GaAs and Al$_x$Ga$_{1-x}$As taken from reference [8]

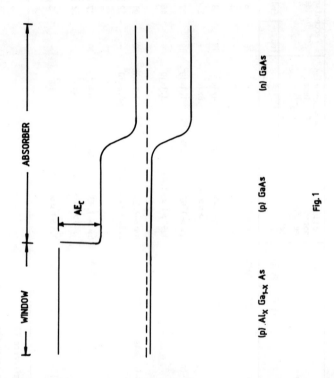

Fig.1

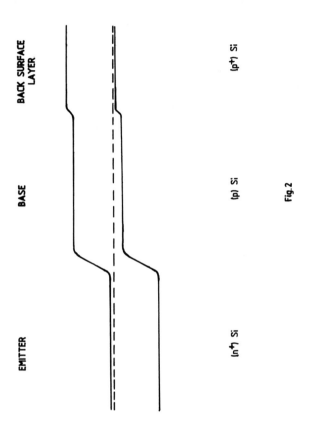

EMITTER BASE BACK SURFACE LAYER

(n⁺) Si (p) Si (p⁺) Si

Fig. 2

456

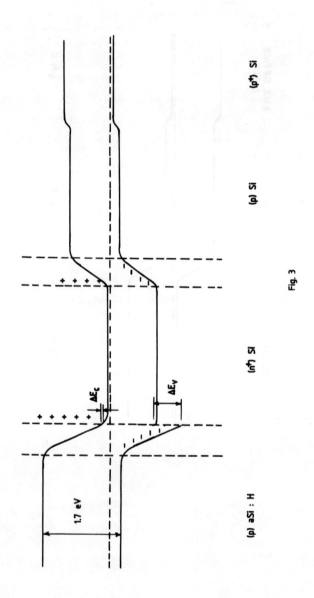

Fig. 3

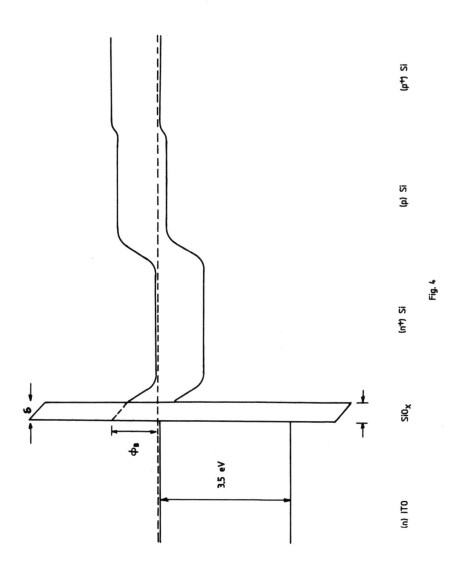

Fig. 4

458

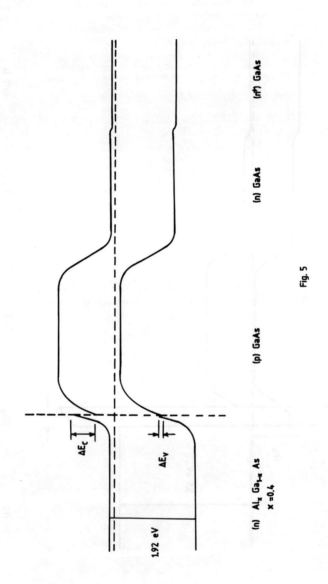

(n) Al$_x$ Ga$_{1-x}$ As
x =0.4

1.92 eV

ΔE$_C$

ΔE$_V$

(p) GaAs

(n) GaAs

(n⁺) GaAs

Fig. 5

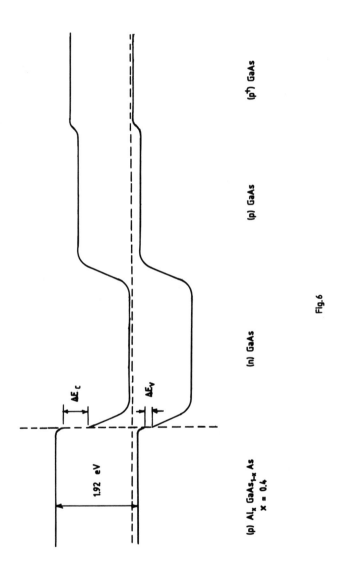

459

Fig.6

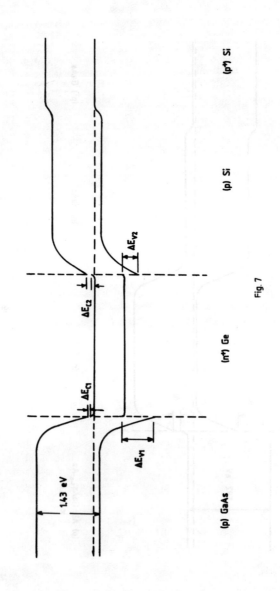

Fig. 7

461

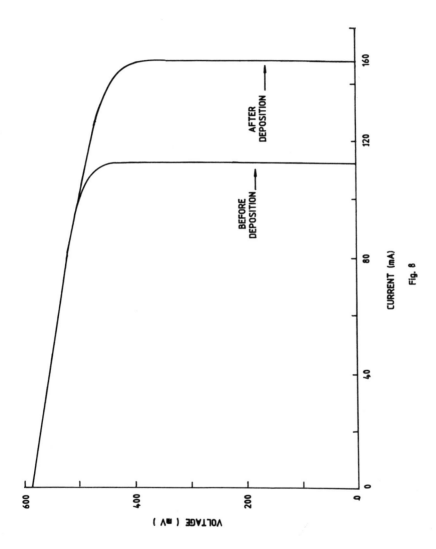

CURRENT (mA)

VOLTAGE (mV)

AFTER DEPOSITION

BEFORE DEPOSITION

Fig. 8

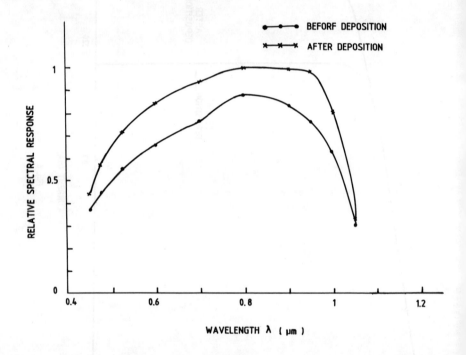

Fig. 9

Considerations for Optimizing Light Coupling to Solar Cells
in a Flat-Plate Photovoltaic Module

B. L. Sopori

Solar Energy Research Institute
1617 Cole Boulevard
Golden, CO 80401

Abstract

This paper describes a procedure to maximize performance of a cell
in a module by optimizing optical coupling to the cell for a given
incident spectrum. An optical model of a module is described
which takes into account absorption in various module media and
the reflections at all interfaces, including scattering at non-
planar surfaces. This model is based on easily measurable parameters. A brief description of the methods to make the measurements of optical parameters of the module materials is given.

Introduction

The optics of a flat-plate photovoltaic (PV) module has received relatively
little attention as compared to the cell design and the mechanical design of a
module. This lack of activity in the area of the optical design of the module
may, in part, be attributed to the fact that various components of the module
such as glass, pottant, and back-skin are limited in variety and that most
design parameters are predetermined by the mechanical considerations. However, because of the constraints placed by the mechanical design, it becomes
increasingly important to understand the optical processes that take place
within the module and, concomitantly, arrive at the proper selection of other
operational parameters so as to optimize the light coupling to the cell.
Since parameters such as the thickness of the glass and the pottant thickness
are predetermined by the mechanical design of the module, the selection of the
parameters for the photovoltaic module designer involves:

a. type of AR coating material for the cell (dictated by the suitable refractive index),

b. thickness of the AR coating,

c. surface characteristics of the cell,

d. surface characteristics of the module glass.

Ideally, the PV module designer would like to reduce reflections at all the
interfaces and the absorption in all the media to zero. However, this cannot
be accomplished in practice; but it is possible to design the module optics so
as to reduce the optical reflections to zero in small ranges of wavelength and
thereby readjust the total average reflectance. However, such a reduction, in
general, can enhance reflectance in other ranges of the spectrum. This is
essentially where the optical design of the module can have the most influence
on the cell performance. In order to accomplish these features a good analyt-

ical model of a PV module, which can be based on easily measurable parameters, is required. This model can then be applied to a cell of known spectral response and the reflectance (measured in air) to determine short circuit current density for various operational parameters. Such a procedure can lead to conditions for optimized light coupling to a cell for module operation.

The optimization of light coupling requires a variety of optical processes that occur in a module to be considered. These include reflections at interfaces, including scattering due to surface roughness of the cell, absorption in various media, contributions to the cell current from the light trapped due to scattering by cell metallization and back-skin. Additionally, it is necessary to have accurate values of the optical parameters of various module materials.

In this paper we first introduce an optical model of a PV module suitable for such calculations followed by a brief description of procedures to measure accurately the optical parameters. Finally, a discussion of how to use this model for optimization of light coupling is given. Additionally, some experimental results, illustrating changes in cell current due to encapsulation as a function of a major coupling parameter, are included.

Optical Model of a Cell in a PV Module

In this section we will describe an optical model which is applicable not only to planar interfaces but also to rough surfaces such as obtained by chemical etching and/or texturing often used for solar cells. Figure 1 illustrates the main features of this model. A photon flux of intensity spectrum I_o (λ) is incident on the module at the air-glass interface. A part of this light is reflected at various interfaces. Some is absorbed within various media and the rest is transmitted into the cell as a result of a number of complex processes. We define R_{ij} (λ) as the reflection coefficient for the light going from medium i to medium j where i = 1,2,3,4 and j = 1,2,3,4 refer to the various media of the module as shown in Figure 1. Correspondingly, α_n (λ) refers to the absorption coefficient of medium n where n = 1,2,3,4. In a module there is generally a gap between the cells; the light reflected from this region and from the cell (including from the metal) suffer multi-reflections within the module and will add to the cell flux. These contributions are represented by R_{inv}.

From the model of Figure 1 the intensity transmission coefficient, representing the fractional flux that is transmitted into the cell, is given by

$$T (\lambda) = (1 - R_{10}) (1 - R_{21}) (1 - R_{2cell}) \exp [(- \alpha_1 t_1 + \alpha_2 t_2)] \quad (1)$$

and the net reflection at the glass-air interface

$$R (\lambda) = R_{10} + (1 - R_{inv}) (1 - R_{10}) R_{2cell} \exp [-2(\alpha_1 t_1 + \alpha_2 t_2)] \quad (2)$$

where R_{2cell} is the reflection coefficient of the cell in the encapsulant medium.

The reflection coefficients $R_{ij} = |r_{ij}|^2$ (where i,j = 0,1,2) are related to the indices of refraction in the media i and j by well known Fresnel expressions. It should be pointed out that even though Fresnel relations are strictly valid for planar interfaces and that, in general, one side of the glass used for encapsulation has textured surface, this approximation is still valid. The reason for its validity is that the textured surface faces the encapsulant material which has a refractive index very close to that of glass.

In equations (1) and (2), all parameters except R_{2cell} and R_{inv} are known. However, R_{2cell}, reflection coefficient of the cell in the encapsulant is quite involved since it cannot be measured directly. We will now describe the procedures to calculate R_{2cell} and R_{inv} from measurements that can easily be made.

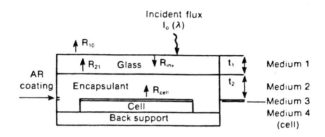

Figure 1. Optical model of a photovoltaic module

Calculation of R_{2cell}

R_{2cell} is the power reflection coefficient of the AR coated cell in the encapsulant medium. Notice that for planar interfaces R_{2cell} can be calculated in a straight forward manner from known data of silicon parameters, AR coating, and the pottant material. However, solar cell surfaces generally are rough and/or textured. Consequently, a planar approximation is not valid. Our approach to circumvent this problem is to measure the reflection coefficient of the AR coated cell in air and use this information to calculate the reflection coefficient in the pottant material. Notice that such a procedure takes into account the surface scattering characteristics of the silicon surface.

In the most practical cases the surface characteristics fit into category of small surface roughness. In this approximation we can write down the expression for R_{2cell} as

$$R_{2cell} = \frac{r^2_{32} + r^2_{43} + 2r_{32}\,r_{43}\,\cos\phi}{1 + r^2_{32}\cdot r^2_{43} + 2r_{32}\,r_{43}\,\cos\phi} \tag{3}$$

where r_{43} is the calculated reflection coefficient of the silicon surface, embedded in medium 3, and can be determined from the reflectance (R) of the AR coated cell measured in the air.

$$r_{43} = \frac{-2r(R-1)\cos\phi \pm \left[4r^2\cos^2\phi\,(R-1)^2 - 4(Rr^2-1)(R-r^2)\right]^{1/2}}{2(R\,r^2 - 1)} \tag{4}$$

R = measured reflection coefficient of the cell in air

$r = (1-n_3)/(1+n_3)$, $\phi = 4 \pi \cdot n_3 t_3/ \lambda$

In equation (4) the positive and negative signs are used when $r^2 < R$ and $r^2 > R$ respectively. This choice of the appropriate roots of equation (4) is dictated by the fact that r_{43} should be continuous and, in the case of AR coated cells, r_{43} should also be negative.

Determination of R_{inv}

R_{inv} depends not only on the optical parameters of the various media but also on the characteristics of the metallization pattern of the cell and the gap between individual cells. We have developed a procedure to determine R_{inv} by measuring the response of the encapsulated cell to a scanning light beam. Using this technique, it has been determined that about 20% of the light incident on the metal and/or back skin is reflected back to the cell via scattering and trapping effects.

Equations developed in this section can be applied to determine the photon flux transmitted into the cell for a given incident flux at the air-glass interface. It may be recognized that some of the design parameters may be easily estimated and then introduced into the model for fine tuning. These include reflections at air-glass interface, for consideration of anti-reflection coating on glass, and at the glass-pottant interface. The validity of these approximations is based on the fact that the media across the interfaces are low index semi-infinite media.

Measurement of Some Optical Parameters

The characteristics of the module materials viz glasses and pottants, play an important role in the overall performance of a cell in the module. There is, however, relatively little information available regarding their optical properties, presumably due to the fact that they are primarily aimed at applications which do not require detailed information of the optical properties (for example, for windows, wind shields, etc.). Additionally, it is difficult to make measurements from which the optical constants can be determined accurately.

In this section, we will describe methods that have been developed to make accurate measurements of optical properties of commonly used module materials i.e. glasses, pottants, and back-skin. The properties measured are: refractive index, reflection coefficient, absorption coefficient and optical scattering characteristics which are important to determine contributions of back-scattered light on the cell performance.

Optically flat surfaces of glasses and pottants were produced to measure specular reflection from the air-sample interface. In order to accomplish this, glass samples were polished in a way that leaves no residual stress at the surface. The pottant samples of uniform thickness were prepared as a sandwich between a polished NaCl window and an optically-matched back support. After pottant curing process the NaCl window was dissolved in water and the sample was dehydrated, leaving one surface of the pottant with an optical finish.

The reflectance and transmission measurements were made in a Cary 17 spectrometer fitted with a reflectance accessory. Diffuse reflectance measurements were made with an integrating sphere. During the reflectance measurements the back reflections were suppressed and a silicon sample of well known reflection coefficient was used as a reference sample.

Absorption measurements of glass were done as follows: The glass samples were polished on both sides and the transmission coefficient measured. The absorption coefficient was determined by the following relation.

$$\alpha (\lambda) = \frac{1}{t} \left\{ \ln \left[1-R(\lambda) \right]^2 / T(\lambda) \right\} \qquad (5)$$

where $\alpha (\lambda)$ and $R(\lambda)$ are the absorption and reflection coefficients respectively at wavelength λ, t is the thickness and $T(\lambda)$ the transmission coefficient. This data was supplemented by measuring transmission coefficients for two different thicknesses and determining $\alpha (\lambda)$ by

$$\alpha = \ln(T_1/T_2)/(t_2 - t_1) \qquad (6)$$

The absorption coefficients of the pottants were determined by the following procedure. The samples consisting of glass/pottant/glass sandwich configuration were prepared with a highly uniform thickness of the pottant. The glasses were selected to match the refractive indices of pottants. The absorption coefficient was determined from equation (6) from the transmission data of samples of different thickness of the same pottant.

The refractive indices were determined from the reflectance data as well as using Brewster angle method (at 6328 Å wavelength).

Typical Results

Table 1 shows refractive indices of different materials obtained from reflectance data and by Brewster angle method at 6328 Å HeNe laser wavelength with a comparison of the published data. Figure 2 shows a comparison of the reflection coefficients of Sunadex and AR coated Solite; also shown in this figure is the reflection coefficient of PVB. Figure 3 shows diffuse reflectance of white Tedlar as a function of wavelength. Figure 4 shows the absorption coefficient of PVB as a function of wavelength.

Similar measurements are made with the following materials.

Glasses: Sunadex, Solite, AR Coated Solite
           chemically tempered glass, Quartz
Pottants: PVB, EVA, SSP
Tedlar

Application of these results to material selection for photovoltaic modules will be discussed in detail elsewhere.

Discussion of the Design Features

As described earlier, the major objective in the module design is to optimize the optical flux transmitted into the cell in a manner which will maximize the short circuit current. Alternately, this requires

468

$$\int_{AM1.5} \frac{SR(\lambda)}{[1 - R(\lambda)]} \cdot T_e(\lambda) \, d\lambda = \text{maximum} \qquad (7)$$

where SR (λ) and R(λ) are the spectral response and the reflectance of the unencapsulated cell respectively (measured in air) and T_e (λ) is the fractional flux transmitted into the encapsulated cell, as given by expression in (1). Notice that the factors which contribute to T_e (λ) are: reflectance at the air-glass interface, absorption in glass and pottant, reflectance at the glass-pottant interface, and reflectance at the pottant cell interface. Additional factor that contributes to T_e (λ) is R_{inv}. These factors are further discussed below.

Table 1. Refractive Index Data

Material	This Data	Previously Published
Sunadex Glass	1.509*/1.515	1.519
Chemically Tempered Glass	1.511*/1.523	-
PVB	1.4826*/1.492	1.4761
EVA	1.476*	1.482
SSP	1.5165*	1.516

*Measured at 6328 Å by Brewster angle.

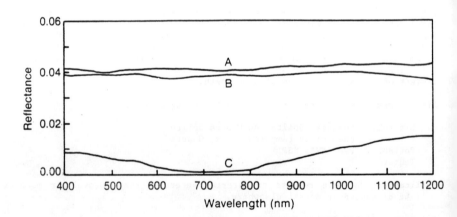

Figure 2. Dependence of the reflection coefficient on wavelength for (A) Sunadex (B) PVB and (C) AR Coated Solite

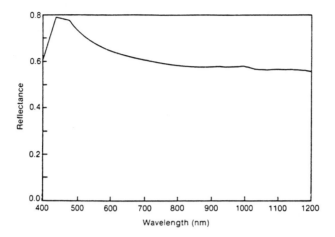

Figure 3. Diffuse reflectance of Tedlar as a function of wavelength

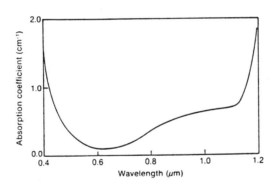

Figure 4. Measured dependence of the absorption coefficient on
wavelength for PVB

To minimize the loss at the air-glass interface, it is necessary to form an AR
coating on the glass surface. A comparison of the reflectance characteristics
of an uncoated module glass (Sunadex) and an AR coated glass (Solite) is shown
in figure 2. The coating on the Solite was a proprietory experimental film.
This figure shows that a reduction in the average reflectance from 0.04 to
0.01 can be obtained by a suitable AR coating. There is considerable research
being done to form an AR coating layer at the glass surface. Several techniques are being tried which include: (i) vacuum deposition of low index
layers such as MgF_2. This is a standard technique for AR coating optical
components; however, due to its high cost this approach is not suitable for
terrestrial modules but is used for concentrator type modules, (ii) acid
leaching, used on soda-lime type of glasses, creates a lower refractive index

film by depleting the surface of alkali-metal radicals, (iii) formation of a porous glass film by dip coating (followed by a low temperature densification) to create a surface layer of low "effective" refractive index, and (iv) ion exchange. To date, none of the approaches for antireflection layer formation have proven to be satisfactory for commercial use. There are questions as to the durability of the surface film, the suitability of these techniques to be applied to low impurity (low absorption) glasses, and the strength of the glasses on which these processes can work to form an antireflection layer.

The refractive index and absorption of light in glass is controlled by impurities in the glass. Impurities also control the mechanical strength of the glass. Glasses currently used for PV modules have been previously developed for other applications and are often tempered to give the glass an increased strength as well as safety features (e.g., these glasses shatter and collapse on a powerful impact). Typical low iron glasses can have total transmittance ~92% in air over most of the useful spectrum. When one side of glass is bonded to a pottant, there is a close optical match between glass and the pottant (see figure 2 for reflectance of glass and PVB) resulting in a small reflection at the glass-pottant interface. Consequently, the total transmittance into the pottant can be quite high — nearly 96% for uncoated and about 99% for AR coated glass. Obviously SSP is better matched for glass than PVB; however, the improvement is very little and the choice between PVB, EVA, and SSP would be dictated by other considerations.

The absorption losses in the pottants are primarily caused by the near IR absorption bands whereas near UV absorption is introduced by absorption tail of UV inhibitors. Additional loss in the pottant can come from inappropriate processing which leads to formation of air bubbles and tubidity of the pottant.

The reflection at the pottant-cell interface can become quite important and can be minimized by an appropriate choice of an AR coating. Although equation (4) for R_{43} (λ) can be used to determine the minimum reflectance, it is important to recognize that minimizing the average reflectance does not necessarily represent the best-match condition as determined by maximizing J_{sc} in equation (7). This procedure is partially illustrated by an example: we start with spectral response of the cell measured in air. Figure 5 shows the SR of a textured polycrystalline silicon solar cell which will be used in this example. This cell had an AR coating consisting of 950 Å of Si_3N_4. Also shown in figure 5 is the reflectance of the cell measured in air. The SR of the cell after encapsulation is determined by equations 1-5 and is also shown in figure 5. Notice that the SR after encapsulation shows an increase in the entire wavelength range. The parameters used in calculation of the SR of encapsulated cell correspond to:

Glass: Sunadex, t_1 = 0.3 cm
Pottant: PVB, t_2 = 0.1 cm
R_{inv} =   0.2 × (fractional area of metal
    and spacing between the cells in
    the final module)
Shadow fraction due to metal = 14%
Cell spacing = 200 mils
t_3 = 950 Å

471

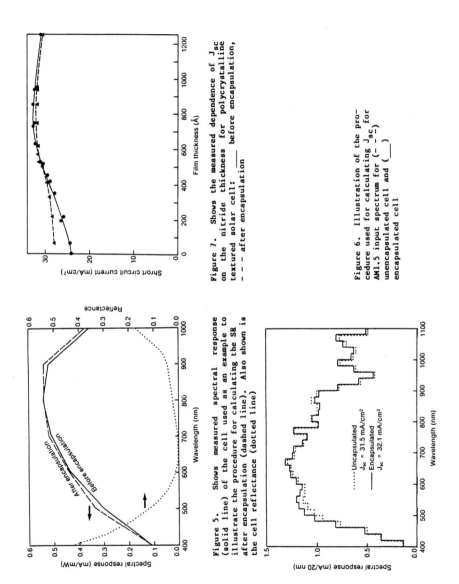

Figure 7. Shows the measured dependence of J_{sc} on the nitride thickness for polycrystalline textured solar cell: ——— before encapsulation, – – – after encapsulation

Figure 6. Illustration of the procedure used for calculating J_{sc} for AM1.5 input spectrum for (– . –) unencapsulated cell and (___) encapsulated cell

Figure 5. Shows measured spectral response (solid line) of the cell used as an example to illustrate the procedure for calculating the SR after encapsulation (dashed line). Also shown is the cell reflectance (dotted line)

472

These responses can now be integrated with respect to the AM1.5 spectrum, as shown in figure 6, to give short circuit currents of 31.5 mA/cm^2 and 32.16 mA/cm^2 for unencapsulated an encapsulated cells respectively. One can use this procedure to calculate J_{sc} for various AR coating thicknesses [by changing $\emptyset = 4\pi \, n_3 t_{3/\lambda}$ in equation (3)] and determine the value of t_3 for maximum J_{sc}. It should be emphasized that the increase in the J_{sc} upon encapsulation comes about because of the particular cell design; the optical losses after encapsulation, due to added reflection at the air-glass interface and absorption in glass and pottant, are over compensated by the light trapping effect and reduction in reflection at the cell surface. Clearly this cannot always be the case. Figure 7 shows measured values of J_{sc} as a function of AR coating thickness before and after encapsulation for cells fabricated on commercially available polycrystalline silicon (4 inch × 4 inch) substrates.

The cells were fabricated on textured polycrystalline p-type substrates using PH_3 diffusion for junction formation. The AR coating consisted of $Si_3N_4/(100 \text{ Å}) \, SiO_2$; the thickness of nitride film was changed for different batches of cells. The metallization consisted of electroless Ni followed by solder coating, and the metal fraction was intentionally kept high at 14%. The encapsulation was done using 5 inch × 5 inch glass/PVB/cell/tedlar configuration.

Conclusion

In this paper we have described various optical processes that need to be considered for optimizing cell performance in a module. The optimum coupling is obtained not by minimizing the average reflectance but by maximizing the integrated spectral response weighted by the incident spectrum. It is shown that the J_{sc} of a cell can be higher after encapsulation compared to that measured in air (before encapsulation). Such a situation occurs whenever added optical loss due to reflection at air-glass interface and absorption in the glass and the pottant are less than the gain in optical coupling due to (i) reduced reflection at pottant-cell interface and (ii) collection of a part light scattered by the metal and back-skin of the cell (light-trapping).

The paper has outlined a procedure to determine an optimum antireflection coating based on an optical model which is applicable to polished as well as rough surfaces. Furthermore, we have presented experimental results showing the dependence of cell performance on the thickness of a single-layer AR coating, before and after encapsulation.

Bibliography

1.   M. Born and E. M. Wolf, Principles of Optics, Pergammon, Oxford, 1979.

2.   F. A. Jenkin and H. E. White, Fundamentals of Optics, McGraw Hill, New York, 1957.

3.   B. L. Sopori and R. A. Pryor, Solar Cells, 8, 249 (1983).

4.   R. Shiffer, Appl. Optics, 26, p. 704 (1987).

5.   B. L. Sopori, "Optical Evaluation of Some Glasses and Encapsulants Used in Terrestrial Photovoltaic Modules," to be published.

6.   Standard for Solar Spectral irradiance tables at air mass 1.5 for 37° tilted surface, ASTM Stand. E892-82.

7.   B. L. Sopori, "A Spectral Response Measurement System for Large-area Solar Cells," Solar Cells, to be published.

EFFECT OF ANNEALING ON ELECTRICAL ACTIVITY OF
SURFACES IN SILICON

Rajesh Kumar, R.K. Kotnala, N.K. Arora and B.K. Das
Materials Division
National Physical Laboratory,
New Delhi-110012
INDIA

ABSTRACT

Using the electron beam induced
current mode of a scanning electron
microscope, electrical activity of
the grain boundary in oxygen rich
poly-silicon has been studied as a
function of heating time and temperature. The minority carrier surface
recombination velocity at the grain
boundary has been found to increase
significantly with heat treatment
due to segregation of oxygen to the
grain boundaries from the intragrain
region. A theoretical model has been
proposed whereby kinetics of oxygen
diffusion to the grain boundary can
be studied and diffusivity of oxygen
in silicon can be obtained. The
intragrain region in poly-silicon
containing fast diffusing elements
such as Cu, Fe and Ni shows enhanced
electrical activity at the surface
due to heat treatment. Minority
carrier trap center density at the
surface, calculated from minority
carrier surface recombination
velocity data, varying from 8.9×10^{11}
to 6.37×10^{13} cm^{-2} have been measured. A finite variation in the
minority carrier trap center density
indicates that metallic impurities
diffuse to the surface from the
bulk of the sample. The activation
energy of impurity atom diffusion is
found to be 1.1 ± 0.1 eV which is
very close to the activation energy
of diffusion of Cu in silicon.

Key words: Silicon, Heat Treatment, Impurity
Redistribution, Electrical Activity

INTRODUCTION

The technological importance of polycrystalline silicon has increased considerably in
recent years due to its application in very
large scale integeration (VLSI), bipolar and
metal oxide semiconductor (MOS) integerated
circuits such as silicon gates, isolation layers,
interconnections, varistors, etc and its potential for low cost alternative to single crystal
for terrestrial solar cells. For an effective
utilization of poly silicon in these devices, a
clearer understanding of its electrical and
electronic properties is essential. The electrical properties of grain boundaries show drastic modifications on heat treatment.

Jain et al[1] established on basis of photo
conductivity studies that the grain boundary
activity of silicon increases by annealing. It
was also observed by Kumari, Jain, Das and
Jain[2] that defects present in poly silicon are
rich in elements like C, O, Fe, Cr, etc. Based
on extensive study with the help of techniques
like EBIC, AES and EELS, Kazmerski[3] concluded
that (i) there is increased electrical activity
at the GB's of polysilicon on heating above $400^{O}C$
(ii) the increased electrical activity is due to
segregation of oxygen to the GB's. and (iii) the
passivation of GB's by the hydrogen is due to a
reduction of activity of oxygen at the GB's.
Courcelles et al[4] using the DLTS technique observed the growth of a level at E_{v} + 0.33 eV and
argued that this is due to formation of a vacancy-
carbon-oxygen complex. Sastry[5] in a recent study
using the DLTS technique has shown that Fe, Ti
and Cr segregate towards the grain boundary. A

detailed study of the migration of various impurities present inside silicon only can explain the change in the photovoltaic properties of such materials during thermal processing.

OXYGEN DIFFUSION TOWARDS THE GRAIN BOUNDARY IN OXYGEN RICH POLY-SILICON

Oxygen is incorporated in Si at the slightly off-centre interstitial position between two neighbouring atoms. Due to thermal treatment, oxygen diffuses from the intragrain region to the grain boundary as a result of which minority carrier recombination velocity at the grain boundary increases[3]. The point defects in silicon have little effect on oxygen diffusion and oxygen diffuse by hopping between interstitial sites.

Consider an oxygen rich silicon bicrystal specimen with its intragrain region free from defects such as stacking faults, dislocations and clusters. If such a specimen is heat treated at temperature T for time t, the oxygen diffuses from the intragrain region to the grain boundary as reported by Kazmerski[3]. The oxygen concentration $C(x,t)$ at a distance x from the grain boundary in the intragrain region is given by

$$C(x,t) = (C_o - C_T) \; \text{erf} \; (\frac{x}{2\sqrt{Dt}}) \qquad (1)$$

where C_o is the intragrain oxygen concentration prior to heat treatment, C_T is the solid solubility of oxygen is silicon at temperature T, D is the oxygen diffusivity and t is the annealing time. The diffusion of oxygen to the grain boundary increases the oxygen concentration $C_g(T,t)$

at the grain boundary after annealing. This may be written as

$$C_g(T,t) = (C_o - C_T) \int_0^\infty \text{erfc}\left(\frac{x}{2\sqrt{Dt}}\right) dx + C_g \qquad (2)$$

In eq. (2) C_g is the grain boundary oxygen concentration prior to heat treatment. Further substituting $y = x/2\sqrt{Dt}$ in equation (2), it reduces to

$$C_g(T,t) = (C_o - C_T)\, 2\sqrt{Dt} \int_0^\infty \text{erfc}(y)\, dy + C_g \qquad (3)$$

or

$$C_g(T,t) = (C_o - C_T)\, 2\sqrt{Dt}/\sqrt{\pi} + C_g \qquad (4)$$

Equation (4) shows that $(C_o - C_T)\, 2\sqrt{Dt}/\sqrt{\pi}$ is the increase in the oxygen concentration at the grain boundary after annealing at temperature T for time t. Oxygen atoms diffused to the grain boundary increases trap density at the grain boundary. Assuming the trap density at the grain boundary, N_{is}, to be proportional to the concentration of oxygen atoms diffused to the grain boundary, N_{is} (T,t) as a function of heating time and temperature can be expressed as

$$N_{is}(T,t) = n\,(C_o - C_T)\, 2\sqrt{Dt}/\sqrt{\pi} + N_{is}^o \qquad (5)$$

In equation (5), N_{is}^o is the grain boundary interface state density prior to heating, n is the number of interface states created by one oxygen atom at the grain boundary. If the grain boundary is not electrically active before heating, then N_{is}^o can be neglected in Eq. (5) without any significant error, hence

$$N_{is}(T,t) = (C_o - C_T)\, 2\sqrt{Dt}/\sqrt{\pi} \qquad (6)$$

The slope of the plot of $\left[N_{is}(T,t)/(C_o-C_T)\right]^2$ vs t will give the diffusivity D and the Arrhenius plot of D will give the activation energy of diffusion of oxygen in silicon.

The interface trap density (N_{is}) at the grain boundary may be calculated from grain boundary minority carrier recombination velocity (S_{gb}) and barrier potential (V_d) measurement. Qualid et al[6], have shown that for small grain boundary barrier potential and low level injection conditions.

$$S_{gb} = \frac{1}{2}\sigma V_{th}\, N_{is}\, \exp\left(\frac{qV_d}{kT}\right) \tag{7}$$

where σ is the minority carrier capture cross-section, V_{th} is the thermal velocity and k is the Boltzman's constant, q is the electronic charge and V_d is the grain boundary potential barrier. S_{gb} can be determined by analysing the electron beam induced current line scan profile of the grain boundary. The grain boundary barrier height can be obtained from zero bias high frequency capacitance measurement using the relation[7].

$$V_d = \frac{q\, N_D \epsilon \epsilon_0\, A^2}{8\, C^2} \tag{8}$$

where ϵ is the relative permitivity of the silicon, ϵ_0 is the permittivity of free space, N_D is the donor concentration in the surrounding grain, A is the area of grain boundary capacitor and C is the zero bias high frequency capacitance of the grain boundary.

DETERMINATION OF DIFFUSIVITY OF OXYGEN IN SILICON

The electron beam induced current mode of
a scanning electron microscope (Cambridge S-600)
was used to measure the minority carrier surface
recombination velocity at the depletion layer
edge of the grain boundary as a function of heating time (30 to 240 min) and temperature
(600 - 900°C) in oxygen rich directionally cast
silicon bicrystal (7-10 Ω-cm , N-type). For
EBIC analysis, Schottky barrier was made on one
side of highly polished silicon surface. Schottky
barrier was formed between thermally evaporated
chromium metal layer (30 nm thick) onto a highly
polished silicon surface. The other side of the
sample was plated with nickel for making an ohmic
contact. The advantage of Schottky barrier over
p-n junction formation is to avoid any heat treatment of the sample so that intrinsic characteristics of the material are not disturbed. The minority carrier recombination velocity at the
depletion layer edge of the grain boundary was
determined from the EBIC profile of the grain
boundary using the method suggested by Burk[8].
A 25 KeV electron beam was used for all the EBIC
work. Once EBIC measurements were over, the
chromium dot was etched away from the silicon
bicrystal and grain boundary capacitance was
measured by making nickel contacts on either side
of the grain boundary. The zero bias high frequency (1MHz) capacitance measurements were carried out at room temperature using a Boonton Capacitance Meter (Model 72 BD). The resistivity of
the sample was determined by the four point probe
technique. The value of N_D was calculated from

the resistivity data. The oxygen concentration
in the intragrain region of the samples was
determined by the Fourier Transform Infrared
Spectroscopy (FTIR) of silicon samples at 9 μm
and found to be 1.7 x 10^{17} atoms/cm^3. The different values of solid solubility of oxygen C_T,
at different temperature were taken from the
solubility curve of oxygen in silicon reported
by R.A. Craven[9].

Figure 1 shows the variation of S_{gb} with
t and T. It should be noted from Fig.1 that
the rate of change of S_{gb} with t is increasing
with increasing T. This variation in the rate
of change of S_{gb} with t for different T is due
to dependence of D upon T. Fig.2 shows the
variation of V_d with t and T. V_d increases
slightly with increasing T but the variation of
V_d with t at constant T is not appreciable.
Plots of $\left[N_{is} (T,t)/(C_o-C_T) \right]^2$ vs t at various
heating temperatures has been found to be
straight lines. Diffusivity of oxygen at various
temperatures were obtained from slope of such
straight lines. The Arrhenius plot of D, for
determining the activation energy of diffusion
of oxygen in silicon is shown in Fig.3. The
computed value of diffusivity of oxygen in
silicon in the temperature range (600-900°C) is
$$D=0.27 \exp \left[- \frac{(2.56 \pm 0.06)\ eV}{kT} \right] cm^2\ sec^{-1},$$
which is in good agrement with other measurements[10,11].
These results show that S_{gb} and V_d increase with
increasing t and resulting in change in $N_{is}(T,t)$
and this change in $N_{is}(T,t)$ correlates with
oxygen diffusion to the grain boundary from intragrain region as a result of annealing.

MIGRATION OF IMPURITIES TO THE SURFACE DURING HEAT TREATMENT

Cu, Fe and Ni create deep levels in silicon and act as very efficient recombination centers and adversely affect minority carrier lifetime. Davis Jr. et al[12] have studied the deteriorating effects of these impurities on the performance of silicon solar cells. It has been observed using electron beam induced current mode of a scanning electron microscope that surface recombination velocity in a region far from the influene of the grain boundary increases with heat treatment in poly-silicon containing fast diffusing elements such as Cu, Fe and Ni. For all the EBIC measurements the samples (Schottky diodes) were mounted on header inside a Cambridge S-600 SEM in such a way that the incident electron beam was normal to the charge collecting barrier. Minority carrier surface recombination velocity was determined by measuring the diffusion length of minority carriers at different accelerating voltages for electron beam[13]. A method suggested by Ioannou and Dimitriadis[14] was used for determination of the minority carrier diffusion length. Fig.4 shows the variation of minority carrier surface recombination velocity with the heating time and temperature. The minority carrier trap density at the surface was determined from surface recombination velocity using the relationship

$$S = \sigma V_{th} \, N$$

Where S = minority carrier surface recombination velocity

       σ = minority carrier capture cross-section

       V_{th} = minority carrier thermal velocity

       N = minority carrier trap centre density

The values of trap center density were calculated which ranged from 8.9×10^{11} to 6.37×10^{13} cm^{-2}. The increase in the minority carrier trap center density at the surface of the sample after giving heat treatment reveals increased carrier recombination process at the surface of the sample. The change in the minority carrier recombination activity of the surface after heat treatment may be due to the segregation of fast diffusing metallic impurities towards the surface from the bulk of the sample. Further assuming that one impurity atom after diffusing towards the surface creates one minority carrier trap center, the rate of change of minority carrier trap center density with heating will give the rate of impurity atoms segregating towards the surface.

Trap center density has been found to increase linearly with heating time at each annealing temperature. The slope of each line represents the rate of change of minority carrier trap centre density, $\frac{dN}{dt}$ with heating time at each temperature. The Arrhenius plot of $\frac{dN}{dt}$ shown in Figure 5 gives an activation energy of 1.1 ± 0.1eV, which is the activation energy of impurities diffusing to the surface. This is very close to the activation energy of diffusion of Cu in silicon Copper, iron and nickel are fast diffusers in silicon. The diffusivity of copper in the temperature range 800-1100°C is controlled by impurity point defect interactions and is higher than iron. Nickel has a higher diffusivity compared to copper and iron. However after diffusion, nickel gets precipitated and only 0.1% remains electrically active. Thus it is possible that more copper atoms diffuse to the surface than iron and nickel.

Hence copper atoms play a vital role in creating
minority carrier trap centers on the surface
of silicon due to heat treatment.

REFERENCES

1. G.C. JAIN, B.C. CHAKRAVARTY and A.
    PRASAD, J. Appl. Phys., 52 (1981) 3700.

2. S. KUMARI, S.K. JAIN, B.K. DAS and
    G.C. JAIN, Solar Cells, 9 (1983) 209.

3. L.L. KAZMERSKI, Proc. of 5th EC Photo-
    voltaic Solar Energy Conf., 1983 p.40.

4. E. COURCELLE, M. MESLI, J.C. MUELLER,
    D. SALLES and P. SIFFERT, Proc. of 16th
    IEEE Photovoltaic Specialists Conf.,
    1982 p. 417.

5. O.S. SASTRY, Ph.D. Thesis, Indian
    Institute of Technology, New Delhi,
    1984.

6. J. OUALID, C.M. SINGAL, J. DUGAS, J.P.
    CREST and H. AMZIL, J. Appl. Phys., 55
    (1984) 1195.

7. G.E. PIKE and C.H. SEAGER, Adv. Ceram.,
    1 (1981) 53.

8. D.E. BURK, IEEE Trans. on Electron
    Devices, ED-29 (1982) 1887.

9. R.A. CRAVEN, in Semiconductor Silicon
    1981, edited by H.R. HUFF, R.J. KRIEGLER
    and Y. TAKEISHI (Electrochemical
    Society, Princeton, 1981) p 254.

10. C. HASS, J. Phys. Chem. Solids, 15,
    (1960) 108.

11. J.W. CORBETT, R.S. MCDONALD and G.D.
    WATKINS, J. Phys. Chem. Solids, 25
    (1964) 873.

12. J.R. DAVIS, JR., A. ROHATGI, R.H. HOPKINS,
    P.D. BLAIS, P. RAI CHOUDHARY, JR
    McCORMICK and H.C. MOLLENKOPF, IEEE Trans.
    on Electron Devices, ED-27 (1980) 677.

13. L. JASTRZEBSKI, J. LAGOWSKI and H.C.
    GATOS, Appl. Phys. Lett., 27 (1975) 537.

14.   D.E. IOANNOU and C.A. DIMITRIADIS, IEEE
      Trans. on Electron Devices, ED-29 (1982)
      445.

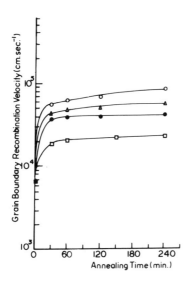

Fig. 1 - Variation of grain boundary recombination velocity with heating time and
temperature, □ 600°C, ● 700°C, △ 800°C,
○ 900°C.

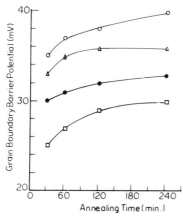

Fig.2 - Dependence of grain boundary barrier
potential on heating time and temperature, □ 600°C, ● 700°C, △800°C, ○
900°C

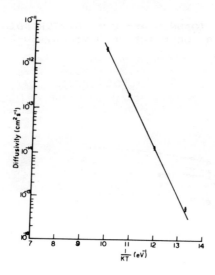

Fig.3 - Arrhenius plot of diffusivity as
a function of reciprocal of heating
temperature.

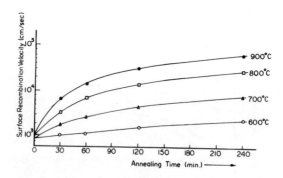

Fig.4 - Variation of minority carrier surface recombination velocity in an
intragrain region with heating time
for different heating temperatures.

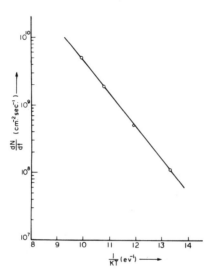

Fig.5 - Arrhenius plot of $\frac{dN}{dt}$ as a function of the reciprocal of heating temperature.

Physics and Technology of Thin Film Polycrystalline Solar Cells

Pratul K. Ajmera
Department of Electrical and Computer Engineering
Louisiana State University
Baton Rouge, LA   70803,   USA
(504) 388-5620

ABSTRACT

The status of the technology of thin film polycrystalline solar cells in the USA along with the physics of their operation are reviewed. The two principal systems based on $CuInSe_2$ and CdTe materials are discussed. The major problems faced by each of the above two solar cell technologies are described. The status of two-junction tandum cells is also discussed.

1. Introduction

Every fifteen minutes, the sun delivers enough radiant energy to the earth to meet mankind's power needs for a full year. The main goal of photovoltaics for large scale terrestial applications is to harness this energy efficiently in a clean, convenient way at an affordable market cost. Among a variety of alternatives to renewable energy sources such as solar-thermal, solar-wind, solar ponds and energy from sea-waves; the photovoltaic approach is the most direct. It involves no moving parts to wear out (other than any slow moving sun tracking mechanism, if used), no intermediary medium (such as the working fluid used in solar thermal and certain other technologies), and no noise or chemical pollution during its normal operation.

Moreover, photovoltaic technology has two advantages that are quite unique. First, the employment of photovoltaic technology is relatively safe and non-obtrusive to the surroundings in which it is employed whether that be a roof top of a house, a hand-held calculator or a central power station. The second advantage stems from the fact that photovoltaic technology is truly modular in that it can be used for generation of electric power from microwatts level needed to power a wrist watch to tens or hundreds of megawatts level needed for a power generating station or to any other level in between. A major drawback of the terrestial photovoltaic technology is that it can produce electricity only when the sun is shining thereby needing a capacity to store energy for night-time usage. Even then, the advantages mentioned above make photovoltaics attractive in the near term energy picture for a large variety of applications.

The terrestial photovoltaic technology is at an emerging state. There are a number of competing technologies such as 1) Single crystal silicon, 2) Amorphous silicon, 3) Polycrystalline thin films, 4) High-efficiency concepts in compound semiconductors and 5) Photoelectro-chemical cells. Each of these competing technologies has its own characteristic advantages and shortcomings. Currently, no single technology dominates the field. In this paper, status of the poly-crystalline thin film solar cells is described.

The research and development effort in the U.S.A. is guided by the U.S. Department of Energy Five-Year Research Plan targets. In order to be competitive with electricity costing about 5 ¢/kWh, the module cost should be below $30/m^2 for 15% efficient flat plate modules and below $50/m^2 for 20% efficient flat plate modules [1]. The former may be achievable in future by a single junction thin film module. Currently, there are two polycrystalline thin film systems that show promise in meeting these future target goals. These are based on semiconductor materials 1) CuInSe$_2$ and 2) CdTe. The status of solar cells from each of these two materials will be described in detail in sections 3 and 4.

2. Polycrystalline Thin Film Solar Cells

The basic structure of a polycrystalline thin film solar cell consists of a thin film polycrystalline "absorber" material with appropriate energy gap and a high value for absorption coefficient for the incident light. The thickness of the absorber typically varies from a fraction of a micrometer to a few micrometers. Generally, the absorber is covered with a "window" material that is a wide band gap polycrystalline semiconductor largely transparent to the incident radiation and is made highly conductive to reduce series resistance losses. This "window-absorber" configuration is the most commonly employed configuration forming a heterostructure cell even though other structures such as a homojunction, Schottky barrier and MIS configurations have also been fabricated in the polycrystalline thin film technology. The defect chemistry of the absorber and its interface are frequently not well understood which precludes formation of reliable homojunctions. Moreover, heterostructure interfaces may be conveniently located to avoid inevitable adverse front surface losses present in shallow homojunctions. A brief report of the polycrystalline thin film technology status is given in reference [1]. The progress in the field during the past two to three years is contained in the papers given in references [2-4]. The highlights of these results along with current problems are described below.

3. CuInSe$_2$ Cells

The ternary CuInSe$_2$ material is a member of the I-III-VI$_2$ chalcopyrite family with a direct energy gap of 1.01 eV. It has a high absorption coefficient approaching a miximum value of 6 x 10^4 cm^{-1} even though this value is about an order of magnitude smaller than the one previously reported [5]. The basic configuration is a CdS(n)/CuInSe$_2$(p) heterostructure. CdS with an energy gap of 2.42 eV acts as a window for the CuInSe$_2$ absorber. Molybdenum is almost exclusively used as a contact for the CnInSe$_2$ layer. A variety of fabrication techniques have been employed to improve upon this basic structure.

The "Boeing" structure is shown in Fig. 1 [6]. Boeing uses elemental vacuum evaporation of Cu, In and Se to deposit CuInSe$_2$. Alumina or glass is used as the substrate. Sputtered molybdenum provides the back contact to the CuInSe$_2$. The resistivity and type of CuInSe$_2$ obtained depends on the Cu:In flux ratio [7]. The near stoichiometric films are p-type as are the Cu-rich films. The In-rich films tend to be n-type. In the Boeing cell, the bottom layer is stoichiometric or slightly Cu-rich making the film more conductive and p-type. The high resistivity top $CuInSe_2$ layer is Cu-deficient. The two layers inter-mix during processing leaving only a very thin high resistive $CuInSe_2$ layer near the heterostructure interface. The Cu:In flux ratio and the substrate temperature play a crucial role in determining the film properties. The window material used is typically n-type CdS also vacuum deposited in two steps. A high resistivity layer followed by a In-doped low-resistivity layer to reduce series resistance and to facilitate ohmic contact formation. Boeing uses coevaporation of CdS and ZnS for the window rather than only CdS. (Zn,Cd)S has a higher band gap and transmits a larger fraction of the incident photons to the absorber. After Al grid metalization, the cell is baked in air at about 200°C from 10 to 30 minutes. This baking step is crucial in obtaining improved open circuit voltage and fill factor. Finally the anti-reflection (AR) coating is applied to complete the cell fabrication. The cells developed by the Institute of Energy conversion at Delaware also use vacuum evaporation technique for depositing $CuInSe_2$ cells [10]. However, they use knudsen cells rather than open boat evaporation as carried out by Boeing.

The role of the air or oxygen bake is quite intriguing as its effects can be reversed by treating the cell in a reducing agent such as hydrazine or an electron beam. It is proposed that oxygen diffuses to the defects where indium atoms are located in the copper sites. The oxygen atom ties up the available electron pair at these defect sites thereby reducing the number of free electrons and making the material more p-type [8,9].

A somewhat different approach is taken by ARCO [11]. They use a (Cd,Zn)S or CdS(n)/$CuInSe_2$(p) interface. The thin undoped (Cd,Zn)S or CdS layer is less than 50 nm thick and is coated with about 1.5μm thick conducting ZnO. The 3.2eV bandgap of ZnO coupled with its high conductivity permits higher energy photons to reach the absorber thereby increasing both the short circuit current and the fill factor. The ZnO layer also acts as a partial anti-reflection coating. The $CuInSe_2$ layer on molybdenum coated glass is about 2μm thick. The exact method used by ARCO to deposit $CuInSe_2$ layer is not disclosed but it does not require a separate air or oxygen bake to optimize its properties as in the case of Boeing like cells.

International Solar Electric Technology (ISET) of Inglewood, CA uses a different approach which shows promise for low cost production. They use electron-beam deposited or sputtered molybdenum on glass substrates to electroplate first a 200nm thick copper layer from an acidic aqueous solution of $CuSO_4$ at room temperature [12]. This is followed by a 440nm thick In layer electoplated from an aqueous solution of $In(SO_3NH_2)_3$. These metal layers are then selenized under H_2Se in argon at 400°C. The $CuInSe_2$ thickness after selenization is about 2 μm. CdS is deposited by evaporation in two steps on the $CuInSe_2$ with a 0.6 μm high resistive layer followed by an In-doped 1.2 μm thick low resistivity layer. The cell structure is completed with Al grid lines and AR coating. The electroplating process is very sensitive to the surface preparation, morphology and composition of the molybdenum coated glass substrate [13]. The problem of film peeling from the substrate needs to be solved by investigating alternatives to the molydenum bottom contact.

Planar magnetron reactive co-sputtering of Cu and In in H_2Se in Ar gas is being carried out at the University of Illinois [14]. The two-layer Boeing structure cells have been fabricated from these films [15]. These is a small problem of In rejection in depositing the top $CuInSe_2$ layer which needs special consideration.

The status of the technology in each of the above mentioned cases may be judged by the best reported active-area solar conversion efficiences in single junction cells based on $CuInSe_2$ films. The ARCO cells have achieved 12.5% [11]. The Boeing cell has achieved 12.5%, ISET electroplated cells have achieved nearly 8% efficiency without the AR coating and the Illinois cells from the sputtered films have achieved nearly 6% efficiency [1]. The ARCO cell has an open-circuit voltage of 487 mV, short-circuit current of 41.1 mA/cm^2 and a fill factor of 0.7 for a cell of 3.6 cm^2 area at AM 1.5 Global Spectrum at 25°C. However, the typical reproducible efficiencies that can be achieved routinely are somewhat lower for each case. In order to achieve 15% module efficiency, the cell active area efficiency should approach about 17% if the module efficiency is projected to be about 90% of the active area efficiency. This would require about 6-10% increase in open-circuit voltage and about 4-6 mA/cm^2 increase in the short-circuit current densities along with an increase in the fill factor to 0.75 [1].

Research is underway along a number of directions to increase the efficiency in the single junction $CuInSe_2$ cells. These include two major approaches. The first is to increase the band-gap of the transparent conducting window layer to transmit the short wavelength photons to the absorber. Both the open circuit voltage and the short circuit current can be increased by this approach. Jet Propulsion Laboratory has developed a technique to deposit In doped n-type ZnSe with energy gap of 2.7eV by reactive magnetron sputtering at low temperatures for this purpose [16]. The second approach is to increase the absorber band gap slightly by using a quaternary rather than a ternary material. This should increase the open circuit voltage. Absorber compounds such as $CuIn_{1-x}Ga_xSe_2$ and $CuInSe_{2x}S_{2-2x}$ have been suggested. A 10.2% efficiency has been achieved for x=0.23 for a $CuIn_{1-x}Ga_xSe_2$ cell [17]. Though higher open circuit voltage has been achieved for this higher energy gap of 1.14eV, the fill-factor in these devices is still poor.

Some of the strongest points in favour of the $CuInSe_2$ technology are its proven stability, its high absorption coefficient and high short-circuit current density. The vacuum growth technique used in the fabrication process involves challanging system design problems to achieve both composition uniformity and thickness uniformity in a large area scaled up process. The film uniformity also demands accurate substrate temperature control on a large substrate area. Fortunately, there is a narrow but distinct composition window permissible in the Cu-In-Se phase diagram from which the high efficiency cells can and have been made [18].

The physics of the CuInSe$_2$ cells is not well understood and currently there does not exist a good working model for these various structures. Explanations on various device aspects have been provided to explain high diode factor ($\approx$ 1.4 to 2), low open circuit voltage and crossover between the dark and the illuminated current voltage curves by various authors especially for the Boeing type cells. The electron beam induced current (EBIC) measurements show that in these two-layer cells the response moves towards the heterostructure interface in the oxidized state after the air bake and moves inwards into the CuInSe$_2$ bulk after reducing treatment in hydrazine [8]. The recent work speculates that in the Boeing structure, the response may not be entirely due to the heterostructure but may also be due to a buried homojunction in CuInSe$_2$ as well [18]. This is further supported by the recent results carried out at low temperatures on the Boeing structure where EBIC peak shifts from near the heterostructure at room temperature into the CuInSe$_2$ bulk at 110K [20]. The low resistivity bottom CuInSe$_2$ layer is not affected at this temperature [21]. The shift in Fermi level with temperature in the high resistivity CuInSe$_2$ top layer near the heterostructure interface is speculated to give rise to an additional space charge region. An increase in the open circuit voltage of 4 to 30 mV with illumination time is also observed in the CdS/CuInSe$_2$ cells [22]. A model involving tunneling of charges from deep level traps in the CdS layer is proposed to explain this light soaking effect [23]. The understanding on the physics of these structures will improve as more data on the device behaviour and the defect chemistry in the material are made available.

4. CdTe Cells

CdTe with an energy gap of 1.45 eV provides an excellent match for absorption of the solar spectrum. The material has a high absorption coefficient with a maximum value of about 10^5 cm^{-1} and can be made either p-type or n-type. The basic structure of the CdTe thin film cells usually involves the CdS(n)/CdTe(p) heterostructure even though homojunction, MIS and schottky barrier cells have also been reported using CdTe material. Several different fabrication techniques have been employed. Some of the more important ones are described below.

The Southern Methodist University team in Dallas employs closed space sublimation to deposit thin layers of CdTe on a substrate separated from a CdTe source by about 1mm [24]. The reaction tube pressure, source-substrate temperature difference and the source temperature are among the critical parameters that need to be controlled. The CdTe layer is deposited on undoped 100-300 nm thick CdS evaporated on fluorine-doped SnO$_2$ coated glass substrate. A key step in the process is the in-situ cleaning of the CdS surface in a reducing atmosphere prior to the CdTe deposition process. This significantly reduces the saturation current density of the cells. Following the CdTe deposition, the surface is etched in K$_2$Cr$_2$O$_7$-H$_2$SO$_4$ solution to give a Te rich surface prior to contacting with Cu doped graphite or Au. The best device produced by this process showed 10.5% efficiency under Global AM 1.5 conditions for a 1.22 cm^2 device with open circuit voltage of 0.748 V, short circuit current of 22.21 mA/cm^2 and fill factor of 0.64.

Ametek Applied Materials Laboratory of Harleysville, PA has developed techniques to electroplate large area sheets of uniform CdTe [25]. They have achieved 10.4% efficient devices based on CdS/CdTe heterostructures and expect to increase the efficiency to above 12% in the near future. They first deposit 150 nm CdS at 450°C in a pyrolytic reactor from an aerosol containing $CdCl_2$ and thiourea followed by electrodeposition of 2 μm thick CdTe. The structure is then heat treated in air for 20 min. at 400°C. Like the $CuInSe_2$ case, air bake is frequently used in CdTe technology to obtain a p-type CdTe layer.

Photon Energy Inc. of El Paso, TX has developed approximately 1 sq. ft. CdS/CdTe panels with an active area efficiency of 5.7%. Matsushita of Japan has developed CdS/CdTe solar cell structure by a screen printing technique and has reported 12.8% efficiency [26]. The process consists of successive printing and firing of CdS, Cd and Te compounds, C, Ag-In, and Ag to form CdS/CdTe structure with 50 PPM Cu in graphite as the electrode to CdTe and Ag-In as the contact to CdS. The CdTe layer near the CdS interface is dense while the layer further aways from the interface is somewhat porous. These cells have better photoresponse in the near infrared region compared to CdS/CdTe cells fabricated by other techniques. This is attributed to a mixed Cd-S-Te compound formation near the heterostructure interface.

The CdS/CdTe heterostructure interface is not ideal. A wider band gap conducting window is preferable in place of CdS. One of the possible candidates currently being investigated for this purpose is (ZnCd)S [27]. There have been two persistent problems with polycrystalline CdTe technology. In general, it is difficult to heavily dope the p-type polycrystalline CdTe. Higher doping is desirable as it reduces series resistance and increases the fill factor. Efforts to increase the doping density have been met with only a limited success [28]. This is frequently referred to as the self-compensating nature of the material. Recently, some progress has been made in this regard. Stanford University researchers have successfully achieved doping levels of 6 x 10^{16} cm^{-3} for As and 10^{17} cm^{-3} for P in p-type CdTe by low energy ion assisted doping of the epitaxial films [29]. Carrier density of 2 x 10^{17} cm^{-3} in epitaxial film is obtained by OMCVD technique [30] and carrier density of 5 x 10^{16} cm^{-3} has been obtained by photon-assisted doping of MBE epitaxial CdTe films [31].

The second problem concerns ohmic contact formation to the p-type CdTe which has a work function higher than most metals. A tunneling contact requires high doping which has been a problem for p-type CdTe films as described above. In past, the contacts to the p-type CdTe layers have resulted in stability problems for the solar cells. Surface etching is found to play a crucial role. A Te-rich surface prior to contacting is desirable. Though acceptable contacts at room temperature have been obtained, their low-temperature performance is often poor. One of the approach being undertaken to avoid the contact problem is to utilize a conducting wide gap material such as ZnTe or In-Sn oxide next to the CdTe. The ohmic contacts are then made to this conducting layer [25]. The cell may now be described by a n-i-p structure with the high resistive CdTe layer being completely depleted. HgTe has also been suggested as a contacting layer [27].

Typically the CdS/CdTe cell has a high diode quality factor (>1.5). The current is modeled to be due to thermally activated recombination processes at room temerpature and above with recombination at and near the interface playing an important role [32]. At low temperatures, tunneling current through the interface states is the dominant mechanism in these devices. The effect of intrinsic and extrinsic states on the device operation is quite significant which emphasizes the importance of controlling the processing parameters.

5. Tandem Cells

The overall efficiency of a photovoltaic module can increase significantly if two or more semiconductors with appropriate different energy gaps are used to better utilize the solar spectrum. A two-junction configuration provides the simplest tandem structure which can be operated in a two-terminal or a four-terminal mode. The later configuration avoids the complicated problem of matching currents through the individual cells. The basic structure is shown in Fig. 2. Efficiencies above 23% have been predicted for top cell energy gap in the range of 1.6 - 2.3 eV and the bottom cell energy gap in the range of 0.95 - 1.4 eV [33]. Institute of Energy Conversion has fabricated two-terminal CdS/CdTe: CdS/CuInSe$_2$ cells by a sequential deposition process [34]. They have obtained only 3% efficiency in these cells primarily due to a contact problem in the top cell. Hybrid tandem modules with 13.7% efficiency have been made out of thin film amorphous Si:H top cell and CuInSe$_2$ bottom cell [35]. The efficiency of these structures are soon projected to reach 17.4%.

Various research is currently being carried out to obtain an efficient top cell from 1.6 - 1.7 eV absorber to be used in tandem with the 1.0 eV CuInSe$_2$ bottom cell. Research in this area includes 1.6 - 1.7 eV energy gap (Cd,Zn)Te absorber structure with ZnTe (p$^+$)/ITO/glass lower contact and CdS (or ZnO) top window [36]. Other materials include (HgZn)Te and (Cd,Mn)Te. In our laboratory we have developed vacuum growth technique to deposit near stoichiometric p-type layers of ZnSnP$_2$ which has an energy gap of 1.66 eV [37]. The deposited ZnSnP$_2$ films on fused quartz, molybdenum and GaAs substrates have interesting optical and electrical properties for fabricating top cell in a tandem structure and are being further investigated [38].

The photovoltaic technology interacts with social and economical factors in the society in a complex way. Various concerns regarding its large scale terrestial application have been addressed including the one related to the health and the safety issues in a large scale manufacturing process [39].

6. Acknowledgements

The author would like to acknowledge the assistance of Dr. Harin S. Ullal of Solar Energy Research Institute in providing a number of technical reports on the subject matter.

7. References

1. K. Zweibel, R. Mitchell and H. Ullal, "Polycrystalline Thin Film: FY 1986 Annual Report", Solar Energy Research Institute SERI/PR-211-3073, NTIS, Springfield, VA, Feb. 1987.
2. Proceedings of the Polycrystalline Thin Film Program Meeting, Lakewood, CO, USA, Solar Energy Research Institute SERI/CP-211-3171, July 20-22, 1987.
3. Solar Cells, Vol. 21, June-Aug. 1987.
4. Proceedings of the 19th IEEE Photovoltaic Specialists Conference, New Orleans, May 1987.
5. J.R. Tuttle, R. Noufi and R.G. Dhere, In Ref. [2] above, pp. 127-132.
6. SERI Photovoltaic Advanced Research & Development: An Overview, SERI/SP-281-2235, pp. 12-15, Feb. 1984.
7. R. Noufi, R. Axton, R. Powell and S.K. Deb. In Abstracts of Presentation: Polycrystalline Thin Film Review Meeting, Golden, CO, Solar Energy Research Institute SERI/CP-211-2548, pp. 25-32, Oct. 1984.
8. R. Noufi, R.C. Powell and R.J. Matson, pp. 55-63, in Ref. [3] above.
9. P. Zurcher, pp. 137-141, in Ref. [2] above.
10. J.D. Meakin, R.W. Birkmire and J.E. Phillips, pp. 17-23, in Meeting Abstracts cited in Ref. [7] above.
11. K.W. Mitchell, pp. 89-96, in Ref. [2] above.
12. V.K. Kapur, B.M. Basol and E.S. Tseng, pp. 65-72, in Ref. [3] above.
13. V.K. Kapur, B.M. Basol, N.L. Nguyen and R.C. Kullberg, pp. 97, in Ref. [2] above.
14. T.C. Lommasson, A.F. Burnett, M. Kim, L.H. Chou and J.A. Thornton, In Proc. 7th Intl. Conf on Ternary and Multinary Compounds, Snowmass, Co., Sept 1986, pp. 207-212, Materials Research Soc. 1987.
15. J.A. Thornton, pp. 71-79, in Ref. [2] above.
16. R.J. Stirn, pp. 173-180, Ref. [2] above.
17. R.A. Mickelsen, J.A. Avery, W.S. Chen, W.E. Devaney, R. Murray, B.J. Stanbery and J.M. Stewart, pp. 61-69, in Ref. [2] above.
18. R.W. Birkmire, J.D. Meakin, J.E. Phillips and R.E. Rocheleau, pp. 81-86, in Ref. [2] above.
19. a) A. Rothwarf, pp. 71-82, b) K.W. Boer, pp. 83-92, C. Goradia and M. Ghalla-Goradia, pp. 95-95-104, d) J.R. Sites pp. 105-111; in Meeting Abstacts cited in Ref. [7] above.
20. R. Noufi, V. Ramanathan and R.J. Matson, pp. 109-114, in Ref. [2] above.
21. T. Datta, R. Noufi and S.K. Deb, Appl. Phys. Lett., 47, pp. 1102-1104, 1985.

494

22. K. Amery and C. Osterwald Solar Cells, 21, pp. 313-327, 1987.
23. M.N. Ruberto and A. Rothwarf, J. Appl. Phys., 61, pp. 4662, 1987.
24. T.L. Chu, S.S. Chu, S.T. Ang and M.K. Mantravadi, Solar Cells, 21, pp. 73-80, 1987.
25. P.V. Meyer, pp. 9-11 in Ref. [2] above.
26. H. Matsumoto, K. Kuribayashi, H. Uda, Y. Komatsu, A. Nakano and S. Ikegami, Solar Cells, 11, pp. 367-373, 1984.
27. T.L. Chu and S.S. Chu, pp. 1-7, in Ref. [2] above.
28. A.L. Fahrenbruch, Solar Cells, 21, pp. 399-412, 1987.
29. A.L. Fahrenbruch, A. Lopez-Otero, K.F. Chien, P. Sharps, and R.H. Bube, Poster I9, 19th Photovoltaic Specialists Conference, New Orleans, LA, May 4-8, 1987.
30. S.K. Ghandi, N.R. Taskar and I.B. Bhat, Appl, Phys. Lett., 50, pp. 900-902, 1987.
31. R.N. Bicknell, N.C. Giles and J.F. Schetzina, Appl. Phys. Lett., 49, pp. 1735-1737, 1986.
32. C.M. Fortmann, A.L. Fahrenbruch and R.H. Bube, J. Appl. Phys., 61, pp. 2038-2045, 1987.
33. K.W. Mitchel, Solar Cells, 21, pp. 127-134, 1987.
34. J.E. Phillips, R.W. Birkmire and J.D. Meakin, pp. 197-203 in Meeting Abstracts cited in Ref. [7] above.
35. K. Mitchell, R. Potter, J. Ermer, R. Wieting, D. Tanner, K. Knapp and R. Gay, Proc. 19th Photovoltaic Spec. Conf., May 4-8, 1987.
36. a) B.M. Basol and V.J. Kapur, pp. 156 in Ref. [2] above.
    b) K. Zanio, pp. 157-159 in Ref. [2] above.
37. P.K. Ajmera, H.Y. Shin and B. Zamanian, Solar Cells, 21, pp. 291-299, 1987.
38. H.Y. Shin and P.K. Ajmera, Materials Letts, 5, pp. 211-214, 1987.
39. P.D. Moskowitz, V.M. Fthenakis, L.D. Hamilton and J.C. Lee, Solar Cells, 19, pp. 287-299, 1986/87.

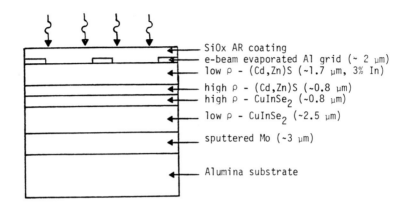

Fig. 1. Boeing (Cd,Zn)S (n)/CuInSe$_2$ (p) structure (not drawn to scale).

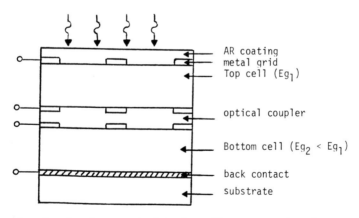

Fig. 2. Example of a four-terminal tandem cell structure (not drawn to scale).

GALLIUM ARSENIDE DEVICES

ELECTRON AND HOLE INJECTIONS INTO SEMI-INSULATING GaAs AND
IMPLICATIONS FOR SIDE-GATING OF LSI CIRCUITS

Kurt Lehovec and Henry Bao
University of Southern California
Los Angeles, California 90089

One-dimensional laminar electron flow from an injecting n+ cathode through semi-insulating (SI) GaAs with EL2 donor centers and shallow acceptors is modelled analytically. The electrically neutral bulk of the SI layer between two n+ contacts assumes a uniform negative charge in proportion to the current density when the drift velocity saturates. We suggest that electron-hole pair generation via partially ionized donors may then occur near the anode at comparatively low fields of several 10^4 V/cm. We suggest that this electron-hole pair generation is fundamental to the side-gating effect with the transistor channel the anode of the side-gating structure. This interpretation differs from the present one which is based on the rapid rise of the space charge limited electron current which occurs when the space charge density of the charged acceptors penetrates to the transistor channel.

1. INTRODUCTION

Semi-insulating (SI) GaAs substrates have room temperature resistivities of about $10^9 \Omega$ cm which eliminates the need for p-n junction isolation in microcircuits and contributes to their superiority over silicon-based circuits by higher speed and improved radiation resistance. However the reduced spacing of subcircuits with increasing integration may result in an undesirable electric interaction known as "side-gating" [1,2,3]. The effect has been traced to carrier injection into the SI substrate from electrically biased conducting regions. Related effects are the occurance of low frequency oscillations [3,4,5,6], instabilities of the I-V characteristics of some structures in the form of a switchback of the voltage at an enhanced current [7], and collection of radiation-generated carriers from the SI substrate [8,9]. Section 2 is a brief review of some of the evidence for these phenomena. Section 3 describes our modeling of the current flow through the SI substrate at carrier injection. In section 4 a new interpretation of the side-gating phenomenon is presented. Section 5 is a summary of the principal results.

2. BRIEF REVIEW OF PHENOMENA RELATED TO CARRIER INJECTION INTO SI GaAs

2.1 Side-gating

Application of a sufficiently large negative bias voltage to a laterally spaced n+ region, the "side-gate", causes a drop-off of the drain current in an n-channel field effect transistor. The drop-off coincides with an abrupt increase of current flow from the side-gate electrode [1,2]. The phenomenon has been explained by space charge limited current through the SI substrate. The resulting high field between the channel and the adjacent substrate requires a positive charge at the channel/substrate interface which arises by removal of conduction electrons from the channel; hence the drop-off of the channel current.

2.2 Low frequency oscillations

Low-frequency oscillations have been observed in transistor circuits used for analogue operation [3-5]. The oscillation appears to be a consequence of a periodic time-variation of the field distribution in the SI-substrate. Such a time variation modulates (i) the carrier injection from the side-gate; (ii) the electron- hole pair generation at the anode; and (iii) the occupancy of electron traps in the substrate [10]. Associated time constants are the transit times of electrons and holes and the trap-release time constant for electrons. The source of the transistor can also act as a "side-gate" electrode for electron injection into the SI substrate underlying the channel [11].

2.3 I-V instability

A switch-back of the I-V characteristics has been observed on GaAs resistors operated at drift velocity saturation [7]. Channel length and dopant concentration of the device were chosen to prevent migrating Gun domains. The switch-back has been modelled [12] assuming generation of electron-hole pairs in the high field near the anode. The holes exit at the cathode and enhance the electron concentration in the bulk of the device by space charge compensation. The increased bulk conductivity generates an inhomogenous field distribution with an anodic high field region. A similar switch back of the I-V characteristic has been observed also on SI GaAs samples with spaced n+ regions along its surface.

2.4 Radiation effects

Electron-hole pairs generated in the SI substrate by short-wave electromagnetic radiation such as γ rays or by impact of high energy particles are separated by the electric field between the substrate and circuit elements overlaying it. The resulting charge collection can cause circuit malfunctions. Particularly obnoxious is the soft error in memory circuits known as single event upset (SEU). The sensitivity of a circuit to radiation has been probed with a focussed electron beam and sensitive spots have been identified [9].

3. MODELLING OF CARRIER INJECTION INTO SI GaAs

The simplest case is one-dimensional laminar flow with a single carrier type present. This situation is realized in samples with anode and cathode having the same dopant and being located at opposite surfaces of an SI wafer. The d.c. I-V characteristics of n+/SI/n+ structures has been analyzed by computer [13]. We have published analytical approximations neglecting the built-in space charge regions adjacent to anode and cathode, diffusion currents and drift velocity saturation [14]. These neglects have now been removed and the results will be described in what follows.

3.1 Material characterization

"Pure" SI substrates used for GaAs microcircuits contain unintentional donors of energy at about midgap and a much smaller concentration of unintentional shallow acceptors. The donors are the EL2 centers, probably As on Ga sites [15], and the acceptors may be C traces. At room temperature all the acceptors are ionized and an equal number of donors is positively charged to provide space charge neutrality in the bulk of the sample. The Fermi-level is pinned slightly above midgap by the large donor concentration relative to that of the acceptors. The Cr doped substrates previously used for semi-insulation have become obsolete because of Cr migration.

3.2 Built-in space charge layers in absence of current flow

Carriers diffuse from doped regions into the adjacent SI substrate causing space charges and built-in fields (Fig. 1, upper row). The mathematical analysis of the field distribution in the absence of applied potentials uses a combination of the Boltzmann distribution and of Poisson's equation.

The positive space charge adjacent to a p+ contact contains four characteristic regions (Fig. 2, upper row) with dominant contributors to the space charge density being, respectively, (i) holes; (ii) fully ionized donors; (iii) partially ionized donors of a concentration in proportion to the hole concentration; and (iv) an excess of ionized donors over fully ionized acceptors. The negative space charge adjacent to a heavily doped n+ contact has only three distinct regions (Fig. 2, lower row) with dominant contributors to the space charge density being, respectively, (i) electrons; (ii) fully ionized acceptors; and (iv) a small excess of fully ionized acceptors over charged donors. The positive space charge density in region (ii) near a p+ contact is orders of magnitude larger, and thus the region is much shallower, than the negative space charge density in the corresponding region (ii) near an n+ contact. The major part of the negative built-in space charge layer near an n+ contact arises from region (ii), while the major part of the positive space charge layer near a p+ contact arises from region (iii). Figure 3 shows that the negative built-in space charge layer near an n+ contact is wider than the positive space charge layer near a p+ contact.

3.3 Modification of the built-in space charge distribution by current flow between two n+ contacts

When a d.c. voltage is applied between two n+ regions the built-in electric field at the cathode and the bulk field have opposite polarities. Thus there is a position x_m near the grounded cathode located at $x = 0$ where the field vanishes and the potential has a minimum $V_m < 0$ (Fig. 1, middle row). The cathodic space charge layer expands with current while the anodic space charge layer contracts. Figure 4 shows a set of potential distributions for different current densities obtained by the analytic approximation

$$W = \begin{cases} W_m + z(u - u_m)^2/2 & \text{for } u < u_m \\ W_m + z(u - u_m) - z^2[1 - exp(-\frac{u - u_m}{z})] & \text{for } u > u_m \end{cases} \quad (1)$$

The potential W is expressed as multiple of the voltage equivalent of temperature, the distance from the cathode u as multiple of the Debye length L_D based on the acceptor concentration A, and the current density z as multiple of the bulk current density for the drift field v_T/L_D. The minimum potential is

$$W_m - W_A = -\ln \frac{A}{zn_T\sqrt{2\pi}} \quad (2)$$

W_A is the potential at the position $u_A \ll 1$ where the electron concentration equals the acceptor concentration, i.e. at the cross-over from region (i) to (ii) in Fig. 2.

The potential minimum occurs about 3 to 4 Debye lengths from the cathode and changes comparatively little with current density. The electron concentration at the potential minimum is

$$n_m = zn_T\sqrt{\pi/2} \tag{3}$$

where n_T is the thermal equilibrium bulk electron concentration. Our approximation assumes that the potential minimum falls into the space charge density region (ii) which restricts it to current densities $1 < z < A/n_T$. However a more restrictive upper limit arises from the occurance of drift velocity saturation ignored in the derivation of (1).

The potential distribution (1) has been derived by combining Poisson's equation for the regions (ii) and (iv)

$$\frac{d^2W}{du^2} = 1 - \frac{n_T}{n} \tag{4}$$

with the current expressed as low field drift current and neglecting the diffusion component. Since the current is completely due to diffusion at u_m, the neglect of the diffusion current would require an infinite electron concentration at u_m. In order to obtain the diffusion current z at u_m we have shifted the electron distribution pertaining to the entire current being drift current by substituting $u + 1$ for u. This provides $n_m/n_T z = 1$ instead of (3), i.e. an error of 25%. This discrepancy does not affect the space charge density at x_m which is essentially $-qA$. The accuracy of our analytic approximations for the potential and carrier distributions can be assessed by examination of the constancy of the resulting net current with position. The maximum error is -25% and occurs for large z-values. However only a few Debye lengths beyond the potential minimum the current arises completely from drift and the approximations both for potential and electron concentration become extremely accurate.

The position x_A where $n = A$ turns out to be a small fraction of a Debye length and thus can be safely neglected. The carrier distribution over the region $0 < x < x_A$ remains quasi-Boltzmann since the net current is very small compared to the drift and diffusion currents which almost compensate each other. For instance the drift current at u_A is $zA/n_T \gg z$. Thus the voltage drop V_A across this region does not change significantly with current. The voltage drop $V_m - V_A$ does decrease with current as seen in Fig. 4, but so does the voltage drop across the anodic space charge region to be discussed subsequently. Thus the relations of our previous paper [14] which neglected diffusion currents and built-in space charge layers are applicable if the sample length is reduced by x_m and the cathodic boundary concentration is replaced by $n_m = zn_T$.

There is no zero field position in the anodic space charge layer, the field increasing continuously toward the anode. The net current arises from an excess of drift current over diffusion current. If the electron distribution were unchanged and the bulk field were merely superimposed over the unchanged built-in field the drift current would increase toward the anode because of the increasing carrier concentration, while the diffusion current would remain unchanged. In order to obtain a space independent net current the diffusion current has to increase also toward the anode requiring a steeper electron distribution, thus a shallower space charge layer at current flow, resulting in a smaller potential drop across the anodic space charge layer with increasing current density.

3.4 Effect of drift velocity saturation

For spacings between the contacts sufficiently wide that the anodic and cathodic space charge regions do not overlap, i.e. typically at least about five Debye lengths, there exists a quasi neutral bulk region of thermal equilibrium electron concentration n_T. The increase of current arises from the increase of drift velocity with bulk field at constant electron concentration n_T (Fig. 1, middle row). However for bulk fields at which drift velocity saturates the current increases by an increase in the bulk electron concentration resulting in a uniform bulk space charge density (Fig. 1, bottom row)

$$\rho_0 = -Aq[1 - \frac{z_s}{z}] \tag{5}$$

where

$$z_s = v_s L_D / \mu v_T \tag{6}$$

is the current density pertaining to the saturation drift velocity v_s. Assuming for simplicity an abrupt transition from the drift velocity $\mu V'$ of the low field region to v_s at the field $(v_T/L_D)z_s$ we have to restrict the validity of (1) to the range where $dW/du < z_s$. Beyond that range i.e. for $u > u_s$, we have

$$dW/du = z_s + (u - u_s)(1 - \frac{z_s}{z}) \tag{7}$$

Figure 5 shows the resulting field distributions. The lower horizontal dotted line refers to the critical field z_s at which the drift velocity saturates. The upper dotted horizontal line refers to a hypothetical field intensity at which electron - hole pair generation sets in. The intercept of that dotted line with the curve pertaining to a current density z gives the sample length $L = u_s L_D$ at which the critical field W'_{BR} for electron hole pair generation is reached at the anode at the current density z. The corresponding voltage across the sample

$$W_{BR} = W_s + (u_L - u_s)\frac{z_s + W'_{BR}}{2} \tag{8}$$

varies substantially linearly with the sample length corresponding to the average field $E_{BR}/2$. The added potential drop across the bulk region resulting from the space charge (5) causes a slight flattening of the I-V characteristics before its sharp rise when hole generation sets in, as indeed observed [17].

4. Discussion

The presently accepted explanation for side-gating is based on Lampert's model for space charge limited electron current through the SI substrate which contains deep electron traps, the EL2 centers, as well as acceptors. According to this hypothesis the current is expected to rise abruptly when the space charge density $-qA$ of region (ii) of Fig. 2 penetrates to the transistor. The corresponding threshold voltage is $V_T = qAL^2/2\epsilon$, or in our reduced units $W_T = u_L^2/2$. However experiments show that the threshold voltage for sidegating varies linearly, and not quadratically, with the distance of the side-gate electrode from the transistor [2,3].

We propose that side-gating occurs at a critical field E_{BR} in the regime of drift velocity saturation in the bulk of the SI layer preceeding the penetration of region (ii) to the transistor. This provides the observed linear relation between the corresponding voltage and the spacing. At the breakdown field a finite probability less than unity exists for an electron to generate an electron hole pair. The mechanism for the abrupt current rise is analogous to that for the switchback in the saturated drift velocity doped GaAs resistor [12]: The hole of the electron-hole pair drifts to the cathode and induces an added electron for space charge neutralization. The added electron has a finite probability to generate an electron hole pair, the generated hole increases further the electron concentration etc. At a certain small finite probability for generation of an electron-hole pair the initially uniform field distribution switches to a lower field bulk region and a narrow higher field anodic region where the hole generation takes place. Matching theory [12] to experiment [7] has given an ionization probability of about $1.8 \times 10^4/(\text{cm electron})$ at a breakdown field of only 4×10^4 V/cm. The breakdown field in SI GaAs should be even lower since electron hole pair generation can occur as a two step process via the partially charged EL2 centers at mid-gap.

At the side-gate threshold we expect lower average fields than the cited 4×10^4V/cm for several reasons: (i) According to (8) the breakdown field is about twice the average field. (ii) The breakdown process is facilitated by the availability of the partially charged EL2 centers. (iii) The side gate geometry is not strictly laminar, since there is some spreading of the fieldlines from the transistor which enhances the anodic field [3]. (iv) There is a considerable field enhancement at the drain and to a lesser degree under the channel from the built-in fields existing there already without substrate current flow. Observed average side-gate threshold fields are about 2×10^3V/cm [2,3].

The effects of surface treatment [16,17,18] and of a negatively biased electrode interposed between the side-gate electrode and the transistor [19] have been cited as evidence for electron flow in a surface channel. In our model these observations may be accounted for by hole removal.

In addition to the transistor current suppression by the impinging substrate field on the channel, which is the same in the former and present interpretations, the impact generated holes may reduce the drain current by recombination with channel electrons. Furthermore holes diffusing through the narrow channel and sucked into the gate may decrease the drain current by shifting the gate voltage to more positive values unless it is clamped in the circuit.

5. Summary and conclusions

The mathematical analysis of one-dimensional carrier flow through SI GaAs with carrier injection from a heavily doped contact shows the following:

(i) The width of the built-in space charge layer at the contact in absence of current flow is characterized by the Debye length L_D defined in terms of the acceptor concentration A; for $A = 10^{13} cm^{-3}$ this length is 1.25 μm and decreases inversely with A. The built-in space charge region adjacent to an n+ contact is about twice as wide as that adjacent to a p+ contact.

(ii) With increasing applied voltage a potential minimum develops near the n+ cathode at a distance of 3 to 4 L_D. Beyond this point the I-V characteristic can be simulated well by an approximation based on the neglect of diffusion current as was done in Ref.

[14] for drift fields in the electrically neutral bulk region in the range $v_T/L_D < E < v_s/\mu$. For $A = 10^{13}$cm3, $v_T/L_D = 200$V/cm.

(iii) For bulk fields beyond the upper limit the drift velocity saturates and the previously neutral bulk of the SI layer becomes uniformly charged at a density which increases in proportion to the current.

(iv) The transistor is the anode and a spaced n+ region is the cathode of an n+/SI/n+ structure in a side-gate set-up. We suggest that side-gating occurs in the drift velocity saturation regime when the anodic field at the transistor substrate boundary becomes sufficiently large for electron - hole pair generation via the EL2 centers. This occurs before the region of space charge density $-qA$ punches through to the transistor.

REFERENCES

1. C.P. LEE, S.J. LEE, and B.B. WELSH, (1982), IEEE Electron Dev. Lett., EDL-3, 97-98.
2. M.S BIRRITTELLA, W. SEELBACH, and H. GORONKIN, (1982), IEEE Trans. Electron Dev., ED-3, 1135.
3. S. MAKRAM-EBEID and P. MINONDO, (1985), IEEE Trans. Electron Dev., ED-32, 632.
4. S. MAKRAM-EBEID, A. MINTOUNEAU, and G. LAWRENCE, (1982), Proc. 2nd Conference Semi-Insulating III-IV Materials, United Kingdom; Shiva Limited, 336-343.
5. P. CAULFIELD and L. FORBES, (1986), IEEE Trans, Electron Dev., ED-33, 925.
6. M. KAMINSKA, J.M. PASSEY, J. LAGOWSKI, and H.C. GATOS, (1982), Appl. Phys. Lett., 41, 989.
7. J.W. ROACH, H.H. WIEDER, and R. ZULEEG, (1987), Trans. Electron Lett., 34 No. 2, 81-84.
8. P.J. McNULTY, W. ABDEL-KADER, A.B. CAMPBELL, A.R. KNUDSON, R. SHAPIRO, F. EISEN, and S. ROOSILD, (1984), IEEE Trans. on Nucl. Sc., NS-31, No. 6, 1128.
9. L.D. FLESNER, (1985), IEEE Tans. Nucl. Sc., NS-32, 4110-4120.
10. B.K. RIDLEY, (1974), J. Phys. C: Solid State Phys., 7, 1169.
11. Personal Communication cited as [12] by CAULFIELD and FORBES [5].
12. K. LEHOVEC, (1986), Proc. of the International Conference, Stockholm, Sweden, August 7-9, HIGH SPEED ELECTRONICS, B. Källbäck and H. Beneking, Editors, Springer-Verlag, 127-131.
13. K. HORIO, T. IKOMA, and H. YANAI, (1986), IEEE Trans. Electron Div., ED-33, No. 9, 1242.
14. K. LEHOVEC and H. BAO, (1987), Solid State Electronics, 30, 479.
15. D.C. LOOK, W.M. THEIS, and P.W. YU, (1987), J. Elec. Mat., 16, 63.
16. D.C. D'AVANZO, (1982), IEEE Trans. Electron Dev., ED-29, 1051.
17. W.M. PAULSON, (1983), IEEE Trans. Nucl. Sc., NS-30, No. 2, 1713.
18. M.F. CHANG, C.P. LEE, L.D. HOU, R.P. VAHRENKAMP, and C.G. KIRKPATRICK, (1984), Appl. Phys. Lett., 44(9), 869-871.
19. C.P. LEE and M.F. CHANG, (1985), IEEE Electron Dev. Lett., EDL-6, 169.

506

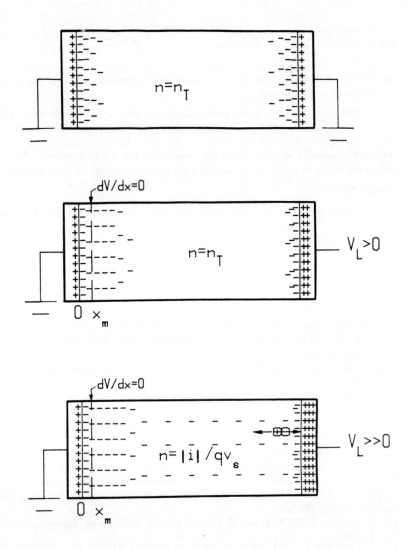

Fig. 1 Schematic of space charge layers in SI GaAs between n+ contacts.
Upper row: Built-in space charge layers at zero current flow.
Middle row: The anodic space charge layer contracts, the cathodic space charge layer expands and a potential minimum arises at x_m. The bulk of the SI layer remains electrically neutral and the electron concentration unchanged for sufficiently small currents that the bulk field does not cause drift velocity sarturation.
Bottom row: The current is sufficiently large that drift velocity saturates in the bulk. The electron concentration increases there in proportion to the current and a uniform space charge density results. At a critical anodic field electron-hole pair generation sets in which we believe to cause the side-gating.

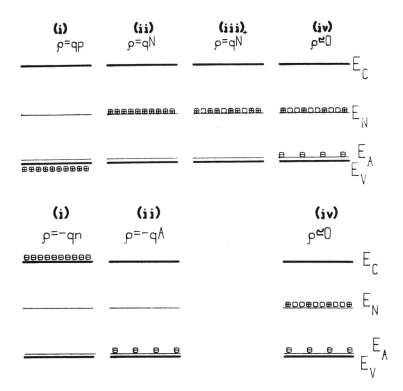

Fig. 2 The dominant contributors to the space charge density in SI GaAs adjacent to a p+ layer (upper row) and to an n+ layer (lower row). The donors are completely charged in regions (i) and (ii) of the upper row, but they are dominant only in region (ii). The acceptors are completely charged in all three regions of the lower row. Region (iii) near the p+ contact has no analogue near the n+ contact.

508

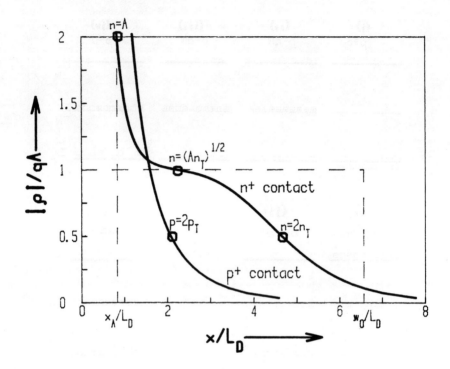

Fig. 3 Built-in space charge density distributions in SI GaAs adjacent to n+ and p+ doped regions. Abscissa expressed as multiple of the Debye length $L_D = [\epsilon \, v_T/Aq]^{1/2}$ for the acceptor concentration $A = 10^{13} cm^{-3}$. The electron and hole concentrations in the neutral bulk are $n_T = 1.28 \times 10^9 cm^{-3}$ and $p_T = 2.5 \times 10^3 cm^{-3}$; w_o is the extension of an equivalent uniform space charge layer with abrupt edge [13]. The electron concentration $n = A$ occurs at $x_A = u_A L_D$.

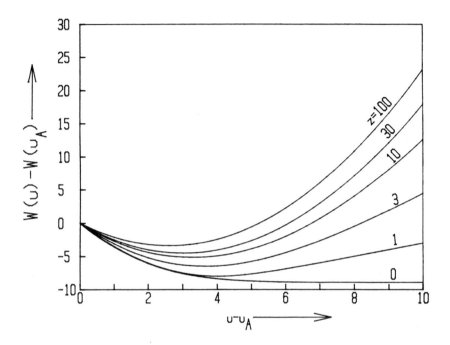

Fig. 4 Potential distributions for the n+ SI GaAs junction of Fig. 3 for various cathodic current densities expressed as multiples of $n_T\ q\ \mu_n v_T/L_D$. The potential is expressed as a multiple of $v_T = kT/q$, and the position coordinate as a multiple of L_D.

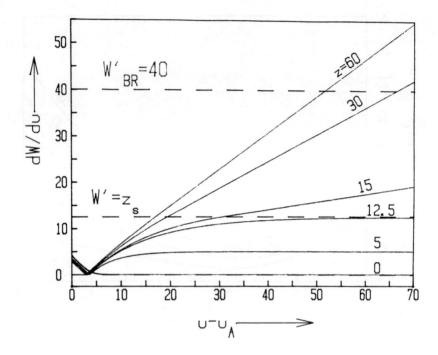

Fig. 5 Normalized field distributions for an n+ SI GaAs junction for various cathodic current densities with drift velocity occuring at $v_s/\mu = 12.5\, v_T/L_D$ (lower dotted line). If electron-hole pair generation would start at $W'_{BR} = 40$, (upper dotted line), the abscissa of the intercept for a given z would provide the sample length for onset of hole generation at the anodic contact at that current density z.

Present Status and Future Trends of High Speed Devices Using GaAs and Related Compounds

Hideki Hasegawa

Department of Electrical Engineering
Faculty of Engineering
Hokkaido University, Sapporo, 060 Japan

Abstract
Present status and future trends of high speed devices using compound semiconductor materials are critically reviewed. Devices include GaAs MESFETs, HEMTs, Heterostructure MISFETs, InP MISFETs, HBTs and quantum electronic devices. Technological issues on speeds and on interface constraints are briefly discussed for future progress.

1. Introduction

It is rather surprising that, of varieties of materials classified as semiconductors, Si had been for decades "the only useful material" for integrated circuits. A major challenge to this, however, has been initiated from early 1980's by GaAs and related compound semiconductors and is currently being actively pursued. Attractive features of these materials for high speed application include 1) high electron mobility or high drift velocity, 2) availability of semi-insulating substrates, 3) high radiation hardness and high operation temperature, 4) capability of material parameter tailoring with the use of mixed crystals, and 5) capability of forming heterojunction barriers and quantum wells with atomic scale controllability .

Varieties of new devices have been invented using these attractive properties, and remarkable progress in performance has been reported in recent years. However, it obviously has stimulated Si device research and Si devices have remarkably increased their speeds in recent years . Thus, it is expected that the competition over speed/power performance among various compound semiconductor and Si device families continues into foreseeable future.

The purpose of the present paper is to review the present status and future trends of the compound semiconductor devices. A particular emphasis is laid on the latest progress achieved in Japan. Technological issues on speeds and on interface constraints are briefly discussed for further successful developments of the compound semicoductor technology.

2. High Speed Integrated Circuits

The merits of high speed devices are best utilized when they are integrated in a large scale to perform complex logic and high density memory functions. This is also where largest market is expected if speeds substantially higher than those of Si devices are realized in LSI/VLSI environments. Owing to recent intensive efforts, three kinds of device technologies, i.e.,

MESFET, HEMT and HBT technologies are at the moment mature enough to fabricate LSIs and test the performance. A brief summary of the recent data on the performance of the compound Semiconductor LSIs is given in Table 1.

Table 1   Performance of compound semiconductor LSIs

circuit/size		device structure	L_g	circuit performance		
				speed	power	year(ref.)
				(access time)		
SRAM	16 K	MESFET SAINT	1.0µm	4.1ns	2.5W	1984(1)
	4K	MESFET SA	1.0µm	2.5ns	0.2W	1986(2)
	4 K	MESFET P$^+$LDD	0.7µm	1-1.5ns	1.6W	1987(3)
	4K	HEMT SA	2.0µm	2.0ns(77K)	1.6W	1984(4)
				(t_{pd}/gate)	(/gate)	
GATE	6K	MESFET LDD	1.0µm	76ps	1.2mW	1987(5)
ARRAY	3K	MESFET SWAT	1.4µm	56ps	4.6mW	1986(6)
	1.5K	HEMT SA	1.2µm	178ps(*)	6.7mW	1986(7)
	4.1K	HEMT SA	0.8µm	108ps(*)	1.6mW	1987(8)
	4K	HBT HI^{2}L	-	300ps		1986(9)

(*)under standard loading condition of FI=FO=3, interconnection
    length of 2 mm
SA = self aligned, LDD = lightly doped drain
SWATT= sidewall assisted

2.1 MESFET Technology

Processing technology of MESFETs is most advanced, and SSI/MSI digital ICs and monolithic microwave ICs (MMICs) are already in production stage.   Recent reliability test data on GaAs SSIs over 10^6 hours has shown that reliability is as good as that of Si devices(10). GaAs MESFETs continues to be major active devices in MMIC as they have been the key discrete device in microwave area.

On the other hand, LSIs are sill in the developmental stage apart from some commercial products. Fabrication yield is still a problem for mass production.   Well-known issue of correlation or non-correlation between FET threshold voltage uniformity and dislocation density seems to have settled in such a way that both are possible depending on the processing conditions, particularly, on the As pressure at implantation annealing(11), which controls V_{Ga} concentration and affects the activation of amphoteric Si(12).

It is also a general consensus that use of submicron gate is essential for GaAs MESFET LSIs in order to overcome the effect of wiring capacitance and achieve speed advantage in LSI/VLSI environments over Si LSIs.   In submicron devices, the so-called short channel effect becomes serious.   In order to avoid or minimize this, buries p-layer structure(13), LDD structure (14) etc. are being investigated.   Examples of such advanced MESFET structures are shown in Fig.1.

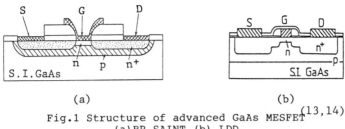

Fig.1 Structure of advanced GaAs MESFET[13,14]
(a) BP SAINT (b) LDD

2.2 HEMT (MODFET) Technology

Establishment of self-alignment technology involving selective dry etching has resulted in a rapid increase of the integration level of HEMT rapidly as shown in Table 1. The key processing step is shown in Fig.2. As compared with MESFETs, HEMT exhibits much less pronounced short channel effect owing to selective doping. With 0.5 μm gate length technology, 10 K gate arrays with t_{pd} of 100 ps and 16K SRAM with access time of 0.5 ns are believed to be feasible(15).

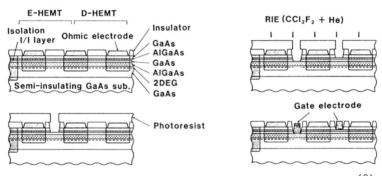

Fig.2 Fabrication process sequence for HEMT LSIs[8]

HEMT is also finding important application as a key device in low-noise amplifiers in microwave and millimeter wave region owing to their superb noise performance.

Recently, volume production capability of MOCVD HEMT wafers has been demonstrated (16), and high performance of MOCVD HEMTs was reported(15).

2.3 HBT Technology

Heterojunction bipolar transistors are characterized by high speeds combined with high drive capability which is essentially required in LSI environments. Although technology is not mature enough to allow high levels of integration primarily due to complex device structures, impressive speed performances recently been reported at SSI levels as summarized in in Table 2. The latest data on Si bipolar device is also included for comparison. f_T of as high as 105 GHz (21) has been obtained. The improvements are mainly due to reduction of parasitic elements such as base resistance and collector capacitance using self-alignment and proton or boron implantation. An example of a self-aligned structure (22) is shown in Fig.3 Future improvement is expected to lead to f_T of above 200 GHz.

Table 2   Performance of HBT SSIs

circuit	gate type	performance	year/ ref.
Ring	ECL	t_{pd} = 5.5 ps	1987(17)
Oscillator	ECL	17.2 ps	1986(18)
	CML	27.6 ps	1986(19)
1/4 Frequency	ECL	f_c = 13.7 GHz	1986(18)
Divider	CML	11 GHz	1986(19)
Si bipolar R.O.	NTL	t_{pd} = 20.5 ps	1987(20)
	ECL	34.1 ps	1987(20)

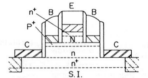

Fig.3   Example of self -aligned HBT (22)

3. Novel High Speed Devices

With the advance of MBE and MOCVD growth technologies and various processing technology combined, which now allow control and manipulation of atomic scale dimensions, possibilities of new devices have emerged. Since such possibilities covers an extremely wide range, only some limited number of examples which look important at the present stag are briefly discussed.

3.1 GaAS Heterostructure MISFETs

Since insulator-semiconductor interface of GaAs is characterized by high-density of interface states which pin the Fermi level, ordinary MISFETs are not feasible. Instead, MISFETs can be constructed by using undoped AlGaAs layer as insulator. Since invention at ETL(23) and at IBM(24), various groups are interested in its LSI applications. Its operation is essentially the same with enhancement type HEMT. But, the advantages are (1)easier V_{th} controllability and uniformity,(2) no DX center problem, no V_{th} shift with temperature, no current collapse (3)feasibility of complimentary circuit.   An example of a complimentary circuit structure(25) is shown in Fig.4(a). High g_m can be achieved by using a thin AlGaAs layer. Maximum g_m of 450 mS/mm (26) was reported for n-channel device, and 50 mS/mm for p-channel device(27).   Recently, an extremely high g_m of 760 mS/mm with ring oscillator t_{pd} of 25 ps has been obtained by further doping the channel as shown in Fig.4(b)(28).

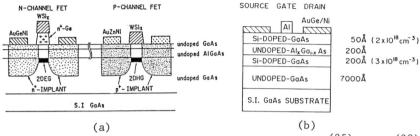

(a)   (b)

Fig.4 Heterostructure MISFET (a) complementary[25](b) DMT[28]

3.2 InP MISFETs

As compared with GaAs, insulator-semiconductor interface
is far easier to control in InP(29) and it is well known that
both enhancement and depletion type MISFETs can be constructed
with various insulators. Early devices suffered from a problem
of severe drain current drift, but it recently has become quite
stable by optimization of processing (for example, within 3 % in
10^5 sec (30)). A recent attempt is to use AlGaAs layer as the
insulator (31). The structure is shown in Fig.5(a) with microwave
performance in (b). This structure allows very stable depletion
mode operation and a much better microwave power performance than
GaAs MESFET is obtained owing to excellent velocity-field
characteristics of InP.

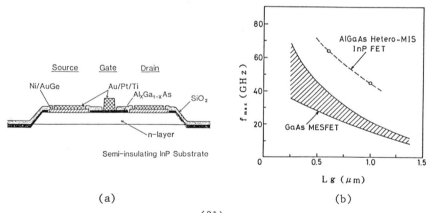

(a)   (b)

Fig.5 AlGaAs hetero MIS InP FET[31] (a) structure (b)performance

3.3 Resonant Tunneling and Other Quantum ELectronic Devices

Advanced crystal growth technology allows control of
electron wave functions. Because of this, resonant tunneling
phenomenon, though conceptually not new, has started to receive
an explosively large revived interest since observation of
negative differential resistance (32). Figure 6 shows the

516

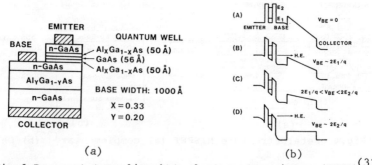

                    (a)   (b)

Fig.6 Resonant-tunneling hot electron transistor (RHET)[33]
      (a) structure   and (b) operation principle

structure and operation principle of RHET where hot electrons are
injected via. energy selective resonant tunneling(33). Exclusive
NOR and memory function can be realized by a single device.
Current gain of 5 obtained in AlGaAs/GaAs/AlGaAs RHET was to
small for integration. However,very encouragingly, current gain
of 25 and the peak-to-valley ratio of the collector current of 15
has recently been achieved at room temperature by changing to
InAlAs/InGaAs/InAlAs system(34).

     Such an effort naturally opens up new possibility of quantum
electronic devices based on confinement in one, two and three
dimensions. Various physical phenomenon and possible device
structures are under investigation. Although nothing definitely
practical and useful has come up yet, it is possible that this
type of efforts may eventually lead us to a "post-silicon" era.

4. Future Trends and Technological Issues

     Obvious future trends are towards higher integration levels,
towards faster speeds with submicron structure and towards new
class of quantum electronic devices with nanometer dimensions.
At the moment, the situation is so much abundant in various
attractive possilibities for future innovations and improvements
that it is difficult to judge which device is really viable.
Figure 7 is an example of computer simulation done at NTT (35)
which tries to place various compound semiconductor device
families on delay-complexity chart. It is not clear at all at
present where new quantum devices are going to be placed on such
a chart.

     There are of course several technological issues that have to
be carefully considered and settled down in order to make
progress along the above mentioned future trends towards larger
scale integration and towards easmaller size. Of these, two of
the fundamental issues are briefly discussed below.

4.1 Speed Issues

     Compound semiconductor high-speed devices pride in their
speeds. Very interestingly, however, it is often not quite clear
what is meant by speed. The first fundamental question is which

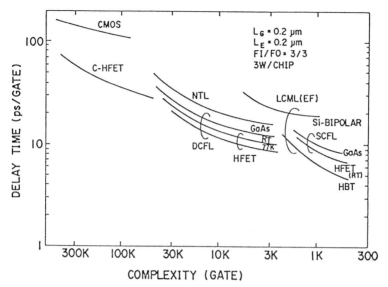

Fig.7 Future trends of semiconductor devices[35]
on delay-complexity chart

quantity (such as carrier transit time in device, g_m or other
quantity) actually determines the device performance in
LSI/VLSI environments. The second question is how this quntity is
related to basic transport quantities such as low-field mobility,
saturation velocity, velocity overshoot etc.

The first question is obviously related to intterconnection
problem and to archtechture of the circuit or system. To answer
the question, or to optimize overall performance, a close
interaction between between systems workers and device workers
are required. In very high speed ICs, microwave properties of
interconnections rather than static capacitance have also taken
into speed considerations(36). As to second question, one often
wonders, for instance, which of mobility, saturation velocity and
velocity overshoot matters most in submicron GaAs MESFETs and
HEMTs, or whether overshoot effect actually exists as many Monte
Carlo simulations predict. If one further comes to quantaum
electronic devices, the situation is even further ambiguous as
exemplified in ambiguity involved in the theory for maximum
ocsillation frequency of a resonnant tunnling diode (37, 38).
Further fundamental theoretical works are obviously required in
these areas. To be able to bridge device properties and
system performance with reasonably accuracy appears, in any case,
to be one of fundamental requirements for choosing viable devices
from almost infinitely abundant possibilities in device
structures and material combinations.

4.2 Interface Constraints

Minuaturization of device implies increased ratio of surface
to volume of the device. Various interfaces including

518

semiconductor-semiconductor (S-S), metal-semiconductor (M-S) and insulator-semiconductor (I-S) interfaces used for device construction, isolation and interconnection, are consequently expected to exert more pronounced effects performance. In particular, as compared with the Si technology, surface passivation technology is not well-established in the compound semiconductor technology. Importance of surface depletion layer control at I-S interface was already indicated in the self-aligned gate structure of GaAs MESFET by which LSI/VLSIs became possible(39). A recent study indicated I-S interface states can degrade device isolation, triggering the so-called side-gating(40).

Understanding and control of interface states at S-S, M-S and I-S interfaces are thus key issues of future device processing. It has recently pointed out that S-S, M-S and I-S interface posses hitherto unnoticed close correlation to each other which cannot be explained by the well known unified defect model(41) for I-S and M-S interfaces or by MIGS model (42) for M-S interface. A novel disorder induced gap state (DIGS) model (43) based on interface disorder, has been proposed, and may provide a guide for atomic-scale control of interfaces.

References

(1) Y. Ishii et al: GaAs IC Symp. (1984) paper no.29.
(2) N. Tanino et al: GaAs IC Symp. (1986) paper no.5-4.
(3) H. Tanaka et al: ISSCC, Dig. Tech. Paper (1987) p.138.
(4) S. Kuroda et al: GaAs IC Symp. (1984) Tech. Dig. p.125.
(5) T. Terada et al: ISSCC Dig. Tech. Paper (1987) p.144.
(6) H. Hirayama et al: ISSCC Dig. Tech. Paper (1986) p.72.
(7) Y. Watanabe et al: ISSCC Dig.Tech. Paper (1986) p.80.
(8) K. Kajii et al: CICC Dig. Tech. Papers (1987) p.199.
(9) H.T. Yuan et al: ISSCC, Dig.Tech. Paper (1986) p.74.
(10) Y. Hosono et al: Proc.1987 IEEE Microwave and
     Millimeterwave Monolithic Circuits Symp. (1987) p.49.
(11) T. Egawa et al: Proc. 4th Conf. Semi-insulating
     III-V Materials, Hakone, (Ohm, 1986) p.567.
(12) S. Miyazawa and K. Wada: Appl. Phys. Lett. **48** (1986) 905.
(13) K. Yamasaki et al: Electron. Lett. **20** (1984) 1029.
(14) S. Asai et al: Ext. Abs. 18th SSDM (1986) p.383.
(15) T. Mimura et al:presented at 1987 Joint Meeting of
     Electricity and Information Related Institutes in Japan.
     Sept. 1987
(16) J. Komeno et al: presented at 1987 Electronic Materials Conf.
     June 24-26, Santa Barbara.
(17) K. Nagata et al: to be published.
(18) T. Ishibashi et al: IEDM, Tech. Dig. (1986) p.809.
(19) F.M. Chang et al: Electron Lett. **22** (1986) 1173.
(20) S. Konaka et al: Ext. Abs. SSDM (1987) p.331.
(21) T. Ishibashi et al: to be published.
(22) N. Hayama et al: IEEE Electron Dev. Lett. **EDL-8** (1987) 246.
(23) K. Matsumoto et al: Electron. Lett. **20** (1984) 463.
(24) P.M. Solomon et al:IEEE Electron Dev. Lett. **EDL-5** (1984) 379.
(25) T. Mizutani et al: Inst. Phys. Conf. Ser. No.79 (1985) p.733
(26) K. Maezawa et al:IEEE Electron Dev.Lett. **EDL-7** (1986) 454.
(27) K. Oe et al: Surface Science, **174** (1986) 378.
(28) H. Hida et al:IEEE Electron Dev. Lett. **EDL-7** (1986) 625

(29) H. Hasegawa and T. Sawada: Thin Solid Films, **103** (1983) 119.
(30) K.P. Pande et al: IEEE Electron Dev. Lett. **EDL-7** (1986) 771.
(31) T. Itoh et al: IEDM 1986, Tech. Dig. p.771.
(32) T.C.L.G. Sollner et al: Appl.Phys.Lett. **43** (1983) 588.
(33) N.Yokoyama et al: Jpn. J. Appl. Phys. **24** (1985) L853.
(34) K. Imamura et al: presented at 45th Device Research Conf. (Santa Barbara, June, 1987): also at 19 SSDM (Tokyo, Aug/1987) Ext. Abs. p.359.
(35) T. Sugeta et al: GaAs IC Symp. Tech. Dig. (1986) p.3
(36) H. Hasegawa and S.Seki: IEEE Trans. Microwave Theory and Tech. MTT-32 (1984) 1721, ibid. 1715.
(37) S. Luryi, Appl. Phys. Lett. 47 (1985) 490.
(38) W.R. Frensley: Phys. Rev.Lett. **57** (1986) 2853.
(39) N. Yokoyama et al: 1981 IEDM, Tech. Dig. p.80.
(40) H. Hasegawa et al: Inst. Conf. Ser.no.74, p.521 (1985).
(41) W.E.Spicer et al: J. Vac. Sci.Technol.16 (1979) 1422.
(42) J. Tersoff: Phys. Rev. Lett. 52 (1984) 465.
(43) H.Hasegawa and H. Ohno:J.Vac.Sci.Technol. **B4** (1986) 1130.

THE PHYSICS OF RESONANT TUNNELLING IN GALLIUM ARSENIDE HETEROSTRUCTURE

D.K. Roy and Amitabh Ghosh
Department of Physics
Indian Institute of Technology, Delhi
Hauz Khas,New Delhi-110 016
INDIA

1.   INTRODUCTION

The effect of size quantization in solids first became experimentally evident when oscillations of tunnelling current as a function
of the applied voltage across a thin insulating film separating a thick
lead and a thin aluminium films were noticed[1]. Observation of such a
phenomenon in Gallium Arsenide heterostructures has currently become an
area of intense research activity with the development of molecular beam
epitaxy. The first such device consisting of a GaAs layer of a few
nanometre thickness sandwiched by slabs of $Ga_{1-x}Al_xAs$ was fabricated by
this technique. Evidence of formation of size quantization in it came
from measurements of optical absorption and emission spectra above the
band gap of $GaAs$[2]. Such structures have ultimately led to the development of quantum well lasers. Introduction of modulation doping in them
has given effect to record carrier mobilities, specially at very low
temperatures[3]. This has subsequently been responsible for the development of modulation doped FETs. Furthermore, when a strong magnetic
field is applied to a two dimensional electron gas confined within the
well normal to the barrier, the remaining two degrees of freedom of
electrons get further quantized leading to the appearance of gaps between
quantized Landau levels[4]. When a large number of similar heterojunctions
are grown along a particular direction, a superlattice is said to be
formed. Such structures made up of InAs and GaSb layers are found to
show some interesting features due to their unusual band edge relationships.
Their characteristics change from a semimetallic to a semiconducting
behaviour as the InAs layer thickness is reduced[5]. Shubnikov deHaas[6]
oscillations in GaAs-GaAlAs superlattices ware observed by Chang et.al
as the orientation of the sample relative to the applied magnetic field
was changed. Magnetophonon resonances in the resistivity vs. magnetic
field plots were also noted subsequently[7]. The infrared radiation
has been detected successfully using such a superlattice. Avlanche photo-
diode-a solid state analogue of a vacuum tube photomultiplier tube has
now been developed using recently introduced computer controlled MBE by
which precise control of potential profiles through compositional grading
has been achieved[8].

In contrast to the lattice matched heteroepitaxy just discussed,
strained layer heteroepitaxy has also been introduced very recently[9].
A very good example of a strained layer superlattice is the Si-Ge heterostructure. Since the lattices of Si and Ge do not match with one another
their thickness are kept very small so that much dislocations are not
generated at the interface. Because the band gap of Si-Ge is smaller
than that of Si, the former is expected to respond to smaller energy
photons. A new type of photodetector to be used in optical communication is thus possible to be developed.

Esaki and Tsu[10] first proposed the negative differential conductivity effect in lattice matched heterostructures due to (i) Bloch
oscillations and (ii) Tunnelling effects. Resonant tunnelling effect in

a double barrier heterostructure was also proposed by them.

They observed a peak to appear in the device I-V characteristic of a double barrier heterostructure (Fig. 1a). This has been found to occur when the energy in the left hand electrode agrees with one of the resonant states within the quantum well about 100 A in thickness (Fig. 1b). When a.d.c. bias is impressed across such a device potential drops are developed over the undoped barrier and well regions. As the materials next to the barriers are generally heavily doped they do not develop any potential drop over them. As a result, a trapezoidal potential profile (Fig. 1b) is formed. This allows the electrons from region 1 to tunnel through the first barrier into the quantized state within the well and finally through the second barrier into the less populated states of region 5. With increasing bias voltages the bound state energy level drops below the conduction band edge of region 1 (Fig. 1c). Electrons must then tunnel through the entire barrier structure to get to the unoccupied states of region 5, giving rise to a drop in the current. The resulting I-V curve has been shown in Fig. 1(d). The negative conductance region in the I-V curve has been used successfully for the generation of micro-waves and other suitable applications.

Although we just discussed the qualitative features of resonant tunnelling in a double barrier heterostructure we would now further elaborate in this paper a rigorous quantum mechanical account of deriving its I-V characteristic.

2.   NEW TUNNELLING THEORY

Before presenting an account of this theory[11], we would first enumerate the various difficulties associated with conventional tunnelling explanations.

a.   Defects with the conventional theory

(i) The primary basis to the explanation of tunnel effect has been the wave-particle duality exhibited by microscopic particles. This, as contained in de Broglie's hypothesis, is presented mathematically by Schrodinger's time-independent wave equation. But, electrons cannot continue to retain their material or wave characteristic inside the barrier and as such they are not in a position to display wave-particle duality during tunnelling. Even then Schrodinger's time-independent wave equation has been used invariably inside the barrier to evaluate their wave characteristics.

(ii) In the conventional approach one regards the electron energy to be conserved during tunnelling. By it, one understands that the electron energy inside the barrier must be the same as the energy outside (E), before tunnelling. If this is so then by virtue of the relation $E = h\nu$, the electron is capable of display its wave nature. On the other hand if one uses $E = \frac{1}{2} mv^2$, the particle characteristics of electrons are exhibited. Thus the energy conservation during tunnelling is equivalent to presuming wave particle duality. Since the electrons cannot acquire the latter characteristic the former premise cannot act as the working principle of the tunnelling theory.

Also, the electron potential energy inside the barrier at a point x is represented by $V(x)$, as described by the barrier shape. It is needless to point out that by doing it one implicitly presumes the particle nature of electrons inside the barrier, once again!

(iii) Conventionally, the quantum mechanical current density is evaluated by employing,

$$J = \frac{q\hbar}{2mi} [\psi^*(x) \frac{d\psi(x)}{dx} - \psi(x) \frac{d\psi^*(x)}{dx}] \tag{1}$$

where the symbols have their usual significance. It should however be remembered that (1) has been evaluated to represent material flow across a conservative field when electrons exhibit wave-particle duality and where it is possible to identify electrons' potential energy exactly. Since the above mentioned conditions are invalid within the barrier, equation (1) cannot be applied to evaluate the tunnelling current density[11].

(iv) On the basis of the time-independent treatment the electron does not lose any phase while tunnelling as the latter turns out to be independent of the barrier width. As such, the process of tunnelling has been regarded to be an instantaneous one.

(v) An alternative but almost equivalent approach of analyzing tunnel effect, based upon the time-dependent perturbation theory, is well established. In it, the tunnelling transition of the electron from the decaying barrier state to the growing one (inside a rectangular barrier) is regarded to be initiated by a small unidentified perturbation inside the forbidden region. Due to the smallness of the latter, the principle of conservation of energy is still considered to be valid. The tunnelling current density then is evaluated by making use of Fermi's golden rule. An approximate expression for the tunnelling time is also possible to be derived by reciprocating the expression for the transition probability per unit time. But contrary to prediction made in (iv) this turns out to be unduly large. But the predictions on tunnelling probability and current density density agree reasonably well with those of the time-independent approach.

b.   Principle of Quantum Measurement and Observation

Owing to various discrepancies of the conventional approaches on tunnelling as elaborated above, a new description of the latter now becomes necessary where the interaction of the electron wave with the potential barrier is required to be considered properly.

Let us first start by considering the case of electron diffraction at a rectangular slit of width 'a'. An electron stream of a well-defined energy E and momentum p_x is presumed to be incident upon it such that its width is comparable to the electron wave length $\lambda (= h/p_x)$. It is also possible to localize electrons before incidence. As these electrons have a well-defined momentum their uncertainties in momenta viz. Δp_x, Δp_y, Δp_z must be zero. Consequently their position uncertainties viz. Δx, Δy and Δz turn out to be unduly large. Since, electrons pass through a narrow slit of width 'a', the y components of their position co-ordinates become uncertain at least by this amount. Consequently, after diffraction one finds

$$\Delta p_y = h/a \tag{2}$$

Hence, the angular spread of the diffraction pattern on the screen is given by

$$\theta = \frac{\Delta p_y}{p_x} = \frac{h/a}{h/\lambda} = \frac{\lambda}{a} \tag{3}$$

Equation (3) is well-known to follow from the laws of physical optics. The expressions for energy uncertainty ΔE and the time of interaction Δt

between the electron wave and the aperture before the effects of diffraction is noticeable may also be calculated, as given below: We know that

$$\Delta E = \frac{\Delta p_y^2}{2m} = \frac{h^2}{2ma^2} \tag{4}$$

Therefore

$$\Delta t \sim \frac{h}{\Delta E} \sim \frac{2ma^2}{h} \sim \frac{2a}{(\Delta p_y/m)} \tag{5}$$

where $\Delta p_y/m$ measures the transverse component of electron's velocity.

Such a phenomenon (e.g. diffraction), where the electron wave interacts with an aperture or similar such objects that imposes on the incident infinite wave train the shape of a wave packet, is known as the Quantum Measurement. The dimensions of the wave packet are however determined by the width of the aperture. The latter which restricts the spatial extension of the wave packet has been termed as the measuring apparatus Before any interaction occurs between the quantum object (e.g. the electron) and the apparatus (e.g. the rectangular slit), the energy of the electron is well defined and is E. Since the interaction process always requires a definite interval of time Δt to produce an observable effect, the electron energy as a consequence of such a quantum measurement process must then be uncertain at least by ΔE, given by (4).

We may now attempt to describe quantum mechanical tunnelling on the basis of the quantum measurement principle discussed above. For that, let us now consider a potential barrier of arbitrary shape (Fig. 2). A stream of electrons of total energy E has been shown to be approaching it and as such get gradually converted into their potential counterparts. At $x = x_1$, electrons come to rest and we may then write $E = V(x_1)$. Thus upto x_1, the classical description of electronic motion can be taken as valid. Beyond this point, electrons face a classically forbidden region of energy height (V_o-E) (Fig. 2). It can now only escape through such a barrier, if the latter behaves as a measuring apparatus. It is then expected to control the position spread as well as the quantum energy fluctuation of the transmitted electrons, like the rectangular slit in the diffraction problem mentioned earlier. After the electrons are incident at $x = x_1$, their where abouts become uncertain at least for a period of τ when they are detected beyond the far side of the barrier. Thus, the latter interacts with the electrons wave for the period τ and imposes on it an energy fluctuation of (V_o-E) over the original energy value E. It follows therefore from the quantum measurement principle that,

$$\tau = \Delta t \sim \frac{h}{(V_o-E)} \tag{6}$$

Thus, electron energy which was E just before tunnelling may now acquire any value between $E+(V_o-E)$ and $E-(V_o-E)$. Due to the superposition of these wavelets at the transmitted end, a wave packet is generated out of an infinite wavetrain of well-defined energy E that is incident upon the barrier. The energy uncertainty $V_2(\tau)$ introduced by the barrier during tunnelling may then be expressed as,

$$-(V_o-E) \leq V_2(\tau) \leq (V_o-E) \tag{7}$$

From (6), equation (7) reduces to,

$$-\frac{\hbar}{\tau} \leq V_2(\tau) \leq \frac{\hbar}{\tau} \tag{8}$$

In the limiting cases (7) may also be written as,

$$\pm|V_2(\tau)| + (V_o-E) = 0$$

i.e.

$$\pm |V_2(\tau)| + V_o = E \tag{9}$$

It should be reasserted that the result expressed in (9) is observable only after tunnelling. But it is difficult to ascertain how do the quantities $V_2(\tau)$ and V_o build up during the tunnelling process. However, during the interaction of the electron wave with the barrier, we may write[12] after a little introspection,

$$V_2(t) + V_1(x) = V(x,t) \tag{10}$$

where $V_2(t)$ and $V_1(x)$ become latent or hidden variables. Immediately after tunnelling one may express by comparing (10) with (9) that,

$$V_1(x) \sim V_o \tag{11}$$

so that $V(x,t)$ approaches E as, the electron tunnels out. The limits for $V_2(t)$ have already been given in (7) and (8).

It should however be noticed carefully that both the height as well as the width of the barrier are intimately related with one another in producing the tunnel effect. This has been made further clear in the following discussion: One may note that a barrier of width $W(= x_r-x_1)$ demands the transmitted wave packet to have the same spatial dimension as itself. Consequently, the momentum spread Δp of the wave packet is given by

$$\Delta p = h/w \tag{12}$$

The corresponding energy spread ΔE of the wave packet then follows to be,

$$\Delta E = \frac{h^2}{2m\ w^2} \tag{13}$$

But, as we have already noticed that $\Delta E \sim (V_o-E)$, one may obtain from (13) that,

$$\chi_o W \sim 2\pi \tag{14}$$

where χ_o is the barrier penetration constant given by,

$$\chi_o = \{ \frac{2m}{\hbar^2} (V_o-E) \}^{1/2} \tag{15}$$

It must however be remembered that (14) is an order of magnitude relation which must be obeyed during tunnelling.

When there is electron transmission high over the barrier, there is no interaction between the electron wave and the latter. The electron potential energy is then described exactly by $V_1(x)$ as given by the barrier shape.

c. **Barrier Electron Wavefunction, Tunnelling Probability and Tunnelling Current Density**

If $\psi(x,t)$ denotes the barrier electron wave function, it must satisfy the equation

$$H\psi(x,t) = i\hbar \frac{\partial \psi(x,t)}{\partial t}$$

But, $\psi(x,t)$ becomes an observable only after tunnelling. By virtue of (10), the barrier electron hamiltonian is given by

$$H = -\frac{\hbar^2}{2m} \frac{\partial^2}{\partial x^2} + V(x,t) = -\frac{\hbar^2}{2m} \frac{\partial^2}{\partial x^2} + V_1(x) + V_2(t) \qquad (16)$$

Next, presuming $\psi(x,t) = X(x)T(t)$, we get from $H\psi(x,t) = i\hbar \frac{\partial}{\partial t}\psi(x,t)$

$$-\frac{\hbar^2}{2m} \frac{1}{X(x)} \frac{d^2 X(x)}{dx^2} + V_1(x) = \frac{i\hbar}{T(t)} \frac{dT(t)}{dt} - V_2(t) = E(\text{say}) \qquad (17)$$

Integrating the time-part of the equation in (17), we get,

$$\ln \frac{T(\tau)}{T(0)} = -\frac{i}{\hbar} \left\{ E.\tau + \int_0^\tau V_2(t)\,dt \right\} \qquad (18)$$

Since, the integral in the exponent of (18) involves the measurement of quantum of action, its value must lie somewhere between $+\hbar$ and $-\hbar$. We may therefore express (18) as follows:-

$$T(\tau) = T(0) \exp\left[-\frac{i}{\hbar} \left\{ E \pm \frac{\hbar}{\tau} \right\} \tau\right] \qquad (19)$$

Equation (19) thus expresses very clearly that as a result of tunnelling there is quantum scattering of electrons in the energy interval $E + \hbar/\tau$ and $E - \hbar/\tau$.

The space part of the differential equation in (17) when isolated and simplified yields,

$$\frac{d^2 X(x)}{dx^2} - K^2(x)\, X(x) = 0 \qquad (20)$$

where

$$K^2(x) = \left[\frac{2m}{\hbar^2} \{V_1(x) - E\}\right]^{1/2} \qquad (21)$$

$V_1(x)$ is a hidden variable and its value is ascertained immediately after tunnelling and is given by (11). If we set $V_1(x) \sim V_o$ in (21), (20) reduces to a very simple form as given below:-

$$\frac{d^2 X(x)}{dx^2} - \chi_o^2\, X(x) = 0 \qquad (22)$$

where χ_o is given by (15). The two allowed solutions of (22) are then given by

$$X_\ell(x) = \alpha \exp(-\chi_o x) ; \quad X_r(x) = \beta \exp(\chi_o x) \tag{23}$$

where α and β are appropriate constants. Now on combining (19) and (23), the general barrier wave function may be represented as,

$$\psi(x,t) = a_\ell(t) X_\ell(x) \exp(- \frac{iE_\ell t}{\hbar}) + a_r(t) X_r(x) \exp(-\frac{iE_r t}{\hbar}) \tag{24}$$

Thus wave function however becomes an observable only after tunnelling. The energies associated with the two parts of the wave function have been designated as E_ℓ and E_r because of the energy uncertainty introduced during tunnelling (eqn. 19). Since, the second part of this wave function builds up at the expense of the first one, as the process of tunnelling continues, they are connected through the time-dependent coefficients. The values of these coefficients are given by[11],

$$a_\ell(t) = \exp[i \{- \frac{W_{11}}{\hbar} - \frac{\eta}{2} \}t] \ [\cos \sigma t + i \frac{\eta}{2\sigma} \sin \sigma t] \tag{25}$$

and

$$a_r(t) = - \frac{iW_{21}\delta}{\hbar} \exp [-i \{\frac{W_{11}}{\hbar} + \frac{\eta}{2} + \omega_{\ell r}\} t] \frac{\sin \sigma t}{\sigma} \tag{26}$$

where

$$W_{11} = W_{22} = \pm \frac{\hbar^2 \chi_o^2}{2m} , \ \pm \frac{\hbar^2 \chi_o^2}{2m}\{ 1+8 \exp(-2\chi_o W)\} \tag{27}$$

$$W_{12}^* = W_{21} = 0 , \ \pm \frac{2\alpha}{\beta} \frac{\hbar^2 \chi_o^2}{2m} \exp(-2\chi_o W) \tag{28}$$

$$|W_{12}|^2 = 16(V_o - E)^2 \exp(-2\chi_o W) \tag{29}$$

$$\sigma = \pm\{\frac{\eta^2}{4} + \frac{|W_{12}|^2}{\hbar^2} \}^{1/2} \ \simeq \frac{\eta}{2} \tag{30}$$

since

$$\eta = \frac{W_{22} - W_{11}}{\hbar} - \omega_{\ell r} = (\omega_o - \omega_{\ell r}) \tag{31}$$

so that

$$(\frac{\eta}{2})^2 >> \frac{|W_{12}|^2}{\hbar^2}$$

also,

$$\delta = \exp(i\theta) \tag{32}$$

accounts for any additional phase difference between $a_\ell(t)$ and $a_r(t)$.

By virtue of (24) the probability density inside the barrier follows to be,

$$P(x,t) = |x_\ell|^2 + \frac{|w_{21}|^2 |x_r|^2}{\hbar^2} \cdot \frac{\sin^2 \sigma t}{\sigma^2}$$

$$+ \frac{2w_{21} X_\ell^* X_r \sin \sigma t}{\hbar \sigma} \sin(\sigma t + \theta_2) \tag{33}$$

The tunnelling current density through the barrier may now be evaluated by using the equation of continuity given below[12]:

$$J = q \int_{x_\ell}^{x_r} \frac{\partial}{\partial t} P(x,t) \Big|_{t = \tau} dx \tag{34}$$

On combining (33) and (34), one obtains,

$$J = J_{01} \frac{\sin \gamma}{\gamma} + J_{02} \sin(\gamma + \theta_2) \tag{35}$$

where

$$\gamma = 2\sigma t \tag{36}$$

$$J_{01} = \frac{4q\hbar^2 \chi_o^3 k_1^2 \tau}{m^2 (k_1^2 + \chi_o^2)} \exp(-2\chi_o W) \tag{37}$$

and

$$J_{02} = \frac{16q\hbar \chi_o^2 k_1^2 W}{m(k_1^2 + \chi_o^2)} \exp(-2\chi_o W) \tag{38}$$

d. RESONANT TUNNELLING IN A DOUBLE BARRIER MODEL

Next, we proceed to evaluate the resonant tunnelling current density across a double barrier model as shown in Fig. 1. Since, in the steady state the current in any region of such a system has to be identical, we would only evaluate the current density across the second barrier for our convenience. The wave function inside the second barrier of Fig. 1 would continue to be represented by (24), but we would now have,

and

$$X_\ell(x) = \alpha \exp\{-\chi_o(x - x_3)\}$$
$$X_r(x) = \beta \exp\{\chi_o(x - x_3)\} \tag{39}$$

The resonant tunnelling current density due to a single electron would continue to be represented by (35) provided we now have,

$$J_{01} = q \cdot \frac{\hbar k_1}{m} Z_{res} \frac{k_5^2 + \chi_o^2}{2k_1 \chi_o} \tag{40}$$

and

$$J_{02} = q \cdot \frac{\hbar k_1}{m} Z_{res} \frac{k_5^2 + \chi_o^2}{2k_1} \tag{41}$$

where Z_{res} is the resonant tunnelling probability approaching unity. When an incoherent stream of electron waves falls upon the barrier, the phase difference among them would vary continuously from $-\infty$ to $+\infty$. The differential tunnelling current density produced by such a wave group having an electron density $\rho_\ell(E) f_\ell(E) dE$ is then given by[13],

$$dJ(E) = \rho_\ell(E)f_\ell(E)dE \sum_{-\infty}^{+\infty} [J_{01} \frac{\sin\gamma}{\gamma} + J_{02}\sin(\gamma+\theta_o)] \tag{42}$$

where

$$\gamma = 2\sigma t = \omega_{\ell r} t \sim \frac{\Delta E t}{\hbar} \tag{43}$$

The minimum observable phase difference at the transmitted end may be obtained to be

$$d\gamma = \frac{\varepsilon_r \tau}{\hbar} \tag{44}$$

where ε_r measures the energy difference between the energy levels there at the energy E. On combining (42) and (44), we may write

$$dJ(E) = \rho_\ell(E)f_\ell(E) \ dE . \frac{\hbar}{\varepsilon_r\tau} [J_{01}\int_{-\infty}^{+\infty} \frac{\sin\gamma}{\gamma} d\gamma$$

$$+ J_{02} \int_{-\infty}^{+\infty} \sin(\gamma+\theta_o)d\gamma]$$

i.e.

$$dJ(E) = \frac{\pi\hbar}{\varepsilon_r\tau} . J_{01} \ \rho_\ell(E)f_\ell(E)dE \tag{45}$$

Equation (45) may also be written as [13],

$$dJ(E) = \frac{\pi\hbar \ \Omega_r}{\tau} \ J_{01} \ \rho_\ell(E) \ \rho_r(E)f_\ell(E)\{ 1-f_r(E)\} \ dE \tag{46}$$

where Ω_r measures the volume of the electrode at the transmitted end.

The net density may now be obtained by integrating (46) over appropriate ranges of incident energy. The resulting expression may then be used in evaluating the I-V characteristic of resonant tunnelling devices.

REFERENCES

1. R.C. Jaklevic, J. Lambe, M.Miktor and W.C. Vassell, Phys.Rev.Lett.
    26, 88 (1971)

2. R. Dingle, in Festokorprobleme, H.J. Quisser ed. Pergamon Press,
    Oxford, England 15, 21,(1975)

3. H.L. Stormer, Surface Science, 132, 519 (1983)

4. K.V. Klitzing, G. Dorda and M. Pepper, Phys. Rev. Lett.,45,494(1980)

5. H. Sakaki, L.L. Chang, G.A. SaiHalasz, C.A. Chang and L. Esaki,
    Phys. Rev. Lett. 26 589 (1978)

6. L.L. Chang, H. Sakaki, C.A. Chang and L Esaki, Phys. Rev. Lett.
    38, 1489 (1977)

7. V.L. Gurevich and Y.A. Firsov, Sov. Phys. J.E.T.P., 13,137 (1961).

8. F.C. Capasso, W.T. Tsang and G. Williams, IEEE Trans. on Electron
    Dev. ED-30, 381 (1983)

9. G.C. Osbourn, J. Vac. Sci. Techl., 21, 469 (1982)

10. R. Tsu and L. Esaki, Appl. Phys. Lett., 22, 562 (1973)

11. D.K. Roy, Quantum Mechanical Tunnelling and its Applications,
    World Scientific, Singapore, 1986.

12. P.N. Roy, P.N. Singh, and D.K. Roy, Phys. Lett. 63A,81 (1977)

13. D.K. Roy, N.S.T. Sai and K.N. Rai, Pramana, 19, 231 (1982)

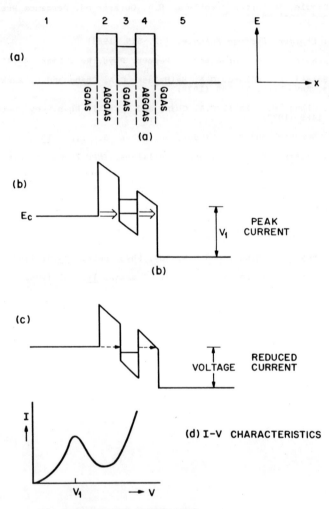

(a)

(b)

(c)

(d) I-V CHARACTERISTICS

Fig.1 RESONANT TUNNELLING

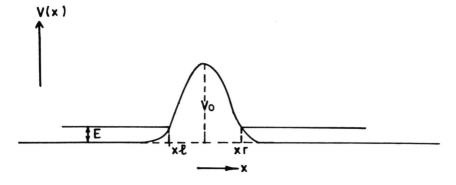

Fig.2. An Arbitrary Potential Barrier